DICTIONARY OF CERAMICS

By A. E. DODD

Information Officer, British Ceramic Research Association

About the Author

DR. DODD has been engaged in ceramic research for more than 30 years, and from his wide knowledge of the field has compiled a comprehensive and accurate dictionary covering the subjects of Pottery, Glass, Vitreous Enamels, Refractories, Clay Building Materials, Cement and Concrete, Electroceramics and Special Ceramics.

DICTIONARY
OF CERAMICS

Pottery, Glass, Vitreous Enamels, Refractories,
Clay Building Materials, Cement and Concrete,
Electroceramics, Special Ceramics

by

A. E. DODD,
M.B.E., Ph.D.,M.Sc., F.R.I.C.,F.I.Ceram.
Information Officer, British Ceramic Research Association

1967

LITTLEFIELD, ADAMS & CO.
Totowa, New Jersey

1967 Edition
By LITTLEFIELD, ADAMS & CO.

Printed in the United States of America

PREFACE

Although in Britain, and in Europe generally, the term *ceramic* is restricted to the products of the pottery, electroceramic, refractories, clay building materials, and abrasives industries, the close relationship of other silicate industries—glass, vitreous enamel and cement—has for long been apparent; in the USA the unity of these industries is acknowledged and the term *ceramic* embraces them all. It seemed that a Dictionary of Ceramics should have this same wide scope. The ceramic technologist specializing in, say, pottery manufacture is more than likely to find that some papers on glass technology are required reading; the building brick manufacturer certainly needs to know something about cement; so, too, the glass technologist should have familiarity with the refractories used in his furnaces.

Mere definitions of terms, however, are not enough; physical and chemical data are often required and references to important original papers are sometimes needed. This dictionary aims to provide such information. The principal sources of the data have been the Mellor Memorial Library of the British Ceramic Research Association and the card-index of the Association's Information Department; I am especially glad to have this opportunity of thanking Dr. N. F. Astbury, Director of Research, for his permission to use these facilities. I also gladly acknowledge the help given by my colleagues at the British Ceramic Research Association and by friends in various branches of the ceramic industry, all of whom readily placed their expert knowledge at my disposal whenever asked. Their kindness did much to make the task of preparing this dictionary enjoyable.

It is also a pleasure to thank the publishers for the meticulous care with which their staff edited the manuscript; they detected omitted cross-references and a number of ambiguities of wording. If errors remain, the responsibility is mine; I shall be grateful to any readers who draw my attention to them so that corrections can be made should this book prove to be sufficiently useful to warrant a further edition.

Stoke-on-Trent, 1964   A. E. DODD.

DICTIONARY OF CERAMICS

A

'A' Glass. A fibre glass containing 10–15% alkali (calculated as Na_2O).

Abbe Number or Abbe Value. A measure of the optical dispersion of a glass proposed by the German physicist E. Abbe. Its usual symbol is ν and it is defined by $\nu = (n_D - 1)(n_F - n_C)^{-1}$ where n_C, n_D and n_F are the refractive indices at wavelengths equivalent to the spectral lines C, D and F. Also known as CONSTRINGENCE and as the ν-VALUE.

Aberson Machine. See SOFT-MUD PROCESS.

Ablation. The removal of a surface skin; the term has been applied to the volatilization of a special protective coating, of ceramic or other material, on space vehicles. The heat absorbed during the ablation process effectively dissipates the heat generated by re-entry of the space vehicle into the Earth's atmosphere.

Abrams' Law. States that the strength (S) of a fully compacted concrete is related to the ratio (R) of water to cement by the equation $S = A/B^R$, where A and B are constants. (D. A. Abrams, *Struct. Mat. Res. Lab. Lewis Inst.*, *Bull.* 1, Chicago, 1918.)

Abrasion. Wear caused by the mechanical action of a solid, e.g. wear of the stack lining of a blast furnace by the descending burden (cf. CORROSION and EROSION).

Abrasion Resistance. The ability of a material to resist mechanical rubbing. Some ceramic products are specially made to have this property, for example blue engineering bricks for the paving of coke chutes.

Abrasive. A hard material used in the form of a wheel or disk for cutting, or as a powder for polishing. The usual abrasives are silicon carbide, corundum, and sand. Abrasive wheels are commonly made by bonding abrasive grains with a ceramic frit and firing the shaped wheel; organic bonds are also used, in which case the wheels are not fired. For the standard marking system see under GRINDING WHEEL.

1

Absolute Temperature

Absolute Temperature (°Abs). See KELVIN TEMP. SCALE.

Absorption. See POROSITY; WATER ABSORPTION.

Absorption Ratio. See SATURATION COEFFICIENT.

Accelerator. A material that, when added to hydraulic cement or to plaster, accelerates the setting. With cement, care must be taken to distinguish between the effect on the initial set and the effect on the longer-term hydration—some additions accelerate setting but inhibit the final development of strength; also, a compound that acts as an accelerator at one level of concentration may become a retarder at some other concentration. In general, alkali carbonates are strong accelerators for portland cement. For plaster, the best accelerator is finely-ground gypsum, which provides nuclei for crystallization; potash alum has also been used.

Accessory Mineral. A mineral that is present in a rock or clay in relatively small quantity, usually as an impurity; e.g. mica is usually found as an accessory mineral in china clay.

Acid Annealing. Dipping of the metal base for vitreous enamelling into dilute HCl followed by annealing; this promotes scaling before the pickling process.

Acid Embossing. The etching of glass with HF or a fluoride.

Acid Frosting. The etching of glass hollow-ware with HF or a fluoride.

Acid Gold. A form of gold decoration for pottery introduced in 1863 by Mintons Ltd, Stoke-on-Trent. The glazed surface is etched with dilute HF prior to application of the gold; the process demands great skill and is used for the decoration only of ware of the highest class. A somewhat similar effect can be obtained by applying a pattern in low-melting flux on the glaze and gold-banding on the fluxed area; this is known as MOCK ACID-GOLD.

Acid Open-hearth Furnace. An OPEN-HEARTH FURNACE (q.v.) used in the refining of hematite iron; little such iron is now made. The particular feature is that the hearth is made of acid refractories—silica bricks covered with a fritted layer of silica sand.

Acid Refractory Material. General term for those types of refractory material that contain a high proportion of silica, e.g. silica refractories ($> 92\%$ SiO_2) and siliceous refractories (78–92% SiO_2). The name derives from the fact that silica behaves chemically as an acid and at high temperatures reacts with bases such as lime or alkalis.

Acid Resistance of Vitreous Enamelware. In USA, the acid resistance of vitreous enamelware at (nominal) room temperature is determined by exposing the enamelled surface to 10% citric acid for 15 min at 80°F. (ASTM – C282). Five classes of enamelware are distinguished according to their subsequent appearance:

AA: no visible stain and passes dry-rubbing test
A: passes blurring-highlight test and wet-rubbing test
B: passes blurring-highlight test; fails wet-rubbing test
C: fails blurring-highlight test; passes disappearing-highlight test
D: fails disappearing-highlight test.

Acid-resisting Brick. A fired clay brick of low water absorption and high resistance to a variety of chemicals. Properties are specified in ASTM – C279.

ACL Kiln or Lepol Kiln. ACL is a trade-mark of the Allis-Chalmers Mfg. Co., USA; Lepol is a trade-mark of Polysius Co., Germany. Both terms refer to a travelling-grate preheater for portland cement batch prior to its being fed to a rotary cement kiln; with this attachment, the length of a rotary cement kiln can be halved.

Actinic Green. An emerald green glass of the type used for poison bottles.

Adams Chromatic Value System. A method for the quantitative designation of colour in terms of (1) lightness, (2) amount of red or green, and (3) amount of yellow or blue. The system has been used in the examination of ceramic colours. (E. Q. Adams, *J. Opt. Soc. Amer.*, **32**, 168, 1942.)

Adams Process. A method for the removal of iron compounds from glass-making sands by washing with a warm solution of acid Na oxalate containing a small quantity of $FeSO_4$; see *J. Soc. Glass Tech.*, **19**, 118, 1935.

Adams–Williamson Annealing Schedule. A procedure, derived from first principles, for determining the optimum annealing conditions for a particular glass; (L. H. Adams and E. D. Williamson, *J. Franklin Inst.*, **190**, 597, 835, 1920).

Adapter. A connecting-piece, usually made of fireclay, between a horizontal zinc-retort and the condenser in which the molten zinc collects.

Adobe. A mud used in tropical countries for making sun-dried (unfired) bricks; the term is also applied to the bricks themselves.

Adsorption. Process of taking up a molecular or multi-molecular film of gas or vapour on a solid surface. This film is not readily removed except by strongly heating the body. The adsorption of moisture on the internal surface of a porous body may cause the body to expand slightly; this is known as MOISTURE EXPANSION (see also GAS ADSORPTION METHOD).

Aerated Concrete. Concrete with a high proportion of air spaces resulting from a foaming process; the bulk density may vary from about 35 to 90 lb/ft³. Aerated concrete is chiefly used for making pre-cast building units. It is also known as GAS CONCRETE, CELLULAR CONCRETE or FOAMED CONCRETE.

Aeration of Cement. The effect of the atmosphere on portland cement during storage. Dry air has no effect, but if it is exposed to moist air both moisture and carbon dioxide are absorbed with erratic effects on the setting behaviour (cf. AIR ENTRAINING).

Aeroclay. Clay, particularly china clay, that has been dried and air-separated to remove any coarse particles.

Aerograph. A device for spraying powdered glaze or colour on the surface of pottery by means of compressed air.

AFA Rammer. Apparatus designed by the American Foundrymen's Association for the preparation of test-pieces of foundry sand; it has also been applied as a method for the preparation of test-pieces of particulate refractory materials. The rammer operates by a 14-lb weight falling through a height of 2 in. on the plunger of a 2-in. dia. mould; normally, the weight is allowed to fall on the mould three times.

AFNOR. Prefix to specifications of the French Standards Association; Association Française de Normalisation, 23 Rue Notre-Dame-des-Victoires, Paris 2.

After Contraction. The permanent contraction (usually expressed as a linear percentage) that may occur if a fired or chemically-bonded refractory product is re-fired under specified conditions of test. Fireclay refractories are liable to show after-contraction if exposed to a temperature above that at which they were originally fired (cf. FIRING SHRINKAGE).

After Expansion. The permanent expansion (usually expressed as a linear percentage) that may occur when a refractory product that has been previously shaped and fired, or chemically bonded, is re-fired under specified conditions of test. Such expansion may take place, for example, if the product contains quartz or kyanite, or if bloating occurs during the test (cf. FIRING EXPANSION).

Afwillite. A hydrated calcium silicate, $3CaO , 2SiO_2 . 3H_2O$; it is formed when portland cement is hydrated under special conditions and when calcium silicate is autoclaved (as in sand-lime brick manufacture).

Agalmatolite. An aluminosilicate mineral that is very similar to PYROPHYLLITE (q.v.).

Agate. A cryptocrystalline variety of silica that has the appearance of a hard, opaque glass; it usually contains coloured Liesegang rings. It is used in the ceramic industry for burnishing gold decoration on pottery and for making small mortars and pestles for preparing samples for chemical analysis.

Agate Ware. Decorative earthenware made, chiefly in the 18th century in Staffordshire, by placing thin layers of differently coloured clays one on another, pressing them together, slicing across the layers and pressing the slab of variegated coloured clay into a mould. When fired, the ware had the appearance of natural agate.

Ageing. (1) A process, also known as SOURING (q.v.), in which moistened clay, or prepared body, is stored for a period to permit the water to become more uniformly dispersed, thus improving the plasticity. A similar effect can be achieved more quickly by TEMPERING (q.v.).

(2) The process of allowing vitreous-enamel slip to stand, after it has been milled, to improve its rheological properties.

Aggregate. See CONCRETE AGGREGATE; LIGHTWEIGHT EXPANDED CLAY AGGREGATE.

Aglite. Trade-name: a lightweight expanded clay aggregate made by the Butterley Co. Ltd, Derby, England, from colliery shale by the sinter-hearth process. The bulk density is: $\frac{1}{2}-\frac{3}{4}$ in., 31 lb/ft^3; $\frac{1}{2}-\frac{3}{16}$ in., 35 lb/ft^3; finer than $\frac{3}{16}$ in., 50 lb/ft^3.

Agricultural Pipes. See FIELD-DRAIN PIPES.

Air Bell. A fault, in the form of an irregularly shaped bubble, in pressed or moulded optical glass.

Air-borne Sealing. A process for the general, as opposed to local (cf. SPRAY WELDING), repair of a gas retort by blowing refractory powder into the sealed retort, while it is hot; the powder builds up within any cracks in the refractory brickwork and effectively seals them against gas leakage. (West's Gas Improvement Co. Ltd., Brit. Pat. 568 159; 21/3/45.)

Air Brick. A fired clay brick, of the same size as a standard building brick, but having lateral holes for the purpose of ventilation, e.g. below floors.

Air Entraining. The addition of a material to portland cement clinker during grinding, or to concrete during mixing, for the purpose of reducing the surface tension of the water so that 4–5% (by vol.) of minute air bubbles become trapped in the concrete. This improves workability and frost resistance and decreases segregation and bleeding. The agents used as additions include: 0·025–0·1% of alkali salts of wood resins, sulphonate detergents, alkali naphthenate, or triethanolamine salts; or 0·25–0·5% of the Ca salts of glues (from hides); or 0·25–1·0% of Ca lignosulphonate (from paper-making).

Air-floated. Material, e.g. china clay, that has been size-classified by means of an Air-separator (q.v.).

Air-hardening Refractory Cement or Mortar. See Chemically Bonded Refractory Cement.

Air Line. A fault, in the form of an elongated bubble, in glass tubing; also known as Hair Line.

Air Permeability. See Permeability.

Air Seal. A method for the prevention of the escape of warm gases from the entrance or exit of a continuous furnace, or tunnel kiln, by blowing air across the opening.

Air Separator. A machine for the size classification of fine ceramic powders, e.g. china clay; the velocity of an air current controls the size of particles classified.

Air-setting Refractory Cement or Mortar. See Chemically Bonded Refractory Cement.

Air-swept Ball Mill. See Ball Mill.

Air Twist. Twisted capillaries as a form of decoration within the stem of a wine glass.

Aired Ware. Pottery-ware that has a poor glaze as a result of volatilization of some of the glaze constituents. The term was used more particularly when ware was fired in saggars in coal-fired kilns, 'air' escaping from a faulty saggar into the kiln while kiln gases at the same time penetrated into the saggar. The term is also sometimes applied to a glaze that has partially devitrified as a result of cooling too slowly between 900 and 700°C.

Alabaster Glass. A milky-white glass resembling alabaster; transmitted light is diffused without any fiery colour.

Albany Clay. A clay found in the neighbourhood of Albany, New York. Because of its fine particle size and high flux content, this clay fuses at a comparatively low temperature to form a greenish-brown glaze suitable for use on stoneware and electrical porcelain.

Alberene Stone. A US stone having properties similar to those of Polyphant Stone (q.v.).

Albite. See Feldspar.

Alcove. A narrow, covered extension to the working end of a glass-tank furnace; it conveys molten glass to a forehearth or revolving pot.

Alginates. Organic binders and suspending agents derived from sea plants. Ammonium alginate and sodium alginate have found some use as additions to suspensions of glazes and enamels.

Ali Baba. Popular name for a large chemical stoneware jar of the type used for the bulk storage of acids; these jars are made in sizes up to 5000 litres capacity.

Alite. The name given to one of the crystalline constituents of portland cement clinker by A. E. Törnebohm (*Tonindustr. Ztg.*, **21**, 1148, 1897). Alite has since been identified as the mineral $3CaO . SiO_2$.

Alkali Neutralizer. See under Neutralizer.

All-basic Furnace. Abbreviation for All-Basic Open hearth Steel Furnace. The whole of the superstructure of such a furnace—hearth, walls, roof, ports, ends—is built of basic refractories. These furnaces were introduced in Europe in about 1935, the object being to make it possible to operate at a higher temperature than that possible with basic O.H. furnaces having a silica roof.

Alligator Hide. A vitreous-enamelware fault in the form of severe Orange Peel (q.v.) and Tearing (q.v.); causes include too-rapid drying and too-heavy application of enamel.

Allophane. An amorphous gel consisting, usually, of 35–40% Al_2O_3, 20–25% SiO_2 and 35–40% H_2O. It is often found in association with halloysite.

Allport Oven. A pottery Bottle Oven (q.v.) in which the hot gases from the firemouths enter the oven nearer to its centre than the usual points of entry around the oven walls; another feature is preheating of the secondary air. (S. J. Allport, Brit. Pat., 570 335; 3/7/45; 606 084; 5/8/48.)

Alluvial Clay. A brickmaking clay of river valleys typified in England by the clays of the Humber estuary, the Thames valley, and the Bridgwater district. The composition is variable.

Alsing Cylinder. A particular type of Ball Mill (q.v.).

Alumina. Aluminium oxide, Al_2O_3. This oxide exists in several forms: principally γ-Al_2O_3, stable up to about 1000°C and containing traces of water or of hydroxyl ions, and α-Al_2O_3, the

Alumina (Fused)

pure form obtained by calcination at high temperature. The so-called β-Al_2O_3 is the compound $Na_2O . 11Al_2O_3$. See also ALUMINA (FUSED); ALUMINA (SINTERED); CORUNDUM.

Alumina (Fused). In spite of its high m.p. (2050°C), alumina can be fused in an oxy-hydrogen flame or in an electric arc. By the former method, large single crystals (boules) can be produced; they are used as bearings, and as dies for wire-drawing, and for other purposes demanding high abrasion resistance. Fused alumina made in the electric arc furnace is usually crushed, bonded with fine alumina powder, shaped and then sintered.

Alumina Porcelain. Defined in ASTM – C242 as: A vitreous ceramic whiteware, for technical application, in which alumina is the essential crystalline phase.

Alumina (Sintered). Alumina, sometimes containing a small amount of clay or of a mineralizer, and fired at a high temperature to form a dense ceramic. Sintered alumina has great mechanical strength and abrasion resistance, high dielectric strength, and low power factor. Because of these properties, sintered alumina is used in thread guides, tool tips, and grinding media; as the ceramic component of sparking plugs, electronic tubes, ceramic-to-metal seals, etc. Sintered alumina coatings can be applied to metals by flame-spraying.

Alumina Whiteware. Defined in ASTM – C242 as: Any ceramic whiteware in which alumina is the essential crystalline phase.

Aluminium Borates. Two compounds have been reported: $9Al_2O_3 . 2B_2O_3$, which melts incongruently at 1950°C; $2Al_2O_3 . B_2O_3$, which dissociates at 1035°C. The compound $9Al_2O_3 . 2B_2O_3$, when made by electrofusion, has been proposed as a special refractory material: R.u.L., > 1500°C; thermal expansion, $4·2 \times 10^{-6}$ (20–1400°C); good thermal-shock resistance.

Aluminium Boride. The usual compound is AlB_2; this dissociates above 980°C to form AlB_{12} and Al.

Aluminium Carbide. Al_4C_3; the dissociation pressure at 1400°C is approx. 1×10^{-5} atm. This carbide is slowly attacked by water at room temperature and has not, as yet, received much attention as a special ceramic.

Aluminium Enamel. A vitreous enamel compounded for application to aluminium. There are three main types: containing lead, phosphate glass, or barium. Lithium compounds have also been used in these enamels.

Aluminium Nitride. AlN; m.p. 2200°C in 4 atm. N_2; dissociates

8

at 1700–1800°C in vacuo. The theoretical density is 3·26. When shaped into crucibles this special ceramic can be used as a container for molten aluminium. It is a good electrical insulator, the resistivity exceeding 10^{12} ohm. cm at 25°C. Modulus of rupture, 38 000 p.s.i. at 25°C and 18 000 p.s.i. at 1400°C.

Aluminium Phosphate. Generally refers to the orthophosphate, $AlPO_4$; m.p. approx. 2000°C but decomposes. It is supplied in various grades, particularly for use as a binder in the refractories industry. The bonding action has been attributed to reaction between the phosphate and basic oxides or Al_2O_3 in the refractory. This type of bond preserves its strength at intermediate temperatures (500–1000°C), a range in which organic bonds are destroyed and normal ceramic bonding has not yet begun.

Aluminium Titanate. $Al_2O_3 . TiO_2$; m.p. 1860°C. This titanate is of potential interest as a special ceramic, being notable for a very low thermal expansion which, however, is non-uniform; stabilization, giving a more uniform expansion, can be achieved by the addition of a small amount of $Fe_2O_3 . TiO_2$.

Alumino-silicate Refractory. A general term that includes all refractories of the fireclay, sillimanite, mullite, diaspore and bauxite types.

Aluminous Cement. See CIMENT FONDU.

Aluminous Fireclay Refractory. This type of refractory material is defined in B.S. 1902 as an alumino-silicate refractory containing 38–45% Al_2O_3.

Alundum. Trade-name: Sintered alumina made by Norton Co.

Alunite. $KAl_3(SO_4)_2(OH)_6$; occurs in Australia, USA (Arizona, Nevada, Utah), Mexico, Italy, Egypt, and elsewhere. The residue obtained after calcination and leaching out of the alkali can be used as raw material for high-alumina refractories.

Amakusa. The Japanese equivalent of CHINA STONE (q.v.).

Amberg Kaolin. A white-firing micaceous kaolin from Hirschau, Oberpfalz, Germany. A quoted analysis is (per cent): SiO_2, 48·0; Al_2O_3, 37·5; Fe_2O_3, 0·5; TiO_2, 0·2; CaO, 0·15; alkalis, 2·6; Loss-on-ignition, 12·2.

Ambetti or Ambitty. Decorative glass containing specks of opaque material; the effect is produced by allowing the glass to begin to crystallize.

Amblygonite. A lithium mineral of variable composition, $(Li,Na)AlPO_4(F,OH)$; sp. gr. 3·0; m.p. 1170°C. It occurs in various parts of southern Africa, but is relatively uncommon and

9

Amboy Clay

therefore finds less use as a source of lithium than do PETALITE and LEPIDOLITE (q.v.).

Amboy Clay. An American siliceous fireclay; it is plastic and has a P.C.E. above 32.

American Hotel China. A vitreous type of pottery-ware. A typical body composition is (per cent): kaolin, 35; ball clay, 7; feldspar, 22; quartz, 35; whiting, 1.

Ammonium Bifluoride. $NH_4F \cdot HF$; used as a saturated solution in hydrofluoric acid (usually with other additions) for the etching of glass.

Ammonium Vanadate. NH_4VO_3; used as a source of vanadium in ceramic pigments, e.g. tin-vanadium yellow, zirconium-vanadium yellow and turquoise, etc.

Ampoule. A small glass container that is sealed (after it has been filled with liquid) by fusing the narrow neck. The glass must resist attack by drugs and chemicals, but must be readily shaped by mass-production methods; normally, a borosilicate glass of fairly low alkali and lime contents is used.

Anaconda Process. A method for the shaping of silica refractories formerly used at some refractories works in USA. The bricks were first 'slop-moulded', then partially dried, and finally re-pressed. The name derives from the town of Anaconda, Montana, USA, where the process was first used, early in the present century, by the Amalgamated Copper Co.

Analysis. See DIFFERENTIAL THERMAL ANALYSIS; PARTICLE-SIZE ANALYSIS; RATIONAL ANALYSIS; ULTIMATE ANALYSIS.

Anatase. A tetragonal form of titania, TiO_2; the other forms are BROOKITE (q.v.) and RUTILE (q.v.). All three structures have similar Ti—O bond distance but differ in the linking of the TiO_6 octahedra. Anatase has been observed in some fireclays and kaolins. It is rapidly converted into rutile when heated above about 700°C.

Andalusite. A mineral having the same composition (Al_2SiO_5) as sillimanite and kyanite but with different physical properties. The principal source is S. Africa; it also occurs in California, Nevada, and New England (USA). When fired, it breaks down at 1350°C to form mullite and cristobalite; the change takes place without significant change in volume (cf. KYANITE). Andalusite finds some use as a refractory raw material.

Andreasen Pipette. An instrument used in the determination of the particle size of clays, by the SEDIMENTATION METHOD (q.v.);

10

it was introduced by A. H. M. Andreasen, of Copenhagen (*Kolloid Zts.*, **49**, 253, 1929).

Andrews' Elutriator. A device for particle-size analysis. It consists of: a feed vessel or tube; a large hydraulic classifier; an intermediate classifier; a graduated measuring vessel. (L. Andrews, Brit. Pat. 297 369; 14/6/27.)

Angle Bead. A special shape of wall tile (see Fig. 6, p. 307).

Angle of Nip. The angle between the two tangents to a pair of crushing rolls at the point of contact with the particle to be crushed. This angle is important in the design of crushing rolls for clay; it should not exceed about 18°.

Angle Tile. A purpose-made clay or concrete tile for use in an angle in vertical exterior tiling.

Anhydrite. $CaSO_4$; there are soluble and insoluble forms. Anhydrite begins to form when gypsum is heated at temperatures above about 200°C. The presence of anhydrite in plaster used for making pottery moulds can cause inconsistent blending times; the moulds tend to be soft, and to give slow casting, slow release of casts from the mould, and flabby casts.

Anionic Exchange. See IONIC EXCHANGE.

Anneal. To release stresses from glass by controlled heat treatment; the process is usually carried out in a LEHR (q.v.). See also ADAMS–WILLIAMSON ANNEALING SCHEDULE; REDSTON–STANWORTH ANNEALING SCHEDULE; TREBUCHON–KIEFFER ANNEALING SCHEDULE.

Annealing Point or Annealing Temperature. The temperature at which glass has a viscosity of 10^{13} poises; also known as the 13·0 TEMPERATURE. When annealed at this temperature glass becomes unstressed in a reasonable time (defined in USA as 15 min).

Annealing Range. The range of temperature in which stresses can be removed from a glass within a reasonable time, and below which rapid temperature changes do not cause permanent internal stress in the glass.

Annular Kiln. A large continuous kiln, rectangular in plan, of a type much used in the firing of building bricks. The bricks are set on the floor of the kiln and the zone of high temperature is made to travel round the kiln by progressively advancing the zone to which fuel is fed. There are two principal types: LONGITUDINAL-ARCH KILN (q.v.) and TRANSVERSE-ARCH KILN (q.v.).

Anode Pickling. See ELECTROLYTIC PICKLING.

Anorthite. See FELDSPAR.

Antimony Oxide

Antimony Oxide. Sb_2O_3; used as an opacifier in enamels and as a decolorizing and fining agent in glass manufacture; in pottery manufacture it is a constituent of Naples Yellow.

Antimony Yellow. See Lead Antimonate.

Antique Glass. Flat glass made by the Cylinder Process (q.v.) and with textured surfaces resembling old glass; it is used in the making of stained-glass windows (cf. Cathedral Glass).

Anti-static Tiles. Floor tiles of a type that will dissipate any electrostatic charge and so minimize the danger of sparking; such tiles are used in rooms, e.g. operating theatres, where there is flammable vapour. One such type of ceramic tile contains carbon. The National Fire Protection Association, USA, stipulates that the resistance of a conductive floor shall be less than 1 megohm as measured between two points 3 ft apart; the resistance of the floor shall be over 25 000 ohm between a ground connection and any point on the surface of the floor or between two points 3 ft apart on the surface of the floor.

A.P. US abbreviation for Annealing Point (q.v.).

Aplite. A rock mined in Virginia for use in glass manufacture; it consists principally of albite, zoisite and sericite.

Apparent Initial Softening. When applied to the Refractoriness-under-Load Test (q.v.), this term has the specific meaning of the temperature at which the tangent to the curve relating the expansion/contraction and the temperature departs from the horizontal and subsidence begins. (For typical curve see Fig. 5, p. 226.)

Apparent Porosity. See under Porosity.

Apparent Solid Density. A term used when considering the density of a porous material, e.g. a fireclay or silica refractory. It is defined as the ratio of the mass of the material to its Apparent Solid Volume (q.v.) (cf. True Density).

Apparent Solid Volume. A term used when considering the density and volume of a porous solid, particularly a refractory brick. It is defined as the volume of the solid material plus the volume of any sealed pores and also of the open pores.

Apparent Specific Gravity. For a porous ceramic, the ratio of the mass to the mass of a quantity of water that, at 4°C, has a volume equal to the Apparent Solid Volume (q.v.) of the material at the temperature of measurement.

Apron. See under Sand Seal.

Arabian Lustre. The original type of on-glaze lustre used by

12

the Moors from the 9th century onwards for the decoration of pottery; the sulphides or carbonates of copper and/or silver are used, the firing-on being in a reducing atmosphere so that an extremely thin layer of the metal is formed on the glaze.

Arbor. US term for the spindle of a grinding machine on which the abrasive wheel is mounted.

Arc-image Furnace. See IMAGE FURNACE.

Arc-spraying. See PLASMA SPRAYING.

Arch Brick. A building brick or refractory brick having the two large faces inclined towards each other. This may be done in two ways: in an END ARCH (q.v.) one of the end faces is smaller than the opposite end face; in a SIDE ARCH (q.v.) one of the side faces is smaller than the opposite side face. Arch bricks are used in the construction of culverts, furnace roofs, etc. (See Fig. 1, p. 37.)

Archless Kiln. Alternative name for SCOVE (q.v.).

Aridized Plaster. Plaster that has been treated, while being heated in the 'kettle', with a deliquescent salt, e.g. $CaCl_2$; it is claimed that this produces a strong plaster having more uniform properties.

Ark. A large vat used in the pottery industry for the mixing or storage of clay slip.

Arkose. A rock of the sandstone type but containing 10% or more of feldspar.

Armouring. Metal protection for the refractory brickwork at the top of the stack of a blast furnace; its purpose is to prevent abrasion of the refractories by the descending burden (i.e. the raw materials charged to the furnace).

Arris. The sharp edge of a building brick or ridge-tile; an ARRIS TILE is a specially shaped tile for use in the ridge or hip of a roof. The ARRIS EDGE on glass is a bevel up to $\frac{1}{16}$ in. wide and at an angle of 45°.

Arsenic Oxide. As_2O_3; sublimes at 193°C; sp. gr. 3·87. Used as a fining and decolorizing agent in glass manufacture.

ASA. Prefix to specifications of the American Standards Association, 70 East 45th Street, New York 17.

Asbestine. A fibrous mineral consisting of mixed hydrated silicates of Mg and Ca; it occurs in New York State.

Asbestos. A group of fibrous minerals, the most important of which is chrysotile, which represents 95% of the worlds' output of asbestos; 60% of this is from Canada, other major sources being

Asbolite

Russia and S. Africa. The formula of pure chrysotile is $3MgO$. $2SiO_2 . 2H_2O$; some of the Mg may be replaced by Fe. Asbestos loses water during prolonged heating above 400°C; at 500°C it rapidly loses strength.

Asbolite. Impure cobalt ore used by the old Chinese potters to produce underglaze blue.

Ashfield Clay. A fireclay associated with the Better Bed coal, Yorkshire, England. The raw clay contains about 57% SiO_2, 27% Al_2O_3, 1·7% Fe_2O_3 and 1·5% alkalis.

ASTM. American Society for Testing and Materials. This Society is responsible for most of the American standard specifications for ceramics. Address: 1916 Race St., Philadeplhia 3, Pa., USA.

Atritor. Trade-name: a machine that simultaneously dries and pulverizes raw clay containing up to 18% moisture; it consists of a feeder, metal separator, pulverizer, and fan. (Alfred Herbert Ltd., Coventry, England.)

Attaclay. See ATTAPULGITE.

Attapulgite. One of the less common clay minerals; it was first found at Attapulgus, USA, and is characterized by a high (10%) MgO content. It has a large surface area and is used in oil refining, as a carrier for insecticides, etc.

Atterberg Test. A method for determining the plasticity of clay in terms of the difference between the water content when the clay is just coherent and when it begins to flow as a liquid. The test was first proposed by A. Atterberg (*Tonindustr. Ztg.*, **35**, 1460, 1911).

Attrition. Particle-size reduction by a process depending mainly on impact and/or a rubbing action.

Attrition Mill. A disintegrator depending chiefly on impact to reduce the particle size of the charge. Attrition mills are sometimes used in the clay building materials industry to deal with the tailings from the edge-runner mill.

Aubergine Purple. A ceramic colour, containing Mn, introduced in the 18th century, when it was used for underglaze decoration.

Auger. (1) An extruder for clay, or clay body, the column being forced through the die by rotation of a continuous screw on a central shaft.

(2) A large 'corkscrew' operated by hand or by machine to take samples of soft material, e.g. clay, during prospecting.

14

Autoclave. A strong, sealed, metal container, fitted with a pressure gauge and safety valve; water introduced into the container can be boiled under pressure, the gauge pressure developed at various temperatures being as follows:

Gauge pressure (p.s.i.):	0	25	50	100	150
Temperature (°C):		100	130	148	170	186

A common method of testing glazed ceramic ware for crazing resistance is to expose pieces of ware to the steam in an autoclave, usually at 50 p.s.i.; the method is unsatisfactory because it confuses the effects of moisture expansion (the prime cause of crazing) and thermal shock. An autoclave test is also sometimes used in the testing of vitreous enamelware. As applied to the testing of portland cement, the autoclave test reveals unsoundness resulting from the presence of any MgO or CaO.

Autocombustion System. An electronically controlled impulse system for the oil firing (from the top or side) of ceramic kilns.

Aventurine. A decorative effect achieved on the surface of pottery or glass by the formation of small, bright, coloured crystals (generally hematite, Fe_2O_3, on pottery bodies, but chrome or copper on glassware).

Ayrshire Bauxitic Clay. A non-plastic fireclay formed by laterization of basalt lava and occurring in the Millstone Grit of Ayrshire, Scotland; there are two types, the one formed *in situ*, the other being a sedimentary deposit. Chemical analysis (per cent) (raw): SiO_2 42; Al_2O_3 38; TiO_2, 3–4; Fe_2O_3, 0·5; alkalis, 0·2.

B

B 25 Block. A hollow clay building block, $25 \times 30 \times 13 \cdot 5$ cm, introduced by the Swiss brick industry in 1953. It is economical to lay (22 blocks per m²), giving a wall 25 cm thick with a U-value of 0·17.

Babosil. Trade-name: a frit for pottery glazes, so-named because it contains barium, boron and silica. The composition is:

$$\left.\begin{array}{l} 0 \cdot 06 \ K_2O \\ 0 \cdot 50 \ Na_2O \\ 0 \cdot 43 \ BaO \end{array}\right\} \quad \begin{array}{l} 0 \cdot 125 \ Al_2O_3 \\ 0 \cdot 68 \ B_2O_3 \end{array} \left\{ \quad 2 \cdot 45 \ SiO_2 \right.$$

Back Ring. See HOLDING RING.

Back Stamp. The maker's name and/or trademark stamped on the back of pottery flatware or under the foot of hollow-ware.

Backing

Backing. The common brickwork behind facing bricks in a brick building, or low-duty refractories or heat insulation behind the working face of a furnace.

Backing Strip. A strip of metal welded to the back of a metal panel prior to its being enamelled; the purpose is to prevent warping.

Bacor; Bakor. A Russian corundum–zirconia refractory for use more particularly in the glass industry; the name is derived from BADDELEYITE (q.v.) and CORUNDUM (q.v.). There are various grades, e.g. BAKOR-20 (62% Al_2O_3, 18% ZrO_2, 16% SiO_2) and BAKOR-33 (50% Al_2O_3, 30% ZrO_2, 15% SiO_2).

Baddeleyite. Natural zirconia, ZrO_2; the only deposits of economic importance are those in Brazil and Russia.

Badging. The application, usually by transfer or silk-screen, of crests, trade marks, etc. to pottery or glass-ware (cf. BACK STAMP).

Baffle Mark. A fault that may occur on a glass bottle as a mark or seam caused by the joint between the blank mould and the baffle plate.

Baffle Wall. See SHADOW WALL.

Bag Wall. A wall of refractory brickwork built inside a down-draught kiln around a firebox, each firebox having its own bag wall. The purpose is to direct the hot gases towards the roof of the kiln and to prevent the flames from impinging directly on the setting (cf. FLASH WALL).

Bait. (1) A tool used in the glass industry; it is dipped into the molten glass in a tank furnace to start the drawing process.

(2) To shovel solid fuel into the firebox of a kiln.

Bakor. See BACOR.

Balance; Balancing. A grinding wheel is said to be in STATIC BALANCE if it will remain at rest, in any position, when it is mounted on a frictionless, horizontal spindle; such a wheel is also in DYNAMIC BALANCE if, when rotated, there is no whip or vibration. The process of testing for balance, and the making of any necessary adjustment to the wheel, is known as BALANCING.

Ball Clay. A sedimentary kaolinitic clay that fires to a white colour and which, because of its very fine particle size, is highly plastic. The name is derived from the original method of winning in which the clay was cut into balls, each weighing 30–35 lb. As dug, ball clays are often blue or black owing to their high content of carbonaceous matter. In England, ball clays occur in Devon and

16

Dorset; in USA, they are found in Kentucky and Tennessee. Ball clays are incorporated in ceramic bodies to give them plasticity during shaping and to induce vitrification during firing. In addition to their use in most pottery bodies and as suspending agents for vitreous enamels, ball clays are used to bond non-plastic refractories such as sillimanite. Current consumption is about 400 000 tons in USA and 100 000 tons in Britain.

Ball Mill. A fine-grinding unit of a type much used in the pottery, vitreous enamel, and refractories industries. It consists of a steel cylinder or truncated cone that can be rotated about its horizontal axis (see CRITICAL SPEED); the material to be ground is charged to the mill together with alloy-steel balls, pebbles, or specially made ceramic grinding media (generally high-alumina). The mill may be operated dry or wet; wet ball-milling is usually a batch process but dry ball-milling may be continuous, the 'fines' being removed by an air current (AIR-SWEPT BALL MILL).

Ball Test. See KELLY BALL TEST.

Balling. (1) The tendency of some ceramic materials or batches, when moist, to aggregate into small balls when being mixed in a machine.

(2) One method of shaping pottery hollow-ware in a JOLLEY (q.v.); a ball of the prepared body is placed in the bottom of the plaster mould and is then 'run up' the sides of the mould as it rotates.

Ballotini. Transparent glass spheres less than about 1·5 mm diameter; presumably a derivative of the Italian *ballotta*, a small ball used for balloting.

Bamboo Ware. A bamboo-like type of CANE WARE (q.v.), somewhat dark in colour, first made by Josiah Wedgwood in 1770.

Banding. The application, by hand or by machine, of a band of colour to the edge of a plate or cup.

Bank Kiln. A primitive type of pottery kiln used in the Far East; it is built on a bank, or slope, which serves as a chimney.

Banks. The sloping parts between the hearth of an open-hearth steel furnace and the back and front walls. They are constructed of refractory bricks covered with fritted sand (Acid O.H. Furnace) or burned-in magnesite or dolomite (Basic O.H. Furnace) cf. BREASTS.

Bannering. Truing the rim of a saggar (before it is fired) by means of flat metal or a wooden board, to ensure that the rim lies in one horizontal plane and will in consequence carry the load of superimposed saggars uniformly.

Baraboo Quartzite. A quartzite of the Devil's Lake region of Wisconsin, USA, used in silica brick manufacture. A quoted analysis is (per cent): SiO_2, 98·2; Al_2O_3, 1·1; Fe_2O_3, 0·2; $Na_2O + K_2O$, 0·1.

Barelattograph. A French instrument for the automatic recording of the contraction and loss in weight of a clay body during drying under controlled conditions; (for description see *Bull. Soc. Franc. Céram.*, No. 40, 67, 1958).

Barium Boride. BaB_6; m.p., 2270°C; sp. gr., 4·32; thermal expansion, $6·5 \times 10^{-6}$; electrical resistivity (20°C), 306 μohm.cm.

Barium Carbonate. $BaCO_3$; decomposes at 1450°C; sp. gr. 4·4. Occurs naturally in Co. Durham as witherite. The principal use in the ceramic industry of the raw material is for the prevention of efflorescence on brickwork; for this purpose it is added to brick-clays containing soluble sulphates. The pure material is used in the manufacture of barium ferrite permanent magnets, and in some ceramic dielectrics to give lower dielectric loss. In the glass industry this compound is used in optical glass and television tubes; it is also used in some enamel batches.

Barium Chloride. $BaCl_2$, a soluble compound sometimes used to replace $CaCl_2$ as a mill-addition in the manufacture of acid-resisting vitreous enamels. It also helps to prevent scumming.

Barium Oxide. BaO; m.p. 1920°C; sp. gr. 5·72.

Barium Stannate. $BaSnO_3$; it is sometimes added to barium titanate bodies to decrease the Curie point. An electroceramic containing approx. 90 mol-% $BaTiO_3$ and 10 mol-% $BaSnO_3$ has a very high dielectric constant (8000–12 000 at 20°C).

Barium Sulphate. $BaSO_4$; m.p. 1580°C; sp. gr. 4·45.

Barium Titanate. $BaTiO_3$; m.p. 1618°C. Made by heating a mixture of barium carbonate and titania at 1300°–1350°C. Because of its high dielectric constant (1350–1600 at 1 Mc and 25°C) and its piezoelectric and ferroelectric properties, it finds use in electronic components; its Curie temperature is 120°C. The properties can be altered by the formation of solid solutions with other ferroelectrics having the perovskite structure, or by varying the $BaO:TiO_2$ ratio on one or other side of the stoichiometric.

Barium Zirconate. $BaZrO_3$; m.p. 2620°C; sp. gr. 2·63. Synthesized from barium carbonate and zirconia and used as an addition (generally 8–10%) to barium titanate bodies to obtain high dielectric constant (3000–7000) and other special properties.

Barker–Truog Process. A process described by G. J. Barker and E. Truog (*J. Amer. Ceram. Soc.*, **21**, 324, 1938; **22**, 308, 1939; **24**, 317, 1941) for the treatment of brickmaking clays with alkali, this being claimed to facilitate shaping and to reduce the amount of water necessary to give optimum plasticity. According to their patent (US Pat. 2 247 467) the clay is mixed with alkali to give pH 7–9 if it was originally acid, or pH 8–10 if originally non-acid; it is also stipulated that the total amount of alkali added shall be limited by the slope of the curve relating the pH to the quantity of alkali added, this slope being reduced to half its original value.

Barratt–Halsall Firemouth. A design for a stoker-fired firemouth for a pottery BOTTLE-OVEN (q.v.); a subsidiary flue system links all the firemouths around the oven wall to assist in temperature equalization. The design was patented by W. G. Barratt and J. Y. M. Halsall (Brit. Pat. 566 838; 16/1/45). Also known as the GATER HALL DEVICE because it was first used at the factory of Gater, Hall & Co., Stoke-on-Trent, where J. Y. M. Halsall was General Manager and which was at the time associated with the Barratt pottery.

Barrelling. The removal of surface excrescences and the general cleaning of metal castings by placing them in a revolving drum, or barrel, together with coarsely crushed abrasive material such as broken biscuit-fired ceramic ware (cf. RUMBLING).

Barton Clay. A clay of the Eocene period used for brickmaking near the coast of Hampshire and in the Isle of Wight.

Barytes. Naturally occurring BARIUM SULPHATE (q.v.) It is the principal source of BARIUM CARBONATE (q.v.) but is also used to some extent in the natural state in glass manufacture.

Basalt Ware. A type of ceramic artware introduced in 1768 by Josiah Wedgwood and still made by the firm that he founded. The body is black and vitreous, iron oxide and manganese dioxide being added to achieve this; a quoted composition is 47% ball clay, 3% china clay, 40% ironstone and 10% MnO_2. The ware has something of the appearance of polished basalt rock (cf. FUSION-CAST BASALT).

Base Code. Alternative name for PUNT CODE (q.v.).

Base Exchange. See IONIC EXCHANGE.

Base Metal. (1) In the phrase 'base-metal thermocouple', this

19

Basic Fibre

term signifies such metals and alloys as copper, constantan, nickel, tungsten, etc.

(2) In the vitreous-enamel industry, the term means the metal (steel or cast-iron) to which the enamel is applied.

Basic Fibre. Glass fibres before they have been processed. A number of fibres may subsequently be bonded together to form a strand.

Basic Open-hearth Furnace. An OPEN-HEARTH FURNACE (q.v.) used in the refining of basic pig-iron. The hearth is built of basic refractory bricks covered with burned dolomite or magnesite.

Basic Refractory. A general term for those types of refractory material that contain a high proportion of MgO and/or CaO, i.e. oxides that at high temperatures behave chemically as bases. The term includes refractories such as magnesite, chrome–magnesite, dolomite, etc.

Basse-taille. Vitreous enamelled artware in which a pattern is first cut in low relief on the metal backing, usually silver; the hollows are then filled with translucent enamel, which is subsequently fired on.

Basset Process. For the simultaneous production of hydraulic cement and pig-iron by the treatment, in a rotary kiln, of a mixture of limestone, coke, and iron ore. (L. P. Basset, French Pat. 766 970; 7/7/34: 814 902; 2/7/37.)

Bastard Ganister. A silica rock having many of the superficial characters of a true GANISTER (q.v.), such as colour and the impression of rootlets, but differing from it in essential details, e.g. an increased proportion of interstitial matter, variable texture, and incomplete secondary silicification.

Bat. (1) A refractory tile or slab as used, for example, to support pottery-ware while it is being fired.

(2) A roughly shaped disk of pottery body—as prepared on a batting-out machine prior to the jiggering of flatware, for example.

(3) A short building brick, either made as such or cut from a whole brick.

Bat Printing. A former method of decorating pottery; it was first used, in Stoke-on-Trent, by W. Baddeley in 1777. A bat of solid glue or gelatine was used to transfer the pattern, in oil, from an engraved copper plate to the glazed ware, colour then being dusted on. The process was still in use in 1890 and has now been developed into the MURRAY–CURVEX MACHINE (q.v.).

Batch. A proportioned mixture of materials. The term is used in most branches of the ceramic industry; in glass-making, the batch is the mixture of materials charged into the furnace.

Batch-type Mixer. See MIXER.

Batt. See BAT.

Batter. In brickwork, the name given to an inclined wall surface or to the angle of such a surface to the vertical.

Battery. Term applied to the large refractory structure in which coke is produced in a series of adjacent ovens (cf. BENCH).

Batting Out. The process of making a disk of prepared pottery body for subsequent shaping in a JIGGER (q.v.).

Battledore. A tool used in the hand-made glass industry for shaping the foot of a wine glass; also known as a PALLETTE.

Bauxite. A sedimentary rock consisting of hydrated alumina together with impurities such as clay, iron oxide, and titania. The principal sources are Jamaica, British Guiana, USSR, Surinam, France, Hungary, and Yugoslavia. It is the raw material for the production of calcined and sintered alumina as used in the abrasives, electroceramic, refractories, and other branches of the ceramic industry.

Bauxitland Cement. See KÜHL CEMENT.

Bayer Process. A process for the extraction of alumina from bauxite, invented by K. J. Bayer in 1888. The bauxite is digested with hot NaOH and the Al_2O_3 is then extracted as the soluble aluminate. Because of this method of extraction, the calcined alumina used in the ceramic industry usually contains small quantities of Na_2O.

Bead. (1) Any raised section round glass-ware.

(2) A small piece of glass or ceramic tubing surrounding a wire.

Beading. (1) The application of vitreous enamel slip, usually coloured, to the edge of enamelware.

(2) Removal of excess vitreous enamel slip from the edges of dipped ware.

Bear. The mass of iron which, as a result of wear of the refractory brickwork or blocks in the hearth bottom of a blast furnace, slowly replaces much of the refractory material in this location. Also known as the SALAMANDER.

Bearer Arch. See RIDER BRICKS.

Beaumé Degrees (°Bé). A system, introduced by a Frenchman, A. Beaumé, in 1768 for designating the specific gravity of liquids. There are two scales, one for liquids lighter than water and one for heavier liquids; only the latter is normally of interest in the ceramic industry, where it is sometimes used in describing the strength of sodium silicate solutions. The precise equivalence between the Beaumé scale and the specific gravity is in some doubt but the following conversion factor has been standardized in USA and may be generally accepted:

$$\text{sp. gr.} = 145/(145 - {}^{\circ}\text{Bé})$$

Bed. Alternative name for a SETTING (q.v.) of gas retorts.

Bedder. A plaster shape conforming to that of the ware to be fired and used to form the bed of powdered alumina on which bone china is fired.

Bedding. The placing of pottery flatware on a bed of refractory powder, e.g. silica sand or calcined alumina, to provide uniform support during the firing process, thus preventing warpage.

Beehive Kiln. See ROUND KILN.

Beidellite. A clay mineral first found at Beidell, Colorado, USA; it is an Al-rich member of the montmorillonite group.

Belgian Kiln. A type of annular kiln patented by a Belgian, D. Enghiens (Brit. Pats. 18 281; 1891: 22 008; 1894: 7 914; 1895). It is a longitudinal-arch kiln with grates at regular intervals in the kiln bottom; it is side-fired on to the grates. Such kilns have been popular for the firing of fireclay refractories at 1200–1300°C.

Belite. The name given to one of the crystalline constituents of portland cement clinker by A. E. Törnebohm (*Tonindustr. Ztg.*, **21**, 1148, 1897). This mineral has since been identified as one form of $2\text{CaO}.\text{SiO}_2$.

Bell. See TRUMPET.

Bell Damper. A damper of the sand-seal type, and bell-shaped. Such dampers are used, for example, in annular kilns.

Belleek Ware. A distinctive type of pottery made at Belleek, County Fermanagh, Northern Ireland. The factory was established in 1857 and the ware is characterized by its thinness and slightly iridescent surface; the body contains a significant proportion of frit.

Belly. (1) The part of a blast furnace where the cross-section is greatest, between the STACK (q.v.) and the BOSH (q.v.); this part of a blast furnace is also known as the WAIST.

(2) The part of a converter in which the refined steel collects when the converter is tilted before the steel is poured.

Belshazzar. A 16-quart wine bottle.

Belt Kiln. A tunnel kiln through which ware is carried on an endless belt made of a wire-mesh woven from heat-resisting alloy. In the pottery industry such kilns have found some use for glost- and decorating-firing.

Bench. (1) A series of chamber ovens or gas retorts built of refractory bricks to form a continuous structure.

(2) The floor of a pot furnace for making glass.

Bend Test. See MODULUS OF RUPTURE.

Beneficiation. US term for any process, or combination of processes, for increasing the concentration of a mineral in an ore; the term is now being used in the UK but the older term 'mineral dressing' is preferred. An example of beneficiation in the ceramic industry is the treatment of certain pegmatites to produce a SPODUMENE (q.v.) concentrate.

Bentonite. A natural clay formed by the weathering of volcanic ash and consisting chiefly of MONTMORILLONITE (q.v.). It occurs as two varieties—sodium bentonite and calcium bentonite; the former swells very considerably when it takes up water. The principal source is USA: Western bentonite (the Na variety from the Wyoming area) and Southern bentonite (the Ca variety from the Mississippi area). The particle size of bentonite is all less than 0.5μ; it has remarkable bonding properties and is used as a binder for non-plastics, particularly for foundry sands.

Berg Method. See DIVER METHOD.

Berkeley Clay. A plastic, refractory kaolin from S. Carolina; P.C.E.34.

Berl Saddle. A chemical stoneware shape for the packing of absorption towers; it is saddle-shaped, size about 1 in., and approx. 2000 pieces are required per ft^3. A packing of Berl Saddles provides a very large contact surface and is a most efficient type of ceramic packing. (Patented in Germany by E. Berl in 1928.)

Berlin Porcelain. A German laboratory porcelain, particularly that made at the Berlin State Porcelain Factory. A quoted body composition is 77% purified Halle clay and 23% Norwegian feldspar, all finer than 10μ. The ware is fired at 1000°C and is then glazed and refired at 1400°C.

Berry Machine. See SOFT-MUD PROCESS.

Beryl. $3BeO.Al_2O_3.6SiO_2$; this mineral is the only economic source of beryllium oxide and other beryllium compounds. It occurs in Brazil, South Africa, India, and Madagascar. Beryl has a low, but anisotropic, thermal expansion.

Beryllia. See BERYLLIUM OXIDE.

Beryllides. A group of intermetallic compounds of potential interest as special ceramics. Cell dimensions and types of structure have been reported for the beryllides of Ti, V, Cr, Zr, Nb, Mo, Hf and Ta (A. Zalkin *et al.*, *Acta Cryst.*, **14**, 63, 1961).

Beryllium Carbide. Be_2C; decomposes at 2150°C with volatilization of Be; sp. gr. 2·44; Knoop hardness (25-g load) 2740. This refractory carbide is slowly attacked by water at room temperature and by O_2 and N_2 at high temperature. It has been used as a moderator in nuclear engineering.

Beryllium Nitride. Be_3N_2; m.p. 2200°C but oxidizes in air when heated to above 600°C. There are two forms: cubic and hexagonal.

Beryllium Oxide or **Beryllia.** BeO; m.p. 2570°C. A special refractory oxide notable for its abnormally high thermal conductivity (125 Btu/h ft^2 deg F ft at 60°C) and mechanical strength (crushing strength 225 000 p.s.i., transverse strength 35 000 p.s.i., tensile strength 14 000 p.s.i.—all at 20°C). Thermal expansion (20–1000°C), $8·8 \times 10^{-6}$. Thermal-shock resistance is good. It acts as a moderator for fast neutrons and is used for this purpose in nuclear reactors; it is also sometimes used as a constituent of special porcelains. Instruments have been mounted on BeO blocks in space-craft. Beryllium compounds are toxic.

Bessemer Converter. See CONVERTER.

Best Gold. See BURNISH GOLD.

Bestowing. The cover of fired bricks (usually three courses) for the setting of a CLAMP (q.v.).

B.E.T. Method. See BRUNAUER, EMMETT and TELLER METHOD.

Better Bed Fireclay. A siliceous fireclay occurring under the Better Bed Coal of the Leeds area, England.

Bevel Brick. A brick having one edge replaced by a bevel (see Fig. 1 p. 37).

Bicheroux Process. A method developed in 1918 by Max Bicheroux, of Aachen, for the production of plate glass; molten glass from a pot is cast between rollers.

Bidet. An item of ceramic sanitaryware designed to facilitate personal hygiene; bidets are used more particularly in France, Spain and South America. (French word.)

Binder. A substance added to a ceramic raw material of low plasticity to facilitate its shaping and to give the shaped ware sufficient strength to be handled; materials commonly used for this purpose include sulphite lye, sodium silicate, molasses, dextrin, starch, gelatine, etc. The term BOND (q.v.) is also sometimes used in this sense but is better reserved to denote the inter-granular material that gives strength to the fired ware.

Bingham Plastometer. A device for the measurement of the rheological properties of clay slips by forcing the slip through a capillary under various pressures; a curve is drawn relating the rate of flow to the pressure. E. C. Bingham, *Proc. A.S.T.M.*, **19**, Pt. 2, 1919; **20**, Pt. 2, 1920.)

Biot Number. The heat-transfer ratio hr/k, where h is the heat-transfer coefficient, r is the distance from the point or plane under consideration to the surface, and k is the thermal conductivity. The Biot Number is a useful criterion in assessing thermal-shock resistance. (M. A. Biot, *Phil. Mag.*, **19**, 540, 1935.)

Biotite. A mica containing appreciable amounts of iron and magnesium; it sometimes occurs as an impurity in such ceramic raw materials as feldspar and nepheline syenite.

Bird's Beak. A special type of wall tile (see Fig. 6, p. 307).

Birdcage. An imperfection occasionally occurring in bottle manufacture, a glass thread (or threads) spanning the inside of the bottle.

Biscuit Firing. The process of kiln firing potteryware before it has been glazed. Earthenware is biscuit-fired at 1100–1150°C; bone china is biscuit-fired at 1200–1250°C.

Biscuit Ware. Pottery that has been fired but not yet glazed. Biscuit earthenware is porous and readily absorbs water; vitreous ware and bone china are almost non-porous even in the biscuit state.

Bismuth Stannate. $Bi_2(SnO_3)_3$; sometimes added to barium titanate bodies to modify their dielectric properties; particularly to produce bodies having an intermediate dielectric constant (1000–1250) with a negligible temperature coefficient.

Bismuth Telluride. Bi_2Te_3; m.p. 585°C. A semiconductor that has found some use as a thermoelectric material.

Bisque. (1) In the vitreous-enamel industry, the dried (but still unfired) coating of enamel slip.

(2) In the pottery industry, the older form of the term BISCUIT (q.v.).

Bit Gatherer. The man whose job is to gather small quantities of glass for use in decorating hand-blown glass-ware.

Bitstone. Broken pitchers or calcined flints (in either case crushed to about $\frac{1}{8}$ in.) formerly used in the bottom of a saggar for glost firing to prevent sticking.

Black Ash. Crude barium sulphide; it has been used as a substitute for barium carbonate as an additive to clay to prevent scumming.

Black Body. As applied to heat radiation, this term signifies that the surface in question emits radiant energy at each wavelength at the maximum rate possible for the temperature of the surface, and at the same time absorbs all incident radiation. Only when a surface is a Black Body can its temperature be measured accurately by means of an optical pyrometer.

Black Core or **Black Heart.** Most fireclays and brick-clays contain carbonaceous matter; if a brick shaped from such clays is fired too rapidly, this carbonaceous matter will not be burned out before vitrification begins. The presence of carbon and the consequent reduced state of any iron compounds in the centre of the fired brick result in a 'Black Core' or 'Black Heart'.

Black Edging. A black vitreous enamel applied over the GROUND-COAT (q.v.) and subsequently exposed, as an edging, by brushing away the COVER-COAT (q.v.) before the ware is fired.

Black Speck. A fault, in glass, particularly in the form of a small inclusion of chrome ore. On pottery-ware the fault is generally caused by small particles of iron or its compounds. In vitreous enamelware also, the fault is caused by contamination.

Blaes. A Scottish name for carbonaceous shales, of a blue-grey colour, associated in the Lothians with oil-shales but differing from these in having a much lower proportion of bituminous matter, in being brittle rather than tough, and in producing when weathered a crumbling mass which, when wetted, is plastic.

Blaine Test. A method for the evaluation of the fineness of a powder on the basis of the permeability to air of a compact prepared under specified conditions. The method was proposed by R. L. Blaine (*Tech. News Bull.*, No. 285, 1941) and is chiefly used in testing the fineness of portland cement.

Blake-type Jaw Crusher. A jaw crusher with one jaw fixed, the other jaw being pivoted at the top and oscillating at the bottom; the output is high but the product is not of uniform size (cf. DODGE-TYPE JAW CRUSHER).

Blakely Test. See TUNING-FORK TEST.

Blank. In the vitreous-enamel industry, a piece cut from metal sheet ready for the shaping of ware. For other meanings see CLOT; OPTICAL BLANK.

Blank Mould. The metal mould in which the PARISON (q.v.) is shaped in glass hollow-ware production.

Blanket Feeding. The charging of batch (to a glass tank furnace) as a broad, thin layer.

Blast Furnace. A shaft furnace, 100–120 ft high, for the extraction of iron from its ore; a smaller type of blast furnace is used for the extraction of lead. In iron-making, ore, limestone and coke are charged at the top and hot air is blasted (hence the name) in near the bottom. The whole furnace is lined with refractory material: fireclay refractories in the STACK (q.v.), aluminous fireclay or carbon refractories in the BOSH (q.v.) and HEARTH (q.v.).

Bleaching Clay. A clay, usually of the bentonite or fuller's earth type, used for decolorizing petroleum products etc.; the liquid is allowed to percolate through a layer of the clay, which adsorbs the colouring matter on its surface.

Bleb. A raised blister on the surface of faulty pottery-ware.

Bleeding or Water Gain. The appearance of water at the surface of freshly placed concrete; it is caused by sedimentation of the solid particles, and can to some extent be prevented by the addition of plasticizers and/or by AIR ENTRAINING (q.v.). See also LAITANCE.

Bleu Persan. A form of pottery decoration in which a white pattern was painted over a dark blue background; the name derives from the fact that the pattern generally had a Persian flavour.

Blibe. A fault, in glass-ware, in the form of an elongated bubble intermediate in size between a seed and a blister. GREY BLIBE consists of undissolved sodium sulphate.

Blinding. (1) The clogging of a sieve or screen.

(2) US term for a glaze fault revealed by a reduction in gloss, and caused by surface devitrification.

Blister. A large bubble sometimes present as a fault in ceramic ware. In glass-ware, if near the surface it is a SKIN BLISTER and if on the inside surface of blown glass-ware it is a PIPE BLISTER (q.v.). The common causes of blisters in vitreous enamelware are flaws in the base-metal, surface contamination of the base-metal, and too high a moisture content in the atmosphere of the enamelling furnace.

Bloach (US). A blemish on plate glass resulting from stopping the grinding before all the hollows have been removed; a 'bloach' is an area of the original rough surface.

Bloating. The permanent expansion exhibited by some clays and bricks when heated within their vitrification range; it is caused by the formation, in the vitrifiying clay, of gas bubbles resulting either from entrapped air or from the breakdown of sulphides or other impurities in the clay (cf. EXFOLIATION and INTUMESCENCE).

Block Handle. A cup handle of the type that is attached to the cup by a solid bar of clay (which is, of course, integral with the handle) (cf. OPEN HANDLE).

Block Mill. See PAN MILL.

Block Rake. A surface blemish, having the appearance of a chain, sometimes occurring on plate glass.

Blocking. A glass-making term having several meanings:

(1) The shaping of glass in a wooden or metal mould;

(2) POLING (q.v.) with a block of wood;

(3) the removal of surface blemishes from glass-ware by re-processing;

(4) the setting of optical glass blanks in a carrier prior to grinding and polishing;

(5) an additional US meaning is the running of a glass-furnace idle at a reduced temperature.

Bloom. (1) Surface treatment of glass (e.g. lenses) by vapour deposition; this decreases reflection at the air/glass interface.

(2) A surface film on glass caused by weathering or by the formation of sulphur compounds during annealing.

Blotter. US term for a disk of compressible material, e.g. blotting-paper stock, for use between a grinding wheel and its mounting flanges.

Blow-and-Blow Process. A method of shaping glass-ware; the PARISON (q.v.) is blown and then a second blowing process produces the final shape of the ware.

Blow Mould. The metal mould in which blown glass-ware is given its final shape.

Blow-pipe Spray Welding. See SPRAY WELDING.

Blowing. See LIME BLOWING.

Blowing Iron. An iron tube used in making hand-blown glass-ware.

Blown Away. See HOLLOW NECK.

Blowpipe. See BLOWING IRON.

Blue Brick. See ENGINEERING BRICKS.

Blueing. The production of blue engineering bricks, quarries or roofing tiles by controlled reduction during the later stages of firing; this alters the normal state of oxidation of the iron compounds in the material and results in a characteristic blue colour.

Blunger. A machine for mixing clay and/or other materials to form a slip; it usually consists of a large hexagonal vat with a slowly rotating vertical central shaft on which are mounted paddles. The process of producing a slip in such a machine is known as BLUNGING.

Blurring-highlight Test. A test to determine the degree of attack of a vitreous-enamelled surface after an acid-resistance test; (see ASTM – C282).

Blushing. A pink discoloration sometimes occurring during the glost-firing of pottery; it is caused by traces of Cr in the kiln atmosphere arising, for example, from CHROME–TIN PINK (q.v.) fired in the same kiln.

Board. See WORK-BOARD.

Bock Kiln. See BULL'S KILN.

Body. (1) A blend of raw materials awaiting shaping into pottery or refractory products.

(2) The interior part of pottery, as distinct from the glaze.

(3) The condition of molten glass conducive to ready working.

(4) The cylindrical part of a Bessemer converter.

Body Mould. In the pressing of glass, that part of the mould which gives shape to the outer surface of the ware.

Boehme Hammer. A device for the compaction of test-pieces of cement or mortar prior to the determination of mechanical strength; it consists of a hammer, pivoted so that the head falls through a definite arc on the test-piece mould to cause compaction under standard conditions.

Boehmite. A monohydrate of alumina; often a constituent of bauxite and bauxitic clays.

Boetius Furnace. A semi-direct coal-fired pot furnace for melting glass; this was the first pot-furnace to use secondary air to increase the thermal efficiency.

Bogie Kiln or Truck Chamber Kiln. An intermittent kiln of the BOX KILN (q.v.) type distinguished by the fact that the ware to be fired is set on a bogie which is then pushed into the kiln; the bogie has a deck made of refractory material (cf. SHUTTLE KILN).

Bohemian Glass. A general term for Czechoslovakian glass,

29

particularly table-ware and chemical ware; it is generally characterized by hardness and brilliance.

Boiling. A fault in vitreous enamelware (particularly in enamelled sheet-steel) visible as blisters, pinholes, specks, dimples or a spongy surface. The usual cause is undue activity of the GROUND-COAT (q.v.) during the firing of the first COVER-COAT (q.v.), but gases are also sometimes evolved during the firing of the cover-coat (cf. BOILING THROUGH).

Boiling Through. A fault in vitreous enamelware, small dark specks appearing on the surface of the ware, usually in consequence of gas evolution from the base metal.

Bole. A friable earthy clay highly coloured by iron oxide.

Bolley's Gold Purple. A colour that has been used on porcelain. A solution of stannic ammonium chloride is left for some days in contact with granulated tin and is then treated with dilute gold chloride solution. The gold purple is precipitated.

Bond. (1) The arrangement of bricks in a wall; the bond is usually such that any cross-joint in a course is at least one-quarter the length of a brick from joints in adjacent courses. For special types of bond see ENGLISH BOND and FLEMISH BOND.

(2) The intergranular material, glassy or crystalline, that gives strength to fired ceramic ware.

(3) The bond in abrasive wheels may be ceramic, silicate of soda, resin, shellac, rubber, or magnesium oxychloride.

Bond's Law. A theory of grinding based on a consideration of the energy required to propagate a crack through a solid material. (F. C. Bond, *Trans. Amer. Inst. min.* (*metall.*) *Engrs.*, **193**, 484, 1952.)

Bond and Wang Theory. A theory of crushing and grinding: the energy (h) required for crushing varies inversely as the modulus of elasticity (E) and specific gravity (S), and directly as the square of the compressive strength (C) and as the approximate reduction ratio (n). The energy in hp h required to crush a short ton of material is given by the following equation, in which all quantities are in f.p.s. units:

$$h = \left[\frac{0 \cdot 001748 C^2}{SE}\right] \left[\frac{(n+2)(n-1)}{n}\right]$$

The theory is due to F. C. Bond and J. T. Wang (*Trans. Amer. Inst. min.* (*metall.*) *Engrs.*, **187**, 875, 1950).

Bondaroy's Yellow. An antimony yellow developed by Fourgeroux de Bondaroy in 1766: white lead, 12 parts; potassium antimonate, 3 parts; alum, 1 part; sal ammoniac, 1 part.

Bonded Roof. A term for the roof of a furnace when the transverse joints in the roof are staggered (cf. RINGED ROOF).

Bonder. A brick that is half as wide again as a standard square (rectangular or arch); such bricks are sometimes used to begin or end a course of bonded brickwork. (See Fig. 1, p. 37.)

Bondley Process. See under METALLIZING.

Bone Ash. Strongly calcined bone, approximating in composition to tricalcium phosphate. $Ca_3(PO_4)_2$; it is used in the making of CUPELS (q.v.). Less strongly calcined bone constitutes about 50% of the BONE CHINA body. It is occasionally used, to the extent of about 2%, in some vitreous enamels. See also CALCIUM PHOSPHATE.

Bone China. Vitreous, translucent pottery made from a body of the following approximate composition (per cent): calcined bone, 45–50; china clay, 25–30; china stone, 25–30. (Note: In the USA, ASTM – C242 permits the term 'bone china' to be applied to any translucent whiteware made from a body containing as little as 25% bone ash.) Bone china, though delicate in appearance, is very strong. It was first made by Josiah Spode, in Stoke-on-Trent, where by far the largest quantity of this type of high-class pottery is still made (cf. PORCELAIN).

Bonnet Hip. See HIP TILE.

Bonnybridge Fireclay. A fireclay occurring in the Millstone Grit in the Bonnybridge district of Scotland. A typical per cent analysis (fired) is: SiO_2, 56–57; Al_2O_3, 36; Fe_2O_3, 3–4; alkalis, 0·75. P.C.E. 32–33.

Bont. N. Staffordshire term for one of the iron hoops used to brace the outside brickwork of a BOTTLE OVEN (q.v.).

Boost Melting. The application of additional heat to molten glass in a fuel-fired tank furnace by the passage of an electric current through the glass.

Boot. Alternative name, preferred in USA, for POTETTE (q.v.).

Borax. $Na_2B_4O_7 . 10H_2O$; sp. gr. 2·36 (anhydrous), 1·7 (hydrated). Occurs in western USA and is an important constituent of vitreous enamel and glaze frits, and of some types of glass. On heating, the water of crystallization is lost by a series of steps:

$$Na_2B_4O_7 . 10H_2O \xrightarrow{62°C} Na_2B_4O_7 . 5H_2O \xrightarrow{130°C} Na_2B_4O_7 . 3H_2O \xrightarrow{150°C}$$

$$Na_2B_4O_7 . 2H_2O \xrightarrow{180°C} Na_2B_4O_7 . H_2O \xrightarrow{318°C} Na_2B_4O_7.$$

The anhydrous borax, or sodium tetraborate, melts at 741°C.

31

Borazon

Borazon. Trade-name (General Electric Co., USA); cubic form of BORON NITRIDE (q.v.).

Boric Oxide. B_2O_3; m.p. approx. $460°C$; sp. gr. $1·84$. Solid B_2O_3 is commonly available only in the vitreous state; two crystalline forms exists, however, α-B_2O_3 (hexagonal) and β-B_2O_3 (a denser form).

Borides. A group of special ceramic materials. Typical properties are great hardness and mechanical strength, high melting point, low electrical resistivity and high thermal conductivity; impact resistance is low but the thermal-shock resistance is generally good. For the properties of specific borides see under the borides of the following elements: Al, Ba, Ca, Cr, Hf, Mo, Nb, Si, Sr, Ta, Th, Ti, U, V, W, Zr.

Boroaluminate. See ALUMINIUM BORATE.

Boron Carbide. B_4C; made by synthesis at high temperatures. Boron carbide has a Knoop Hardness of 2300–2800, which is second only to that of diamond; for this reason it is used for grinding and drilling. Its transverse strength is approx. 40 000 p.s.i. and compressive strength over 400 000 p.s.i. Boron carbide is also refractory (m.p. $2450°C$), chemically resistant and abrasion resistant; it therefore finds use in nozzles and other high-temperature locations. The sp. gr. is $2·52$ and thermal expansion (25–$800°C$) $4·5 \times 10^{-6}$.

Boron Nitride. BN; m.p. $3000°C$ (sublimes at $2730°C$). A special ceramic existing in two forms that correspond to the graphite (hexagonal) and the diamond (cubic) forms of carbon. The cubic form is as hard as diamond and was originally made (R. H. Wentorf, *J. Chem. Phys.*, **26**, 956, 1957) by the simultaneous application of very high pressure (85 000 atm.) and temperature ($1800°C$). It has since been found that, in the presence of a catalyst, both the temperature and pressure can be reduced. It is made under the trade-name Borazon by General Electric Co., USA. The hexagonal form of BN is readily produced by high-temperature reaction between B_2O_3 or BCl_3 and ammonia. Hot-pressed hexagonal BN is machinable. It oxidizes in air at $800°C$. but its m.p. under a pressure of N_2 is $> 3000°C$. Its electrical resistivity is $1·7 \times 10^{13}$ ohm.cm at $20°C$ and $2·3 \times 10^{10}$ ohm.cm at $500°C$. Uses include dielectic valve spacers, crucibles and rocket nozzles.

Boron Oxide. See BORIC OXIDE.

Boron Phosphate. BPO_4; vaporizes at $1400°C$; sp. gr. $2·81$;

32

related structurally to high-cristobalite. It has been used as a constituent of a ceramic body that fires to a translucent porcelain at 1000°C.

Boron Phosphide. BP; m.p. > 2000°C but readily oxidizes, which limits its potential use.

Boron Silicides. See SILICON BORIDES.

Borosilicate Glass. A silicate glass containing at least 5% B_2O_3 (ASTM definition); a characteristic property of borosilicate glasses is heat resistance.

Bort. Industrial diamond of the type used as an abrasive for cutting and grinding.

Bosh. (1) The part of a blast furnace between the tuyère belt and the lintel; it is usually lined with high-grade fireclay refractory and is water-cooled.

(2) In the glass industry, a tank containing water for cooling glass-making tools.

Boss; Bossing. In the process of pottery decoration known as GROUND-LAYING (q.v.), brush marks are removed by BOSSING, i.e. striking, the ware with a pad, or BOSS, made by stuffing cotton-wool into a silk bag.

Botting Clay. Prepared plastic refractory material for use in the stopping of the tap-holes in cupolas. A typical composition would be 50–75% fireclay, up to 50% black sand, 10% coal dust and up to 5% sawdust.

Bottle Brick. A hollow clay building unit shaped like a bottom-less bottle, 12 in. long, 3 in. o.d., 2 in. i.d. and weighing $2\frac{1}{4}$ lb. The neck of one unit is placed in the end of another to build beams, arches or flat slabs; steel reinforcement can be used. Bottle bricks have been used in France (where they are known as 'Fusées Céramiques'), in Switzerland, the Netherlands, and in S. America.

Bottle Oven. A type of intermittent kiln, usually coal-fired, formerly used in the firing of pottery; such a kiln was surrounded by a tall brick hovel or cone, of typical bottle shape.

Bottom Pouring or Uphill Teeming. A method of teeming molten steel from a ladle into ingot moulds. The steel passes through a system of refractory fireclay tubes and enters the moulds at the bottom; the refractory tubes are of various shapes—TRUMPET, GUIDE-TUBE, CENTRE BRICK and RUNNER BRICKS (all of which q.v.).

Boulder Clay. A glacial clay used in making building bricks, particularly in the northern counties of England.

Boule

Boule. A fused mass of material, pear-shaped, particularly as produced by the VERNEUIL PROCESS (q.v.). Sapphire (99·9% Al_2O_3) boules, about 2 in. long, are produced in this way, and are used, for example, in making thread guides, bearings and gramophone needles.

Bouyoucos Hydrometer. A variable-immersion hydrometer. The original instrument was graduated empirically to indicate the weight of solids per unit volume of suspension; it was subsequently developed for particle-size analysis. (G. J. Bouyoucos, *Soil Sci.*, **23**, 319, 1927; **25**, 365, 1928; **26**, 233, 1928.)

Bowmaker Test. A method of forecasting the durability of refractory glass-tank blocks proposed by E. J. C. Bowmaker (*J. Soc. Glass Tech.*, **13**, 130, 1929). The loss in weight of a sample cut from the tank block is determined after the sample has been immersed for 3 h. in HF/H_2SO_4 at 100°C; the acid mixture is 3 parts by vol. HF (commercial 50–60% HF) and 2 parts by vol. pure conc. H_2SO_4. The test is no longer considered valid.

Box-car roof. Popular name for the KREUTZER ROOF (q.v.).

Box Feeder. A device for feeding clay to preparation machines. It consists of a large metal box, open topped, with the bottom usually formed by a steel-band conveyor or by a conveyor of overlapping steel slats; for plastic clay the feeding mechanism may be a number of revolving screw shafts.

Box Kiln. A relatively small industrial kiln of box-like shape and intermittent in operation.

Boxing. The placing of biscuit hollow-ware, e.g. cups, rim to rim one on another; this helps to prevent distortion during firing.

Boxing-in. A method of setting in a kiln so that, for example, special refractory shapes can be fired without being stressed and deformed; also known as POCKET SETTING.

Bozsin Box. A box, with heat-insulated walls, containing a temperature recorder; it was designed by M. Bozsin to travel with the ware through a vitreous-enamelling furnace.

Brabender Plastograph; Brabender Plasti-Corder. Trade-names: instruments designed in USA to assess the plasticity of clays and other materials on the basis of stress measurement during a continuous shearing process.

Brackelsburg Furnace. A rotary furnace, originally fired by pulverized coal, for the melting of cast iron; the lining of the first such furnaces was a rammed siliceous refractory, but silica brick linings have also been used successfully. (C. Brackelsberg, Brit. Pat. 283 381; 13/4/27.)

Bracken Glass. Old English glass-ware made from a batch in which the ash from burnt bracken supplied the necessary alkali.

Bracklesham Beds. Pale-coloured clays intermingled with glauconite sand occurring in parts of Southern England and worked for brickmaking to the S.W. of London and near Southampton.

Brake Linings. See FRICTION ELEMENTS.

Brasqueing. A process sometimes used for the preparation of the interior of a fireclay crucible prior to its use as a container for molten metal. The crucible is lined with a carbonaceous mixture, it is then covered with a lid and heated to redness. (From French word with the same meaning.)

Bravaisite. A clay mineral containing Mg and K, and of doubtful structure; it has variously been stated to be a mixture of kaolinite and illite or of montmorillonite and illite.

Brazilian Test. A method for the determination of the tensile strength of concrete, ceramic, or other material by applying a load vertically at the highest point of a test cylinder or disk (the axis of which is horizontal), which is itself supported on a horizontal plane. The method was first used in Brazil for the testing of concrete rollers on which an old church was being moved to a new site (cf. BRITTLE-RING TEST).

Break-up of Matt Glaze. The term BREAK-UP is applied more particularly to the glazes containing rutile used on wall tiles. Some of the added rutile dissolves in the glaze, the yellow or brown titanates thus formed subsequently collecting round the undissolved rutile crystals to give the marbled effect known as RUTILE BREAK or BREAK-UP.

Breast Wall. (1) The side-wall of a glass-tank furnace above the tank blocks; also known as CASING WALL, CASEMENT WALL or JAMB WALL.

(2) The refractory wall between the pillars of a glass-making pot furnace and in front of the pot.

Breasts. The sloping parts joining the hearth of an open-hearth furnace to the furnace ends below the ports and adjoining brickwork (cf. BANKS).

Bredigite. The form of CALCIUM ORTHOSILICATE (q.v.) that is stable from about 800–1447°C on heating, persisting down to 670°C on cooling.

Breezing. A thin layer of crushed anthracite or of coarse sand spread on the siege of a pot furnace before setting the pots.

Brenner Gauge. An instrument for the non-destructive determination of the thickness of a coating of vitreous enamel; it depends on the measurement of the force needed to pull a pin from contact with the enamel surface against a known magnetic force acting behind the base metal.

Breunnerite. Magnesite containing 5–30% of ferrous carbonate; some of the Austrian 'magnesite' used as a raw material for basic refractories is, more strictly, breunnerite.

Brewster. Unit of photoelasticity: 1 brewster is equivalent to a relative retardation of $10^{-13}\text{cm}^2\text{dyn}^{-1}$. Named after Sir D. Brewster who, in 1816, demonstrated that glass becomes birefringent when stressed.

Brianchone Lustre. A lustre decoration for pottery-ware distinguished by the fact that the reducing agent necessary to form the thin deposit of metal is incorporated with the colour so that a reducing atmosphere in the kiln is not needed. The usual procedure is to apply the metal as its resinate dissolved in an organic solvent. Although the easiest lustre to apply, it is less durable than lustres produced in a reducing fire.

Brick. See following types: ACID-RESISTING; CLAY BUILDING; CONCRETE; FLOOR; PAVING; PERFORATED; SAND–LIME; SEWER. For Brick Shapes, both for building and for furnace construction, see Fig. 1, p. 37.

Brick Clays. Clays suitable for the manufacture of building bricks occur chiefly in the Carboniferous and more recent geological systems. In the U.K. about 30% of the bricks made are from Carboniferous clays, 30% from the Oxford clays, 10% Glacial clays, 6% Keuper Marl; the remaining 24% are made from Alluvial clays, the so-called Brick-earths, Tertiary, Cretaceous, Devonian, Silurian, and Ordovician deposits. Brick clays are impure and most of them vitrify to give bricks of adequate strength when fired at 900–1100°C.

Brick Earth. An impure loamy clay, particularly that of the Pleistocene of the Thames Valley, used for brickmaking.

Bridge; Bridge Wall. A refractory wall separating two parts of a furnace. The bridge wall (or firebridge) in a boiler furnace terminates the combustion chamber. In a glass-tank, the bridge wall separates the melting end from the working end of the furnace; the wall is in this case usually double, is pierced by the THROAT (q.v.) and is spanned at the top by refractory tiles or bridge covers.

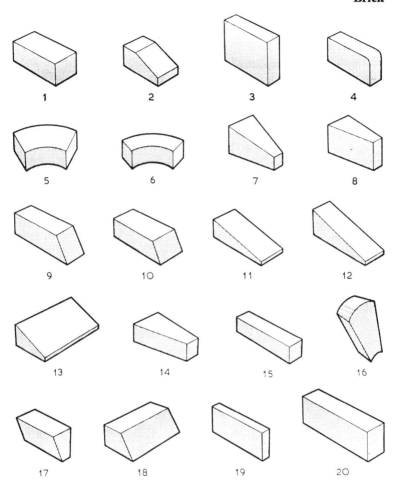

FIG. 1. SOME BRICK SHAPES, PARTICULARLY REFRACTORY-
BRICK SHAPES FOR FURNACE CONSTRUCTION.

1. Standard Square. 2. Bevel Brick. 3. Bonder. 4. Bullnose. 5. Circle Brick. 6. Circle Brick on Edge. 7. Dome Brick. 8. End Arch. 9 End Skew on Edge. 10. End Skew on Flat. 11. Feather End. 12 Feather End on Edge. 13. Feather Side. 14. Key Brick. 15. Pup. 16. Radial Brick. 17. Side Arch. 18. Side Skew. 19. Split. 20. Whelp.

Bridging Oxygen

Bridging Oxygen. An oxygen ion placed between two silicon ions, e.g. in the structure of a silicate glass.

Bright Gold or Liquid Gold. A material for the decoration of pottery-ware; it consists essentially of a solution of gold sulphoresinate together with other metal resinates and a flux (e.g. a bismuth compound) to give adhesion to the ware. This form of gold decoration is already bright when drawn from the decorating kiln (cf. BURNISH GOLD).

Brilliant Cut. Process of decorating flat glass by cutting a pattern with abrasive wheels followed by polishing.

Brimsdown Frit. A lead bisilicate frit for glazes made from 1913 until 1928 by Brimsdown Lead Works, Brimsdown, Middlesex, England. The batch consisted of litharge, silica, and Cornish Stone. Although the frit contained 64% PbO its solubility, when tested by the Home Office method then used, was only 1–2%.

Brindled Brick. A building brick made from a ferruginous clay and partially reduced at the top firing temperature; it has a high crushing strength.

Brinell Hardness. An evaluation of the hardness of a material in terms of the size of the indentation made by a steel ball when pressed against the surface. The test is primarily for metals; it has been applied to clay products, e.g. building bricks, but without great success. (I. E. Brinell, *Comm. Congr. Internat. Math. Essai*, **2**, 83, 1901.)

Bristol Glaze. A feldspathic type of glaze, generally opaque, maturing at 1200–1250°C and suitable for use on stoneware. Both the transparent and the opaque types are compounded from feldspar, whiting, ZnO, china clay and flint; the opaque type contains more feldspar and less whiting. Bristol glazes tend to be rather dull and often show pin-holes.

Britmag. Trade-name: dead-burned magnesia made by the sea-water process in Britain.

Brittle-ring Test. A test to determine the behaviour of a ceramic material under tensile stress; a test-piece in the form of an annulus is loaded along a diameter so that maximum tensile stresses develop on the inner periphery of the annulus in the plane of loading. For theory of this test see *Mechanical Properties of Engineering Ceramics*, p. 383, 1961. (cf. BRAZILIAN TEST.)

Broken-joint Tile. A single-lap roofing tile of a size such that the edge of one tile, when laid, is over the centre of the head of a tile in the course next below.

38

Broken Seed. See SEED.

Brokes. Term used in the English ball-clay mines for clay that will not cut into balls; such clay is generally of low plasticity and poor fired colour.

Brongniart's Formula. A formula relating the weight (W, oz) of solid material in 1 pint of slip (or slop glaze), the weight (P, oz) of 1 pint of the slip, and the specific gravity (S) of the dry solid material:

$$W = (P - 20) \times S/(S - 1)$$

The formula was established for slop glazes by A. Brongniart (*Traité des Arts Céramiques*, Vol. 1, p. 249, 1854).

Brookfield Viscometer. An electrically operated rotating cylinder viscometer in which the drag is recorded directly on a dial; it has been used in the testing of vitreous-enamel slips (*J. Amer. Ceram. Soc.*, **31**, 128, 1948).

Brookite. The orthorhombic form of titania, TiO_2; the other forms are ANATASE (q.v.) and RUTILE (q.v.). It is comparatively rare and is rapidly transformed into rutile at temperatures above about 800°C.

Broseley Tile. An old name for a plain clay roofing tile; such tiles were made in the Broseley area of Shropshire, England.

Brownies. Term sometimes applied to brown spots in white vitreous-enamel ground-coats; more commonly known as COPPER-HEADS (q.v.).

Brownmillerite. A calcium aluminoferrite that occurs in portland and high-alumina cements. It was originally thought to have the composition $4CaO \cdot Al_2O_3 \cdot Fe_2O_3$ (m.p. 1415°C); there is, however, a continuous series of solid solutions in the $CaO-Al_2O_3-Fe_2O_3$ system, the usually accepted Brownmillerite composition being only a particular point in the series.

Brucite. $Mg(OH)_2$; deposits in Ontario and Quebec (Canada) and in Nevada (USA), are used as a source of MgO.

Bruise. A concentration of cracks in the surface of glass-ware caused by localized impact.

Brulax System. An impulse system of oil firing, particularly for the top-firing of annular kilns, developed by A. A. Niesper, in Switzerland, in 1955.

Brunauer, Emmett and Teller Method. A procedure for the determination of the total surface area of a powder or of a porous solid by measurement of the volume of gas (usually N_2) adsorbed on the surface of a known weight of the sample. The mathematical basis of the method was developed by S. Brunauer, P. H. Emmett and E. Teller—hence the usual name B.E.T. METHOD (*J. Amer. Chem. Soc.*, **60**, 309, 1938).

Brunner's Yellow. An antimony yellow recipe given by K. Brunner in 1837: 1 part tartar emetic, 2 parts lead nitrate, 4 parts NaCl. The mixture is calcined and then washed free from soluble salts prior to its use as a ceramic colour.

Brush Marks. A surface imperfection found on the exterior of some bottles; the marks resemble a series of fine vertical laps and are also known as SCRUB MARKS.

Brushing. The removal of bedding material from pottery-ware after the biscuit firing.

BS; BSI. Abbreviations for British Standard and British Standards Institution. The Institution is responsible for the preparation (through industry committees on which interested parties are represented) of national standards for Britain; copies of these standards, and of any foreign standard, can be obtained from the Institution at 2 Park St, London, W.1.

Bubble Cap. A small, hollow, chemical stoneware hemisphere with serrations round the bottom edge; used on stoneware trays in de-acidifying towers in the chemical industry.

Bubble Glass. Glass-ware containing gas bubbles sized and arranged to produce a decorative effect (cf. FOAM GLASS).

Bubble-pressure Method. A technique for the determination of the maximum size of pore in a ceramic product; this size is calculated from the pressure needed to force the first bubble of air through the ceramic when it is wetted with a liquid of known surface tension. The method is used, for example, in the testing of ceramic filters (B.S. 1969).

Bubble Structure. The relative abundance, size and distribution of gas bubbles in a ceramic glaze or in a vitreous enamel. In the latter, it is the dull enamels that contain most bubbles, glossy enamels being relatively bubble-free.

Bubbly Clay. A clay which, because it contains small amounts of organic matter, causes bubbles if used in vitreous enamels.

Buck. A support used in the firing of heavy vitreous enamel-ware.

Buckstave or Buckstay. A steel bracing designed to take the thrust of the brickwork of a furnace.

Bührer Kiln. The ZIG-ZAG KILN (q.v.) invented by J. Bührer (Brit. Pat. 562; 1867).

Building Brick. See CLAY BUILDING BRICK; CONCRETE BRICK; SAND–LIME BRICK.

Building Clays. See BRICK CLAYS.

Bulged Finish. See FINISH.

Bulk Density. A term used when considering the density of a porous solid, e.g. an insulating refractory. It is defined as the ratio of the mass of the material to its BULK VOLUME (q.v.).

Bulk Specific Gravity. For a porous ceramic, the ratio of the mass to that of a quantity of water which, at 4°C, has a volume equal to the BULK VOLUME (q.v.) of the material at the temperature of measurement.

Bulk Volume. A term used relative to the density and volume of a porous solid, e.g. a refractory brick. It is defined as the volume of the solid material plus the volume of the sealed and open pores present.

Bulking. The tendency of fine particles, e.g. sand, to occupy a greater volume when slightly moist; the finer the particles the more pronounced is this effect of surface moisture.

Bull's Eye. (1) A thick round piece of glass or lens.

(2) See ELECTRODE RING.

Bull's Kiln. A CLAMP (q.v.) of a type designed by W. Bull (Brit. Pat. 1977; 31/5/1875) in which the bricks to be fired are set in a trench below ground level; this type of kiln finds some use in India. (Also known as a BOCK KILN.)

Bullers' Rings. Annular rings ($2\frac{1}{2}$ in. dia. with a central hole $\frac{7}{8}$ in. dia.) made by pressing a blend of ceramic materials and fluxes; the rings are not fired. The constituents are so proportioned that the contraction of the rings during firing can be used as a measure of the temperature to which they have been exposed in a kiln; the contraction is measured, after the firing process and when the rings are cold, by means of a gauge. The rings were introduced in about 1900 by Bullers Ltd., Stoke-on-Trent, England, from whom four types are now available for use in various temperature ranges within the overall range 960–1400°C.

Bullhead. See KEY BRICK.

Bullion. Flat glass of uneven thickness made by the hand-spinning of a gob of glass at the end of an iron rod.

Bullnose. A building brick or refractory brick having one end face rounded to join one side face. Such bricks built above one another can be used to form a rounded jamb, hence the alternative name JAMB BRICK (q.v.) (See Fig. 1, p. 37).

Bung. (1) A vertical stack of saggars or of ware.

(2) A refractory brick shape used in the roof of a malleable-iron furnace.

Bunsen's Extinction Coefficient. The reciprocal of the thickness that a layer of glass, or other transparent material, must have for the intensity of transmitted light to be decreased to one-tenth of its intensity as it falls on the layer.

Burgee. Contaminated sand resulting from the grinding of plate glass. (N. England dialect; the same word is used for poor-quality coal.)

Burgos Lustre. A red lustre for porcelain made by suitably diluting a gold lustre with a bismuth lustre; some tin may also be present. (From Burgos, Spain.)

Burley Clay. A refractory clay, intermediate in alumina content between a flint clay and a diaspore clay, that occurs in Missouri, USA; elsewhere it is known as Nodular Fireclay. The name derives from the diaspore oolites, known to the local miners as 'burls', found in these clays. The Al_2O_3 content of the raw clay is 55–63%.

Burning. An alternative (but less appropriate) term for FIRING (q.v.).

Burning Bar; Burning Point; Burning Tool. A support, usually made of heat-resisting alloy, for vitreous enamelware during firing. Defects in the ware caused by contact with these supports are known as BURNING TOOL MARKS.

Burning Off. A fault in vitreous enamelling resulting from the apparent burning-away of the ground-coat; in reality, the fault is due to the enamel having become saturated with iron oxide. To prevent this, the fusion temperature of the ground-coat should be raised by altering its composition.

Burnish Gold or Best Gold. The best gold decoration on pottery is made bright by BURNISHING, i.e. rubbing, usually with a blood-stone or agate. It is applied to the ware as a suspension of gold powder in essential oils, with or without the addition of a proportion of BRIGHT GOLD (q.v.); other ingredients are a flux (e.g. lead borosilicate or a bismuth compound) to promote adhesion to the ware and an extender such as a mercury salt. This type of gold decoration is dull as taken from the kiln, hence the need for subsequent burnishing.

Burnover. An underfired STOCK BRICK (q.v.) from the outside of a clamp; such bricks are usually re-fired.

Burr. (1) A partially fused waste brick from a kiln, or several such bricks fused together.

(2) A rough edge on the base-metal used in vitreous enamelling; it must be removed before the enamel is applied.

Bursting Expansion. In the refractories industry this term has the specific meaning of surface disintegration of basic refractories caused by the absorption of iron oxide. The expansion that leads to this form of failure results from solid solution of magnetite (Fe_3O_4) in the chrome spinel that forms a major constituent of chrome and chrome–magnesite refractories. A laboratory test submits a test-piece cut to the size of a 2 in. cube, to the action of 40 g of mill-scale (crushed to pass a 30 B.S. sieve) for 1 h at 1600°C; the expansion is expressed as a linear percentage.

Bursting Off. Breaking blown glass-ware from the end of the blowing iron.

Bushing. (1) An electric glass-melting unit for the production of glass fibres, which are drawn through platinum orifices in the base.

(2) A liner fitted in the feeder that delivers glass to a forming machine; this liner is also known as an ORIFICE RING.

Bustle Pipe or Hot-blast Circulating Duct. A metal tube of large diameter which surrounds a blast furnace at a level a little above the tuyères; it is lined with refractory material and distributes the hot air from the hot-blast-stoves to the pipes known as GOOSENECKS which in turn carry the air to the tuyères.

Button. Part of a piece of pressed glass designed to produce a hole when knocked out; also sometimes called a CAP or KNOCK-OUT.

Button Test. A test for the fusibility of a vitreous enamel frit, or powder, first proposed by C. J. Kinzie (*J. Amer. Ceram. Soc.*, **15**, 357, 1932) and subsequently standardized (ASTM – C374). Also known as the FUSION-FLOW TEST and as the FLOW BUTTON TEST.

Bwlchgwyn Quartzite. A quartzite from Bwlchgwyn, N. Wales, used as a raw material for silica-brick manufacture. Chemical analyses are:

	SiO_2	Al_2O_3	Alkalis
As quarried:	96·6	1·1	0·4
Washed:	97·4	0·7	0·3

C

°C. Degrees Celsius (formerly, and still more commonly, known as Degrees Centigrade). For conversion table to Fahrenheit see Appendix 3.

C_{80}. See CONTRAST RATIO.

Cabal Glass. A special glass consisting solely of *CA*lcium oxide, *B*oric oxide and *AL*umina, hence its name; it was first made by H. Jackson for the UK Ministry of Munitions in 1917–19.

Cable Cover. A fired clay—in this context generally (but erroneously) known as 'Earthenware'—or concrete conduit for covering underground electric cables; the principal objects of these covers are to give warning of the cable's presence and to protect it from excavating tools. For specification see B.S. 2484.

Cadmium Niobate. $Cd_2Nb_2O_7$; a ferroelectric compound of potential value as a special electroceramic; the Curie temperature is $-103°C$.

Cadmium Oxide. CdO; the vapour pressure is approx. 1 mm at 1000°C and 7 mm at 1100°C. It is occasionally used in ceramic colours. The oxide is poisonous.

Cadmium Selenide. This compound, or the polyselenide, is responsible for the colour in selenium ruby glass. With some CdS in solid solution it forms the basis of Cd–Se red ceramic colours; the presence of up to 3% V_2O_5 is claimed to stabilize the colour; blackening in the presence of a glaze or flux containing Pb is caused by the formation of PbSe.

Cadmium Sulphide. CdS; sp. gr. 4·8; m.p. 1750°C. A constituent of some ruby-coloured glass, of some yellow enamels, and of Cd–Se reds for pottery decoration. Sintered CdS can be used as a photoconductive ceramic.

Cadmium Titanate. $CdTiO_3$; a special ferroelectric ceramic having the ilmenite structure at room temperature; the Curie temperature is approx. $-220°C$.

Cadmium Zirconate. $CdZrO_3$; occasionally used as an addition to barium titanate bodies, the effect being to reduce the dielectric constant and Curie temperature.

Calcite. $CaCO_3$; the mineral constituent of limestone, chalk, and marble. Used as a constituent of soda–lime glass and, as WHITING (q.v.), in some pottery bodies; it is a major component of the batch used in portland cement manufacture.

Calcium Aluminates. The five compounds (q.v.) are: TRICAL-CIUM ALUMINATE ($3CaO.Al_2O_3$); DODECACALCIUM HEPTALUMINATE ($12CaO.7Al_2O_3$); CALCIUM MONOALUMINATE ($CaO.Al_2O_3$); TRI-CALCIUM PENTALUMINATE ($3CaO.5Al_2O_3$); CALCIUM DIALUMINATE ($CaO.2Al_2O_3$); and CALCIUM HEXALUMINATE ($CaO.6Al_2O_3$).

Calcium Boride. CaB_6; m.p. 2235°C; sp. gr. 2·45; thermal expansion, $5·2 \times 6·5 \times 10^{-6}$; electrical resistivity (20°C), 124 μohm.cm.

Calcium Carbonate. See CALCITE; WHITING.

Calcium Chloride. $CaCl_2$; used for FLOCCULATION (q.v.) in the preparation of glazes; about 0·05% is normally sufficient. It is also sometimes used as a mill-addition in the preparation of vitreous enamel slips. Calcium chloride solution accelerates the rate of setting of portland cement; it is also sometimes added to concrete mixes as an integral waterproofer.

Calcium Dialuminate. $CaO.2Al_2O_3$; m.p. 1705°C; thermal expansion, $5·0 \times 10^{-6}$. Present in high-alumina cement but does not itself have cementing properties.

Calcium Ferrite. In the binary system, two ferrites are formed—$CaO.Fe_2O_3$; and $2CaO.Fe_2O_3$; the former may occur in some HIGH-ALUMINA CEMENT (q.v.).

Calcium Fluoride. See FLUORSPAR.

Calcium Hafnate. $CaHfO_3$; m.p. 2470 ± 20°C; sp. gr. 5·73; thermal expansion (10—1300°C), 7×10^{-6}.

Calcium Hexaluminate. $CaO.6Al_2O_3$; melts incongruently at 1850°C to form corundum and a liquid.

Calcium Metasilicate. See WOLLASTONITE.

Calcium Monoaluminate. $CaO.Al_2O_3$; m.p. 1605°C. A principal constituent of HIGH-ALUMINA CEMENT (q.v.).

Calcium Orthosilicate or Dicalcium Silicate. $2CaO.SiO_2$; m.p. 2130°C. Occurs in four crystalline forms: α, stable above 1447°C; α', bredigite, stable from about 800—1447°C on heating, 1447°—670°C on cooling; β, larnite, stable or metastable from 520–670°C; γ, stable below 780–830°C. Material in which a considerable amount of $2CaO.SiO_2$ has been formed by high-temperature reaction, falls to a powder—'dusts'—on cooling because of the inversion (accompanied by a 10% increase in volume) to the γ-form at 520°C. The inversion can be prevented by the addition of a stabilizer, e.g. B_2O_3 or P_2O_5. Calcium orthosilicate is a constituent of portland cement and may be formed in dolomite refractories.

45

Calcium Oxide

Calcium Oxide. See LIME.

Calcium Phosphate. $3CaO.P_2O_5$ or $Ca_3(PO_4)_2$; m.p. 1820°C. Crucibles made of this phosphate withstand FeO melts at 1600°C and permit the refining of steel to a phosphate content of 0·025%. See also BONE ASH.

Calcium Silicates. The four compounds (q.v.) are: WOLLASTONITE ($CaO.SiO_2$); RANKINITE ($3CaO.2SiO_2$); CALCIUM ORTHOSILICATE ($2CaO.SiO_2$); and TRICALCIUM SILICATE ($3CaO.SiO_2$).

Calcium Stannate. $CaSnO_3$; sometimes used as an additive to barium titanate bodies, one effect being to lower the Curie temperature.

Calcium Sulphate. See ANHYDRITE; GYPSUM; PLASTER OF PARIS.

Calcium Sulpho-Aluminates. Two compounds exist: the 'high' form, $3CaO.Al_2O_3.3CaSO_4.30–32H_2O$; and the 'low' form, $3CaO.Al_2O_3.CaSO_4.12H_2O$. Both forms may be produced by reaction between $3CaO.Al_2O_3$ and gypsum during the hydration of portland cement. The 'high' form is also produced when cement and concrete are attacked by sulphate solutions.

Calcium Titanates. Three compounds exist: $CaO.TiO_2$, m.p. 1915°C; $4CaO.3TiO_2$, melts incongruently at 1755°C; $3CaO.2TiO_2$ melts incongruently at 1740°C. Calcium titanate refractories slowly deform under load at high temperature; they resist attack by basic O.H. slag and portland cement clinker but react with materials containing SiO_2 and/or Al_2O_3. $CaO.TiO_2$ has a dielectric constant of 150–175; it finds use as a component of more complex titanate dielectrics.

Calcium Zirconate. $CaZrO_3$; a material having useful dielectric and refractory properties. As a dielectric it is normally used in minor amounts (3–10%) in titanate bodies; as a refractory it can be used, even under reducing conditions, up to 1700°C provided that it is not in contact with siliceous material. M.p. approx. 2350°C; sp. gr. 4·74; thermal expansion 8–11 $\times$ 10^{-6} (25–1300°C).

Calgon. Proprietary name for a complex SODIUM PHOSPHATE (q.v.) sometimes used as a deflocculant for clay slips.

Callow. A term, of localized use, for the overburden of a clay-pit.

Calorific Value (CV). The quantity of heat released when a unit weight (or volume) of fuel is completely burned. Units: 1 Btu/lb = 1·8 cal/g. Typical values: coal, 11 000—15 000 Btu/lb; liquid fuels, 17 000—21 000 Btu/lb; town gas, 450—520 Btu/ft³.

Calorite. Trade-name: a pyroscope similar to a pyrometric cone but cylindrical; they are made for use between 500 and 1470°C. (Wengers Ltd., Stoke-on-Trent, England.)

Cameo. (1) Glass jewellery, etc., made of two or more layers of glasses of different colours and carved in relief.

(2) Similar ornamental pieces made of JASPER WARE (q.v.)

Campaign. The working life of an industrial furnace between major repairs; the term is particularly applied to open-hearth furnaces and to glass-tank furnaces.

Canal. The section of a flat-glass tank-furnace through which molten glass flows to the drawing chamber.

Candle. See CERAMIC FILTER.

Cane. Glass rods of small or medium diameter.

Cane Marl. Local name for one of the low-quality fireclays associated with the Bassey Mine, Littlerow and Peacock coal seams of N. Staffordshire, England.

Cane Ware. Eighteenth-century English stoneware of a light brown colour; it was a considerable advance on the coarse pottery that preceded it but, for use as tableware, cane ware was soon displaced by white earthenware. During the 19th and the earlier part of the present century, however, cane ware continued to be made in S. Derbyshire and the Burton-on-Trent area as kitchen-ware and sanitary-ware; it had a fine-textured cane-coloured body with a white engobe on the inner surface—often referred to as CANE-AND-WHITE.

Cank. An indurated clay with cemented particles, or a useless mixture of clay and stone; the term is chiefly used in the Midlands and North of England.

Cannock. South Staffordshire term for a ferruginous nodule occurring in a fireclay; the name derives from the town of Cannock in that area.

Cannon. A small ($\frac{1}{2}$–2 fl. oz) thick-walled glass bottle of the type used to contain flavouring essences.

Cannon Pot. A small POT (q.v.) for glass melting.

Cant. The bottom outside edge of a SAGGAR (q.v.); it is important that this should be rounded to help the saggar to resist thermal and mechanical shock.

Cantharides Lustre. A silver lustre for artware similar in appearance to the lustre on the wings of the cantharides beetle; it is yellow and has usually been applied to a blue glaze.

Canton Blue. A violet-blue ceramic colour made by the addition of barium carbonate to cobalt blue. A quoted recipe is (per cent): cobalt oxide, 40; feldspar, 30; flint, 20; $BaCO_3$, 10.

Caolad Flint. A form of cryptocrystalline silica occurring at Cloyne, Co. Cork, Eire. It has a sp. gr. of 2·26 and is readily ground, without the need for pre-calcination, for use in pottery bodies.

Cap. See BUTTON.

Carbides. A group of special ceramic materials; see under the carbides of the following elements: Al, Be, Cr, Hf, La, Nb, Si, Ta, Th, Ti, U, V, W, Zr.

Carbon. See CARBON REFRACTORIES; GRAPHITE.

Carbon-Ceramic Refractory. See PLUMBAGO.

Carbon Dioxide Process. A method of bonding refractory grains by mixing them with a solution of sodium silicate, moulding to the required shape and then exposing the shape to CO_2. The process was first mentioned in Brit. Pat. 15 619 (1898), but did not come into general use until about 1955, when it began to be employed in the bonding of foundry sands and cores. The process has been tried for bonding rammed linings in small ladles.

Carbon Monoxide Disintegration. The breakdown of refractory materials that sometimes occurs (particularly with fireclay refractories) when they are exposed, within the temperature range 400–600°C, to an atmosphere rich in carbon monoxide. The disintegration is due to the deposition of carbon around 'iron spots' in the brick, following the well-known dissociation reaction:

$$2CO \rightleftharpoons CO_2 + C.$$

Carbon Refractories. These refractories, consisting almost entirely of carbon, are made from a mixture of graded coke, or anthracite, pitch and tar; the shaped blocks are fired (packed in coke). The fired product has an apparent porosity of 20–25%; crushing strength 7000–10 000 p.s.i.; R.u.L. (50 p.s.i.), > 1700°C; thermal expansion (0–1000°C), 0·65%. The principal use is in the lining of blast furnaces, particularly in the hearth and bosh (cf. PLUMBAGO).

Carborundum. Trade-mark of the Carborundum Co., Niagara Falls, USA, and Trafford Park, Manchester, England. This firm pioneered the industrial synthesis of silicon carbide and their trade-mark is often used as a synonym of SILICON CARBIDE (q.v.)

Carboxymethylcellulose (CMC). An organic compound that finds use in the ceramic industry as an additive to glazes and engobes to prevent friability before the coating is fired; CMC has been added to vitreous-enamel slips to prevent settling.

Carboy. A glass container for acids, etc.; it has a narrow neck and its capacity is 5 gallons or more. (See B.S. 678.)

Carburettor. The chamber of a water-gas plant, lined with refractory material and often filled with CHECKERS (q.v.) on which oil is sprayed to enrich the gas (cf. SUPERHEATER).

Carder Tunnel Kiln. A tunnel-kiln designed in about 1928 by Carder and Sons, Brierley Hill, England, for the firing of stoneware at 1200°C.

Carlton Shape. A tea-cup the top half of which is cylindrical, the bottom half being approximately hemispherical but terminating in a broad, shallow, foot. For specification see B.S. 3542.

Carman Equation. A relationship, derived from KOZENY'S EQUATION (q.v.), permitting determination of the specific surface, S, of a powder from permeability measurements:

$$S = 14\sqrt{[p^3/KV(1 - p^2)]}$$

where p is the porosity of the bed of powder, V is the kinematic viscosity of the flowing fluid, and K is a constant. In his original application of this equation, Carman used a simple apparatus in which liquids were used as permeating fluids. (P.C. Carman, *J. Soc. Chem. Ind.*, **57**, 225, 1938.)

Carnegieite. $Na_2O.Al_2O_3.2SiO_2$; m.p. 1526°C. It is formed when NEPHELINE (same composition) is heated above 1248°C and is sometimes found in fireclay refractories that have been attacked by Na_2O vapour.

Carolina Stone. A CHINA STONE (q.v.) used to some extent in the US pottery industry.

Carrousel. A four-wheeled bogie fitted with a rotating framework which carries two sets of STILLAGES (q.v.) for the handling of bricks from a dryer to a Hoffmann type of kiln. (From the French word for a merry-go-round.)

Carry. Term sometimes used in Scotland for OVERBURDEN (q.v.).

Case Mould. See under MOULD.

Cased Glass. Glass-ware with a superimposed layer of another glass having a different composition and usually coloured. The thermal expansions of the two glasses must be carefully matched (cf. PLY GLASS).

Casement Wall. See BREAST WALL.

Casher Box. A metal box used to catch a glass bottle after it had been severed from the blow-pipe in the old hand-blown process.

Casing Wall. See BREAST WALL.

Cassel Kiln. See KASSEL KILN.

Castable Refractory. A mixture of graded refractory material and aluminous hydraulic cement, e.g. CIMENT FONDU. Such materials are generally cast into place but may be sprayed by means of a cement gun.

Casting. Shaping a fluid material (which subsequently solidifies) by pouring it into a mould. The process is used in pottery manufacture and in glass-making. To cast pottery-ware, a SLIP (q.v.) is poured into a plaster mould which absorbs a proportion of the water so that the body builds up on the walls of the mould. In the production of thin-walled ware, e.g. tableware, excess slip is poured out of the mould when the required thickness of ware has been formed. For thick ware, e.g. sanitary fireclay, the SOLID CASTING process is used, with an inner plaster core; slip is in this case poured into the space between mould and core and the body is allowed to build up without any slip being poured off.

Casting-pit Refractories. Specially shaped refractories (usually fireclay) for use in the casting of molten steel. The individual items included in the term are: LADLE BRICKS, ROD COVERS, STOPPERS, NOZZLES, MOULD BRICKS, TRUMPETS, GUIDE TUBES, CENTRE BRICKS, RUNNERS, CONES and MOULD PLUGS (q.v.).

Casting Spot. A fault that sometimes appears on cast pottery as a vitrified and often discoloured spot on the bottom of the ware or as a semi-elliptical mark on the side. It occurs where the stream of slip first strikes the plaster mould and is attributable to local orientation of platy particles of clay and mica in the body. The fault can be largely eliminated by adjusting the degree of deflocculation of the slip so that it has a fairly low fluidity. The fault is also sometimes called FLASHING.

Castle; castling. Local term for the setting of bricks on a dryer car, two-on-two in alternate directions.

Cat's Eye. (1) Glass tubing of the SCHELLBACH (q.v.) type for use in spirit-levels.

(2) A fault, in glass-ware, in the form of a crescent-shaped blister which may contain foreign matter.

Cataphoresis. The movement of colloidal particles in an electric field; this forms the basis of the purification of clays by the so-called ELECTRO-OSMOSIS (q.v.) method.

Celite

Catenary Arch. A sprung arch having the shape of an inverted catenary (the shape assumed by a string suspended from two points that are at an equal height from the ground). The stress pattern in such an arch is such that there is no tendency for any bricks to slip relative to one another.

Cathedral Glass. Rolled flat-glass textured on one side to resemble old window glass (cf. ANTIQUE GLASS).

Cathode Pickling. See ELECTROLYTIC PICKLING.

Cationic Exchange. See IONIC EXCHANGE.

Cauchy Formula. A formula proposed by A. L. Cauchy, a 19th century French mathematician, relating the refractive index (n) of a glass to the wavelength (λ) of the incident light:

$$n = A + B\lambda^{-2} + C\lambda^{-4} \ldots$$

A closer relationship is the HARTMAN FORMULA (q.v.).

Caveman. An odd-job man around (frequently under) a glass furnace.

Cavity Wall. A wall built so that there is a space between the inner and outer leaves, which are tied together at intervals by metal or other ties. Such a wall has improved thermal insulation and damp-proofness.

C/B Ratio. A term that has been used for the SATURATION COEFFICIENT (q.v.).

C & D Hot Top. A HOT-TOP (q.v.) designed by W. A. Charman and H. J. Darlington (hence the name 'C & D') at the time, about 1925, when they were both employed by Youngstown Sheet and Tube Co., USA. The hot top, which is fully floating, consists of a cast-iron casing lined with fireclay or insulating refractories; a refractory bottom ring is attached to the lower end of the casing to protect the latter from the hot metal.

C–D Principle. The Convergence–Divergence principle used in the FRENKEL MIXER (q.v.).

Celadon. An artware glaze of a characteristic green colour, which is obtained by introducing a small percentage of iron oxide into the glaze batch and firing under reducing conditions so that the iron is in the ferrous state. The name was used by the first Josiah Wedgwood for his self-coloured green earthenware.

Celeste Blue. A ceramic colour made by softening the normal cobalt blue by the addition of zinc oxide.

Celite. The name given to one of the crystalline constituents of portland cement clinker by A. E. Törnebohm (*Tonindustr.*

Cell

Ztg., **21**, 1148, 1897). This constituent has now been identified as a solid solution of $4CaO.Al_2O_3.Fe_2O_3$ and $6CaO.2Al_2O_3.Fe_2O_3$.

Cell. One of the spaces in a hollow clay building block. According to the US definition (ASTM – C43) a cell must have a minimum dimension of at least $\frac{1}{2}$ in. and a cross-sectional area of at least 1 in².

Cell Furnace. A glass-tank furnace in which glass in the melting end and auxiliary chambers is heated electrically.

Cellular Concrete. See AERATED CONCRETE.

Celsian. Barium feldspar, $BaO.Al_2O_3.2SiO_2$; m.p. 1780°C. There are two crystalline forms, resulting in a non-linear thermal expansion curve. Celsian refractory bricks have been made and have found some use in electric tunnel kilns.

Celsius. The internationally approved term for the temperature scale commonly known as Centigrade; for conversion table to Fahrenheit see Appendix 3.

Cement. See CIMENT FONDU; HIGH EARLY-STRENGTH CEMENT; KEENE'S CEMENT; PORTLAND CEMENT; RAPID-HARDENING PORTLAND CEMENT; REFRACTORY CEMENT; SILICATE CEMENT; SOREL CEMENT.

Cement Bacillus. This name has been applied to the compound $3CaO.Al_2O_3.3CaSO_4.31H_2O$, which is formed by the action of sulphate solutions on portland cement and concrete.

Cement-kiln Hood. The head, which may be mobile or fixed, of a rotary cement kiln; through the hood the burner passes and within the hood the clinker discharges from the kiln to the cooler.

Centigrade. For conversion table to Fahrenheit, see Appendix 3.

Centre Brick. A special, hollow, refractory shape used at the base of the GUIDE TUBES (q.v.) in the bottom-pouring of molten steel. The Centre Brick has a hole in its upper face and this is connected via the hollow centre of the brick to holes in the side faces (often six in number). The Centre Brick distributes molten steel from the trumpet assembly to the lines of RUNNER BRICKS (q.v.). It is also sometimes known as a CROWN BRICK or SPIDER.

Centrifuge. A device in which the rate of settling of particles from a liquid suspension is accelerated by subjecting them to a high centrifugal force; this is done by charging the suspension into carefully-balanced 'buckets' at opposite ends of a diameter of a rapidly spinning disk or shaft. Such equipment is used industrially in the treatment of raw materials; in the ceramic laboratory, the centrifuge is used as a method of particle-size analysis, particularly for particles below about 1μ.

Ceramel. Obsolete term for CERMET (q.v.).

Ceramic. The usual derivation is from *Keramos*, the Greek word for potters' clay or ware made from clay and fired; by a natural extension of meaning, the term has for long embraced all products made from fired clay, i.e. bricks and tiles, pipes and fireclay refractories, sanitary-ware and electrical porcelain, as well as pottery tableware. In 1822 silica refractories were first made; they contained no clay, but were made by the normal ceramic process of shaping a moist batch, drying the shaped ware and firing it. The word 'ceramic', while retaining its original sense of a product made from clay, thus tacitly began to include other products made by the same general process of manufacture. There has, in consequence been no difficulty in permitting the term to embrace the many new non-clay materials now being used in electrical, nuclear and high-temperature engineering.

In USA a radical extension of meaning was authorized by the American Ceramic Society in 1920; chemically, clay is a silicate and it was proposed that the term 'ceramic' should be applied to all the silicate industries; this brought in glass, vitreous enamel, and hydraulic cement. In Europe, this wider meaning of the word has not yet been fully accepted.

Ceramic Filter. A ceramic characterized by an interconnected pore system, the pores being of substantially uniform size. Such ceramics are made from a batch consisting of pre-fired ceramic, quartz or alumina together with a bond that, during firing, will vitrify and bind the surfaces of the grains together. The pore size of different grades varies from about 10μ to 500μ. The filters are commonly available as tiles or tubes, the latter sometimes being known as CANDLES; special shapes can be made as required. Uses include filtration, aeration, electrolytic diaphragms and air-slides (cf. FILTER BLOCK; SINTERED FILTER).

Ceramic-to-metal Seal. The joining of a metal to a ceramic is generally accomplished by METALLIZING (q.v.) the ceramic surface and then brazing-on the metal component. Ceramic-to-metal seals are used in electrically insulated and vacuum-tight 'lead-throughs', especially for high-power h.f. devices; components so made are more rugged and resist higher temperatures than those having a glass-to-metal seal, thus permitting a higher bake-out temperature and use in a nuclear environment.

Ceramic Whiteware

Ceramic Whiteware. This term is defined in the USA (ASTM – C242) as: A fired ware consisting of a glazed or unglazed ceramic body which is commonly white and of fine texture; the term includes china, porcelain, semivitreous ware and earthenware.

Ceria. See CERIUM OXIDE.

Cerium Nitride. CeN; produced by the action of N_2 on Ce at 800°C or NH_3 on Ce at 500°C.

Cerium Oxide. CeO_2; m.p. 2600°C; sp. gr. 7·13; thermal expansion (20–800°C), $8·6 \times 10^{-6}$. A rare earth derived from the monazite of India, Brazil, Florida, Australia and Malagasy. Used in the polishing of optical glass; it is also effective both as a decolorizer and as a colouring agent for glass.

Cerium Sulphides. There are three sulphides: CeS, m.p. 2450°C ± 100°C; Ce_3S_4, m.p. 2050°C ± 75°C; and Ce_2S_3, m.p. 1890°C ± 50°C. Special ceramic crucibles have been made of these sulphides, but they can be used only *in vacuo* or in an inert atmosphere; such crucibles are suitable as containers for molten Na, K, Ca and other highly electropositive metals. The thermal-shock resistance of CeS is good, that of Ce_2S_3 poor, and that of Ce_3S_4 intermediate.

Cermet. A material containing both ceramic and metal. Experiments have been made with a wide range of ceramics (oxides and carbides) and metals (iron, chromium, molybdenum, etc.) in an attempt to combine the refractory and oxidation-resistant properties of ceramics with the thermal-shock resistance and tensile strength of metals. Success has been only partial. The most promising cermets are TiC/Alloy (the alloy containing Ni, Co, Mo, W, Cr in various proportions), Al_2O_3/Cr and Al_2O_3/Fe. The principal use for cermets would be for the blades of gas turbines and other high-temperature engineering units; special cermets are also currently being used in brake linings. (See J. R. Tinklepaugh and W. B. Crandall *Cermets*, Reinhold, 1960.)

Chain Hydrometer. A type of hydrometer that is operated at constant depth by a chain loading device similar to that used on some analytical balances. It has been used for the determination of the particle-size distribution of clays.

Chair. (1) A wooden chair of traditional design used by a glass-blower.

(2) A team of workmen producing hand-made glass-ware.

Chairman. See GAFFER.

Chalcedony. A cryptocrystalline, fibrous form of quartz; flint, for example, is chalcedonic.

Chalk. A soft rock consisting of fine-grained calcium carbonate; used as an alternative material to limestone in cement manufacture, and in the production of WHITING (q.v.).

Chalked or Chalky. The condition of vitreous enamelware that, in consequence of abrasion or chemical attack, has lost its gloss and has become powdery.

Chamber Dryer. A type of dryer in which shaped clayware is placed in chambers in which the temperature, humidity, and air flow can be controlled; the ware remains stationary during the drying process. The Keller Dryer is of this type, its distinctive feature being the system of handling the bricks to be dried by means of STILLAGES (q.v.) and FINGER-CARS (q.v.).

Chamber Kiln. An ANNULAR (q.v.) kiln of the type in which the setting space is permanently subdivided into chambers; examples include the SHAW KILN (q.v.) and the MENDHEIM KILN (q.v.).

Chamber Oven. A refractory-lined gas-making unit; the capacity of such an oven may vary from about 1 to 5 tons of coal.

Chamotte. A W. European word (originally German) which has now also been adopted in the UK, denoting refractory clay that has been specially fired and crushed for use as a non-plastic component of a refractory batch; it is generally made by calcination of lumps of the raw clay in a shaft kiln or in a rotary kiln (cf. GROG).

Champlevé Enamelware. One type of vitreous-enamel artware: a pattern is first cut into the base-metal but, where the pattern requires that enamels of different colours should meet, a vertical fin of metal is left so that the colour boundary will remain sharp— the two enamels not running into one another when they are fused. (French word meaning 'raised field'.)

Charlton Photoceramic Process. A positive print is formed on ceramic ware by a process involving the application of emulsion to the ware, exposure in contact with the negative, and development. (A. E. Charlton, *Ceramic Industry*, **56,** No. 4, 127, 1951.)

Charp. Trade-name for Calcined High-Alumina Refractory Powder; it is made from Ayrshire bauxitic clay.

Chaser Mill. Name occasionally used for an EDGE-RUNNER MILL (q.v.).

Chatelier. See LE CHATELIER.

Check, Crizzle, Vent

Check, Crizzle, Vent. A surface crack in glass-ware. (See also SMEAR.)

Checkers; Checker Bricks. Refractory bricks or special shapes set in a regenerator in such a way as to leave passages for the movement of hot gases; waste gases passing from a furnace through the checkers give up heat which, on reversal of the direction of gas flow in the furnace, is subsequently transferred to the combustion air and fuel.

Checking. (1) A defect in cast-iron enamelware, raised lines appearing on the surface as a result of cracks in the ground-coat.

(2) The term has been used in USA to signify surface cracking or crazing of clayware.

Cheeks. The refractory side-walls of the ports of a fuel-fired furnace.

Chemical Stoneware. A vitreous type of ceramic product for use in the chemical industry and in other locations where resistance to chemical attack is sought. A typical body composition is: stoneware clay or plastic fireclay, 25–35%; ball clay, 30–40%; feldspar, 20–25%; grog 20–30%. The grog may be crushed stoneware or porcelain, fused alumina, or silicon carbide, depending on the combination of physical properties sought in the finished product. The firing temperature is usually 1200–1300°C. Chemical stoneware has a crushing strength of nearly 20 tons/in²; the coefficient of linear expansion (20–100°C) is 4×10^{-6}. In the UK, chemical stoneware must comply with B.S. 784. In the USA, this type of ware is termed 'Chemical Porcelain'.

Chemically Bonded Refractory Cement. A jointing cement for furnace brickwork that may be one of two types: *Air-setting refractory cement*—finely ground refractory material containing chemical agents, such as sodium silicate, which ensure that the cement will harden at room temperature; *Air-hardening refractory cement*—finely ground refractory material containing chemical agents that cause hardening at a temperature below that at which vitrification begins but above room temperature.

Chequers; Chequer Bricks. See CHECKERS; CHECKER BRICKS.

Chert. A general term for a cryptocrystalline siliceous rock that may occur in either nodular or tabular form. Chert stones were used as pavers and runners in the old paddle type of mill for grinding potters' materials; trimmed chert blocks are now used as a lining material for ball mills.

Chest Knife. A tool used in hand-blown glass-making for

56

removing the MOIL (q.v.) from the blowing-iron; the moils are allowed to crack off while the blowing-irons are in a receptacle called a 'chest'.

Chevenard Dilatometer. An apparatus for the measurement of thermal expansion; it depends on the recording, by means of an optical lever, of the differential expansion of the test-piece and that of a standard. It finds use in Western Europe for the testing of ceramic products. (P. Chevenard *Compte Rend.*, **164**, 916, 1917.)

Chill Mark. A surface defect on glass-ware characterized by its wrinkled appearance; also known as FLOW LINE (q.v.). It is caused by uneven contact in the mould prior to the forming.

China. Within the UK pottery industry this term refers to BONE CHINA (q.v.). an essential feature of which is translucency. In the USA, however, ASTM – C242 defines the word as any glazed or unglazed vitreous ceramic whiteware used for non-technical purposes, e.g. dinnerware, sanitary-ware, and art-ware, provided that they are vitreous (cf. PORCELAIN).

China Clay (Kaolin). A white-firing clay consisting essentially of KAOLINITE (q.v.). Major deposits occur in Cornwall (England); Georgia, S. Carolina and Florida (USA); Germany, France, Czechoslovakia, and elsewhere. Annual output in USA is 2 500 000 tons; in England, 1 500 000 tons. Of these tonnages only about 20% is used in the ceramic industry (pottery and refractories). Typical composition (per cent): SiO_2, 47; Al_2O_3, 38; Fe_2O_3, 0·8; Alkalis, 1·7; loss-on-ignition, 12.

China Process. A US term defined (ASTM – C242) as the method of producing glazed whiteware in which the body is biscuit fired and glaze is then applied followed by glost firing. (In the UK this process is used in the manufacture both of earthenware and of bone china.)

China Sanitary-ware. US term for vitreous ceramic sanitary-ware.

China Stone. Partly decomposed granite, consisting of feldspathic minerals and quartz; it is used as a flux in pottery bodies. Examples in the UK are Cornish Stone and Manx Stone. The former is available in various grades, e.g. Hard Purple, Mild Purple, Hard White and Soft White; the feldspars are least altered in the Hard Purple, alteration to secondary mica and kaolinite being progressively greater in the Mild Purple, Hard White, and Soft White; the purple stones are so coloured by the small amount of fluorspar present. Manx Stone (from Foxdale, Isle of Man) is virtually free from fluorine.

Chinese Blue or Mohammedan Blue. The mellow blue, ranging in tint from sky-blue to greyish-blue, obtained by the early Chinese and Persian potters by the use of impure cobalt compounds as colorants.

Chipping. The chipping of vitreous enamelware is often attributable to the enamel coating being too thick, or to the curvature of enamelled edges being too sharp. Chipping of the edges of ceramic tableware is also accentuated by poor design.

Chittering. A fault that may appear as a series of small ruptures along the edge or rim of pottery-ware. True chittering is caused by incorrect fettling.

Choke. A glass-making fault appearing as a constriction in the neck of a bottle.

Choke Crushing. The method of running a pair of crushing rolls with the space between the rolls kept fully charged with clay; with the rolls operated in this way much of the crushing is achieved by the clay particles bearing on each other—this gives a finer product than with FREE CRUSHING (q.v.).

Chrome–Alumina Pink. A ceramic colour consisting principally of Cr_2O_3, Al_2O_3 and ZnO; when used as a glaze stain, the glaze should contain ZnO and little, if any, CaO. It is recommended that, for use under-glaze, the glaze should be leadless. The colour depends on diffusion of Cr into the insoluble Al_2O_3 lattice and is normally stable up to 1300°C.

Chrome Green. See VICTORIA GREEN.

Chrome–Magnesite Refractory. A refractory material made from chrome ore and dead-burned magnesite, the chrome ore preponderating. Such refractories may be fired or they may be chemically bonded; they are frequently metal-cased and find most use in the steel industry. A typical chemical analysis is (per cent): SiO_2, 3–7; Fe_2O_3, 10–14, Al_2O_3, 12–18; Cr_2O_3, 25–30; CaO, 1–2; MgO, 35–40. In Western Europe, P.R.E. define a chrome–magnesite refractory as containing 25–55% MgO.

Chrome Ore. An ore consisting primarily of chrome spinels, (Fe, Mg) O.(Cr, Fe, Al)$_2$O$_3$; it occurs in ultrabasic igneous rocks and in rocks derived from them by alteration, e.g. serpentines. The chief sources are in S. Africa, Rhodesia, Turkey and Russia; there are other important deposits in Yugoslavia, Greece, India, Philippines and Cuba. The composition varies widely; a typical

58

refractory-grade ore contains (per cent) Cr_2O_3, 40–45; SiO_2, 3–6; and Al_2O_3, FeO, and MgO, 15–20 of each. The principal use in the refractories industry is in the manufacture of chrome–magnesite bricks for the steel industry.

Chrome Refractory. A refractory brick made from chrome ore without the addition of other materials. Chrome refractories are neutral, chemically, and find use as a separating course of brickwork between silica refractories and basic refractories; their R.u.L. (q.v.) is considerably lower than that of chrome–magnesite refractories. Chemical composition (per cent): SiO_2, 4–7; Fe_2O_3, 12–15; Al_2O_3, 16–20; Cr_2O_3, 38–42; CaO, 1–2; MgO, 15–20. In Western Europe, P.R.E. define chrome refractories as containing $\leqslant 25\%$ MgO and $\geqslant 25\%$ Cr_2O_3.

Chrome Spinel. A SPINEL (q.v.) in which the trivalent metal is Cr.

Chrome–Tin Pink. A colour for ceramic glazes; it was first used in England in the 18th century and was originally called 'English Pink'. The colour is probably produced by the precipitation of fine particles of chromic oxide on the surface of tin oxide in an opaque glaze; lime must also be present.

Chrome–Zircon Pink. About 70% of the SnO_2 used in CHROME–TIN PINK (q.v.) can be replaced by zircon without impairing the colour or stability.

Chromic Oxide. Cr_2O_3; m.p. 2275°C; sp. gr. 5·21; thermal expansion (100–1000°C) $7·3 \times 10^{-6}$. A source of green colours for the ceramic industry but usually added in the form of a chromate, e.g. potassium dichromate. The oxide is highly refractory and has been used to a limited extent to improve the bond in CHROME–MAGNESITE REFRACTORIES (q.v.)

Chromite. Ferrous chromite, $FeO.Cr_2O_3$; the principal constituent of many CHROME ORES (q.v.).

Chromium Borides. At least seven chromium borides have been reported. The compound that has received most attention is the diboride, CrB_2; m.p. 2000°C (approx.); sp. gr. 5·4; thermal expansion $4·6 \times 10^{-6}$. When heated in air, a coating of B_2O_3 is formed and prevents further oxidation up to the limiting temperature at which B_2O_3 itself ceases to be stable. It has been used for the flame-spraying of combustion-chamber linings.

Chromium Carbide. Several Cr carbides exist, e.g. Cr_3C_2, m.p. 1890°C; sp. gr. 6·68. They are hard, refractory, and chemically resistant. As such they have found limited use in bearings and corrosion-resistant nozzles, etc.

Chromium Oxide

Chromium Oxide. See CHROMIC OXIDE.

Chromium Silicides. Several compounds exist: Cr_3Si, Cr_2Si, $CrSi$, and $CrSi_2$. The m.p. varies from 1570°C for $CrSi_2$ to 1710°C for Cr_3Si. Hardness varies from 76–89 Rockwell A. These silicides resist oxidation up to about 1000°C; they have good resistance to thermal shock but low resistance to impact.

Chrysotile. The principal mineral of the ASBESTOS (q.v.) group. When pure it has the composition $3MgO.2SiO_2.2H_2O$.

Chum. A shaped block of wood or plaster on which a bat of pottery body can be roughly shaped before it is placed in the mould for jolleying in the making of large items of hollow-ware; a piece of flannel is placed over the chum before it is used and this subsequently slips off the chum when the shaped clay is removed.

Chunk Glass. Rough pieces of optical glass obtained when a pot of glass is broken open.

CIE System. The name derives from the initials of the Commission Internationale de l'Eclairage. It is a trichromatic system of colour notation that is being used, for example, in the glass industry.

Ciment Fondu. Trade-name: aluminous hydraulic cement. (Lafarge Aluminous Cement Co. Ltd., 73 Brook St., London, W.1.) For general properties of this type of cement see HIGH–ALUMINA CEMENT.

Cimita. A natural mixture of clay and feldspar occurring in parts of Chile. The composition is not uniform but a typical analysis (per cent) is: SiO_2, 58; Al_2O_3, 33; Fe_2O_3, 1; Alkalis, 4; H_2O, 4.

Cinder Notch. See SLAG NOTCH.

Circle Brick. A brick with two opposite larger faces curved to form parts of concentric cylinders (cf. RADIAL BRICK, and see Fig. 1, p. 37).

Circle System of Firing. See under ROTARY-HEARTH KILN.

Cire-perdue. French words for LOST WAX (q.v.).

Citadur. Trade-name: a HIGH-ALUMINA CEMENT (q.v.) made in Czechoslovakia.

Citric Acid. An organic acid sometimes used for the treatment of steel prior to enamelling.

Cladding. An external non-load bearing facing, e.g. of bricks, faience or tiles, used in frame-type buildings (cf. VENEERED WALL).

Clamming. Local name for the brick and fireclay filling of the

WICKETS (q.v.) of an old-type pottery kiln; sometimes spelled CLAMIN.

Clamp. A kiln constructed, except for the permanent foundations, of the bricks that are to be fired, together with combustible refuse and breeze. STOCK BRICKS (q.v.) in the London area were formerly fired exclusively in clamps but these produced only about 50% of first-quality bricks and many have now been replaced by annular or tunnel kilns.

Clark Circle System. See under ROTARY-HEARTH KILN.

Clash. Local Scottish term for a thin slurry of clay and water.

Clay. A natural material characterized by its plasticity, as taken from the clay-pit or after it has been ground and mixed with water. Clay consists of one or more clay minerals together with, in most cases, some free silica and other impurities. The common clay mineral is kaolinite; most clays consist of kaolinite in various degrees of atomic disorder. See also BRICK CLAYS, CHINA CLAY, FIRECLAY and the clay minerals HALLOYSITE, KAOLINITE, MONTMORILLONITE, etc.

Clay Building Brick. A brick for normal constructional purposes; such bricks can be made from a variety of BRICK CLAYS (q.v.). Relevant British Standards are: B.S. 657 (Dimensions) and B.S. 1257 (Testing). The US Standards are ASTM – C62 (Building Brick); ASTM – C216 (Facing Brick); ASTM – C67 (Sampling and Testing).

Clay Lath. B.S. 2705 describes this as copper-finished steel wire mesh at the intersections of which suitably shaped unglazed clay nodules have been bonded by a firing process. Clay lath provides a stable, well-keyed base to cover the whole of a surface with the minimum number of joints; it is supplied in rolls or mats.

Clay Shredder. A unit for the preliminary preparation of plastic clay. The machine consists of a hopper with a flat or conical base; adjustable knives operate from a vertical, central, rotating shaft. The clay falls from the shredder through slots in the casing.

Clayite. A term proposed by J. W. Mellor (*Trans. Brit. Ceram. Soc.*, **8**, 28, 1909) 'for that non-crystalline variety of the hydrated aluminium silicate—$Al_2O_3.2SiO_2.2H_2O$—which occurs in china clay and in most clays yet examined. Kaolinite is crystalline clayite'. When X-ray analysis was subsequently able to demonstrate the crystallinity of all the clay minerals except ALLOPHANE (q.v.), the term was discarded.

Clayspar. Trade-name for a siliceous raw material occurring in

Cleaner

Scandinavia and containing approx. 95% SiO_2, 2·5% Al_2O_3, 1·5% K_2O, 0·5% Na_2O.

Cleaner. A hot solution of alkalis (strength about 5%) used to remove grease and dirt from the base-metal before it is enamelled.

Clear Clay. A clay such as kaolin that is free from organic matter and so does not give rise to bubbles if used in a vitreous enamel; such clays are used in enamels when good gloss and clear colours are required.

Clinker. (1) Hydraulic cement in the unground state as it issues from the rotary or shaft kiln in which it was made.

(2) Dead-burned basic refractory material, e.g. stabilized dolomite clinker.

(3) Lumps of fused ash from coal or coke as formed, for example, in the fireboxes of kilns.

(4) For the similar word as applied to continental engineering bricks see KLINKER BRICK.

Clinkering Zone. That part of a cement kiln which is in the temperature range (1350–1600°C) in which the constituents react to form the CLINKER (q.v.).

Cloisonné Enamelware. One type of vitreous-enamel artware: a pattern is outlined on the base-metal by attaching wire fillets to the metal; enamels of different colours are applied within the partitions thus made; the ware is then fired and the surface is polished. (From French word meaning 'partitioned'.)

Closed-circuit Grinding. A size-reduction process in which the ground material is removed either by screening or by a classifier, the oversize being returned to the grinding unit. Typical examples are a dry-pan with screens, dry-milling in an air-swept ball mill, and wet-milling in a ball mill with a classifier.

Closed Porosity. See under POROSITY.

Closer. See PUP.

Closet Suite. A suite of ceramic sanitaryware including the closet and the flushing cistern.

Clot. Roughly shaped clay or body ready for a final shaping process.

Clot Mould. The mould, in some types of stiff plastic brick-making machines, into which a clot of clay is extruded and from which it is then ejected prior to the final re-pressing.

Clunch. A name used in some parts of England for a tough, coarse, clay or for a marly chalk.

CMC. Abbreviation for CARBOXYMETHYLCELLULOSE (q.v.).

CO₂ Process. See CARBON DIOXIDE PROCESS.

Coade Stone. A vitreous ware, used for architectural ornament, made in London by Mrs Coade from 1771 until her death in 1796; manufacture finally discontinued in about 1840. The body consisted of a kaolinitic clay, finely ground quartz and flint, and a flux (possibly ground glass).

Coating. See METAL PROTECTION; REFRACTORY COATING.

Cob Mill. A disintegrator for breaking down agglomerates of the raw materials used in the compounding of vitreous-enamel batches.

Cobalt Chloride, $CoCl_2.6H_2O$; Cobalt Nitrate, $Co(NO_3)_2.6H_2O$. Soluble cobalt compounds added to the slip that is made when preparing pottery bodies for the purpose of neutralizing the slightly yellow colour caused by the presence of 'iron' impurities; the principle is the same as the use of a 'blue bag' in the laundry. About 0·01% is sufficient.

Cobalt Oxides. Grey or 'prepared' oxide, CoO; cobaltic oxide, Co_2O_3; black oxide, Co_3O_4. Used to give a blue colour to glass and pottery ware, and added (together with nickel oxide) to ground-coat enamels for steel to improve their adherence. The tint of cobalt oxide colours can be modified by adding other oxides (see MAZARINE BLUE, WILLOW BLUE and CELESTE BLUE). With the addition of the oxides of manganese and iron a black colour can be produced.

Cobalt Sulphate, $CoSO_4.7H_2O$. A soluble cobalt compound used for the same purpose as COBALT CHLORIDE (q.v.). It is also sometimes used in vitreous enamels for mottling single-coat grey-coloured ware.

Coconut Piece. A special shape of ceramic wall tile (see Fig. 6, p. 307).

Cod Placer. See under PLACER.

Coefficient of Scatter. A term used in the testing of vitreous enamelware and defined (ASTM – 286) as: The rate of increase of reflectance with thickness at infinitesimal thickness of vitreous enamel over an ideally black backing; for method of test see ASTM – C347.

Coesite. A form of silica produced at 500–800°C and a pressure of 35 kilobars; sp. gr. 3·01; insoluble in HF. Named from L. Coes who first obtained this form of silica (*Science*, **118**, 131, 1953).

Coil Building. A primitive method of shaping clay vessels by rolling clay into a 'rope', which is then coiled to form the wall of the vessel; the inner and outer surfaces of the roughly shaped ware are finally smoothed.

Coke Oven. A large, refractory-lined structure consisting of a series of tall, narrow chambers in which coal is heated out of contact with air to form coke. Silica refractories are used for most of the coke-oven structure.

Colburn Process. The production of sheet glass by vertical drawing for about 4 ft and then bending over a driven roller so that the cooling sheet then travels horizontally. The process was invented in USA by I. W. Colburn in 1905 and was subsequently perfected by the Libbey–Owens Company.

Cold Crushing Strength. See CRUSHING STRENGTH.

Cold-process Cement. Older name for SLAG CEMENT (q.v.).

Colemanite. Naturally occurring calcium borate, $2CaO.3B_2O_3.5H_2O$; there are deposits in Nevada and California. It is used in some pottery glazes.

Collar. A short fireclay section used to join the main (silica) part of a horizontal gas retort to the metal mouthpiece.

Collaring. The final process in THROWING (q.v.) a vase: both hands are used to compress the clay and thus reduce the size of the neck.

Collet. A flange for the mounting of an abrasive wheel on a spindle.

Colloid. Material in a form so finely divided that, when dispersed in a liquid, it does not settle out unless flocculated by a suitable electrolyte. The finest fraction of plastic clays is in this form. The colloidal size range is approx. $0.001–0.20\mu$.

Colloid Mill. A high-speed dispersion unit yielding a suspension of particles of the order of 1μ in size. The original mill of this type was designed by H. Plauson (Brit. Pat. 155 836 and subsequent patents).

Colorimeter. An instrument for COLOUR MEASUREMENT (q.v.).

Colour Measurement. There are two basic methods of measurement: (1) Spectrophotometer, typified by the Beckman and the Hardy instruments; (2) Tristimulus Filter Method, typified by the 'Colormaster', Hilger, and Hunter instruments. The subject of colour measurement is important in the ceramic industry—particularly for glazed tiles, sanitary-ware and vitreous enamel-ware, which may have to match. For a good introduction to colour measurement see: W. D. Wright, *The Measurement of*

Colour (Hilger, 2nd Ed.); D. B. Judd and G. Wyszecki, *Colour in Business, Science and Industry* (Wiley, 1963).

Colour Twist. Twisted coloured glass rods as a form of decoration within a wine-glass stem.

Comb Rack. A bar of acid-resisting metal, e.g. Monel, used to support and to separate metal-ware in a PICKLING (q.v.) basket, while it is being prepared for the application of vitreous-enamel slip.

Combed. A surface texture of narrowly spaced lines produced on clay facing bricks by fixing wires or plates above the extruding column of clay so that they 'comb' its surface.

Comeback. In the vitreous-enamel industry this term denotes the time that elapses before a batch-type furnace regains its operating temperature after a fresh load of ware has been placed in it.

Comminution. A term covering all methods of size reduction of materials, whether by crushing, impact, or attrition.

Commons. Clay building bricks that are made without attention to appearance and intended for use in the inner leaf of cavity walls or for internal walls. The crushing strength of such bricks varies from about 1500 to 6000 p.s.i., the water absorption from about 10 to 30 wt. %.

Compacting Factor. A factor indicating the workability of any concrete made with aggregate not exceeding $1\frac{1}{2}$ in., but of particular value for assessing concretes of low workability. The factor is the ratio of the bulk weight of the concrete when compacted by being allowed to fall through a specified height, to that when fully compacted. For details see B.S. 1881.

Compaction. The production of a dense mass of material, particularly of a powder, in a mould. Ceramic powders may be compacted by dry-pressing, tamping, vibration, isostatic pressing, or explosive pressing.

Compass and Wedge. Term sometimes used for a brick that has a taper both on the side and on the face, e.g. a 9 in. brick tapered $4\frac{1}{2}/3\frac{1}{2}$ in. and $2\frac{1}{2}/2$ in.

Compo. A siliceous, highly grogged fireclay composition, with crushed coke or graphite sometimes added, for use as a general ramming material in the casting-pit of a steelworks or in a foundry. A typical composition is (per cent): crushed firebrick, 65–70; chamotte, 15–20; siliceous fireclay, 14; foundry blacking, 1. The batch is made up with about 10% water. (The word is an abbreviation for 'Composition'.)

Composition Brick

Composition Brick. Scottish term for a common building brick made by the stiff-plastic process from clay and colliery waste; characteristically, it has a black core.

Compound Rolls. Two or more pairs of CRUSHING ROLLS (q.v.) arranged above one another, the upper pair acting as a primary crusher and the lower pair as a secondary crusher. Compound rolls find use in the size-reduction of brick clays.

Concrete. A mixture of hydraulic cement, fine aggregate (e.g. sand) and coarse aggregate, together with water; it is placed *in situ* or cast in moulds and allowed to set. The ratio of the three solid constituents is usually expressed in the order given above, i.e. '1 : 2 : 5 : concrete' means 1 part cement, 2 parts fine aggregate and 5 parts coarse aggregate. The properties are largely determined by the cement and 'fines', but, for normal constructional purposes, the coarse aggregate must be graded to give dense packing.

Concrete Aggregate. Normal (as opposed to lightweight) concrete aggregate includes sand and gravel, crushed rock of various types and slag. The nomenclature is given in B.S. 812; the mineralogical composition is dealt with in ASTM – C294 and C295.

Concrete Block. The properties required of concrete blocks, both dense and lightweight, are specified in the UK in B.S. 2028. In USA the properties required of a solid concrete building block are specified in ASTM – C145; the properties of hollow concrete blocks are specified in ASTM – C90 and C129; for methods of sampling and testing see ASTM – C140 and C426.

Concrete Brick. A building brick made from portland cement and a suitable aggregate. ASTM – C55 specifies two qualities: Grade A (for use where exposed to frost), crushing strength $\not< 2500$ p.s.i., water absorption $\not> 15$ lb/ft^3; Grade B (for back-up or interior walls), crushing strength $\not< 1500$ p.s.i.

Condenser. The condenser attached to a horizontal zinc retort for the cooling of the zinc vapour and its collection as metal is made of fireclay.

Conditioning Zone. (1) The part of a tank furnace for flat glass where the temperature of the glass is adjusted before it flows into the forehearth or drawing chamber.

(2) That part of the feeder, away from the wall of a glass-tank furnace, in which the temperature of the molten glass is adjusted to that required for working.

66

Conduit (for Electric Cable). See CABLE COVER.

Cone. (1) The usual term for a PYROMETRIC CONE (q.v.) or a TEST CONE (q.v.); for nominal squatting temperatures of Pyrometric Cones see Appendix 2.

(2) A special fireclay shape sometimes used in the bottom-pouring of molten steel; it has a hole through it and conveys the steel from runner brick to ingot mould. The cone is frequently made as an integral part of a runner brick, in which case the latter is known as a RISER BRICK or END RUNNER (q.v.).

Cone Crusher. A primary crusher for hard rocks, e.g. quartzite used in making silica refractories; it consists of a hard steel cone that is rotated concentrically within a similarly shaped steel casing (cf. GYRATORY CRUSHER).

Cone Hip. See HIP TILE.

Cone-screen Test. A works' test for the fineness of milled enamels. A standard volume of enamel slip is washed through a conical sieve; the amount of oversize residue is read from a graduated scale along one side of the sieve.

Cone Wheel. A small cone-shaped abrasive wheel of the type frequently used in portable tools.

Congruent Melting. The melting point of a solid is said to be congruent if the material melts completely at a fixed temperature and without change in composition; alumina melts congruently at 2050°C (cf. INCONGRUENT MELTING).

Consistodyne. Trade-name; a device for attachment to the barrel of a pug for controlling the workability of the clay; (*Ceramic Age*, **77**, No. 11, 34, 1961).

Consistometer. Term used in the vitreous-enamel industry for various instruments designed for the evaluation of the flow properties of enamel slips. The earliest consistometer was that used by R. D. Cooke (*J. Amer. Ceram. Soc.*, **7**, 651, 1924); this was of the capillary type and was developed into an instrument suitable for works' control by W. N. Harrison (*ibid.*, **10**, 970, 1927). Other instruments for controlling the properties of vitreous-enamel slips include the GARDNER MOBILOMETER (q.v.).

Constringence. See ABBE NUMBER.

Continuous Chamber Kiln. See TRANSVERSE-ARCH KILN.

Continuous Kiln. A kiln in which the full firing temperature is continuously maintained in one or other zone of the kiln. There are two types: ANNULAR KILN (q.v.) and TUNNEL KILN (q.v.).

Continuous Vertical Retort. A type of gas retort, built of silica

Contrast Ratio

or siliceous refractories. Coal is charged into the top of the retort, coke is extracted from the bottom, and town gas is drawn off, the whole operation being continuous (cf. HORIZONTAL RETORT). Continuous vertical retorts are also used in the zinc industry, in which case they are built of silicon carbide refractories.

Contrast Ratio (C_{80}). A ratio related to the reflectance of a vitreous enamel coating and defined in ASTM – C347 as: The ratio of the reflectance of an enamel coating over black backing to its reflectance over a backing of reflectance 0.80 (80%).

Contravec. Trade-name: a system for the blowing in of air at the exit end of a tunnel kiln to counteract the normal convection currents. (Gibbons Bros. Ltd., Dudley, England.)

Conversion. A change in crystalline structure on heating that is not immediately reversible on cooling. The most important example in ceramics is the conversion of quartz at high temperature into cristobalite and tridymite (cf. INVERSION).

Converter. A refractory-lined vessel supported on trunnions and used for the production of steel by oxidation of the impurities in the molten pig-iron that forms the charge; this is done by a blast of air or oxygen. In the original BESSEMER CONVERTER the blast passed through the converter bottom, the basic refractory lining being pierced with tuyères for this purpose. In the TROPENAS CONVERTER the blast strikes the surface of the molten metal via tuyères passing through the refractory wall—hence the alternative name SIDE-BLOWN CONVERTER.

Cooke Elutriator. A short-column hydraulic elutriator for sub-sieve sizes designed by S. R. B. Cooke (*U.S. Bur. Mines, Rept. Invest.* No. 3333, 1937).

Cooler. As used in the portland cement industry, the term 'cooler' refers to the ancillary unit of a cement kiln into which hot clinker is discharged to cool before it is conveyed to the grinding plant.

Cooling Arch. A furnace for the annealing of glassware, which is placed in the furnace and remains stationary throughout the annealing (cf. LEHR).

Copacite. Trade-name; a Canadian SULPHITE LYE (q.v.).

Copper Carbonate. The material is the basic carbonate, $CuCO_3 . Cu(OH)_2$; it is used as a source of copper for coloured glazes.

Copper Enamel. A vitreous enamel specifically compounded for application to copper; the composition is essentially lead silicate with small additions of alkalis, arsenic oxide and (sometimes) tin oxide.

Copperhead. A fault (reddish-brown spots) liable to appear in the ground-coat during vitreous enamelling; the spots are exposed areas of oxidized base metal. The usual cause is either boiling from the base-metal or inadequate metal preparation.

Copperlight. A glass window pane, $\frac{1}{4}$ in. thick and up to 16 in^2 size, fitted in a special copper frame and used as a 'fire-stop'.

Copper Oxide. Cupric oxide (black), CuO; cuprous oxide (red) Cu_2O. Used as colouring agents in pottery and glass. Copper oxide normally gives a green colour but under reducing conditions it gives a red due to the formation of colloidal copper, as in rouge flambé and sang de boeuf art pottery and copper ruby glass.

Copper Ruby Glass. See RUBY GLASS.

Copper Titanate. $CuTiO_3$; a compound sometimes added in amounts up to 2% to $BaTiO_3$ to increase the fired density.

Coquille. Thin glass with a radius of curvature of $3\frac{1}{2}$ in.; used in the production of sun glasses; from French word meaning 'shell' (see also MICOQUILLE).

Coral Red. A ceramic colour. One form of coral red consists of basic lead chromate; this compound is unstable and the decorating fire must be at a low temperature.

Corbel. Brickwork in which each course projects beyond the course immediately below.

Cord. A fault in glass resulting from heterogeneity and revealed as long inclusions of glass of different refractive index from that of the remainder of the glass.

Cordierite. A magnesium alumino-silicate $2MgO.2Al_2O_3.5SiO_2$; part of the Mg can be replaced by Fe or Mn. Cordierite exists in three crystalline forms but only the α-form is commonly encountered; it melts incongruently at 1540°C. There is a large deposit of cordierite in Wyoming, USA, but it is usually synthesized by firing a mixture of clay, steatite and alumina (or an equivalent batch) at 1250°C or above. The mineral has a low and uniform thermal expansion ($2\cdot3 \times 10^{-6}$) and cordierite bodies therefore resist thermal shock. The electrical resistance is adequate for many purposes when high thermal shock resistance is also required, e.g. arc chutes, electric fire-bars, fuse cores, etc. The dielectric constant is $5\cdot0$ and the power factor $0\cdot004$, both measured at 1 mc and 20°C.

Core. (1) The central part of a plaster mould of the type used in SOLID CASTING (q.v.).

(2) The central part of a sand-mould as used in foundries.

(3) A one-piece refractory or heat-insulating shape for use at the top of an ingot mould and serving the same purpose as a HOT-TOP (q.v.); this type of core is also sometimes called a DOZZLE.

Corhart. Trade-name for various types of electrically fused refractories: Corhart Standard is a fused mullite–corundum refractory; Corhart Zac is a similar product containing zirconia. These electrocast refractories are used chiefly in glass-tank furnaces. (Corhart Refractories Co., Louisville 2, Kentucky, USA; English agent—British Hartford–Fairmont Ltd, Greenford, Middlesex.)

Cornelius Furnace. A type of glass-melting furnace in which the glass is heated by direct electrical resistance. The design was introduced in Sweden by E. Cornelius (Brit. Pat. 249 554; 23/3/25; 303 798; 8/1/29).

Corner Wear. The wear of an abrasive wheel along one or both of its circumferential edges.

Cornish Crucible. A small (e.g. $3\frac{1}{2}$ in. high, 3 in. diam.) clay crucible of a type used for the assaying of copper; these crucibles are made from a mixture of about equal parts of ball clay and silica sand.

Cornish Stone. See CHINA STONE.

Corridor Dryer. Term sometimes used for a CHAMBER DRYER (q.v.).

Corrosion. Wear caused by chemical action (cf. ABRASION and EROSION (q.v.)).

Corundum. The only form of alumina that remains stable when heated above about 1000°C; also known as α-Al_2O_3; m.p. 2050°C; hardness 9 Moh; sp. gr. 3·97; thermal expansion (20–1000°C) $8\cdot5 \times 10^{-6}$. It occurs naturally, but impure, in S. Africa and elsewhere but is generally produced by extraction from bauxite followed by a firing process at high temperature. Corundum is used as an abrasive and as a special refractory and electroceramic, e.g. in sparking plugs (see also ALUMINA).

Cottle. Term used in the N. Staffordshire potteries for the material, e.g. stiff canvas, used to form the sides of a plaster mould while the plaster is being poured in and until the plaster has set. (Probably from N. country *Cuttle*—a layer of folded cloth.)

Coulter Counter. A high-speed device for particle-size analysis designed by W. H. Coulter (*Proc. Nat. Electronics Conf.*, **12**, 1034, 1956) and now made by Coulter Electronics Inc., Chicago, USA. A suspension of the particles flows through a small aperture having an immersed electrode on either side, with particle concentration such that the particles traverse the aperture substantially one at a time. Each particle, as it passes, displaces electrolyte within the aperture, momentarily changing the resistance between the electrodes and producing a voltage pulse of magnitude proportional to particle volume. The resultant series of pulses is electronically amplified, scaled and counted.

Counter Blow. In the BLOW-AND-BLOW (q.v.) process of shaping glass-ware, the operation during which the parison is blown out.

Course. By convention, a course of brickwork includes one layer of mortar as well as the bricks themselves.

Cove Skirting. A special shape of ceramic wall tile (see Fig. 6, p. 307).

Cover. An item of KILN FURNITURE (q.v.). The cover is the flat refractory shape forming the top of a CRANK (q.v.); it protects the top piece of ware and at the same time holds the PILLARS (q.v.) in position.

Cover-coat. The final coat applied to vitreous enamelware, resulting in the top surface. Normally there is a ground-coat and a cover-coat, but some enamels are now sufficiently opaque for single-coat application to appropriate grades of base-metal.

Covered Pot. See under POT.

Covering Power. The ability of a glaze or vitreous enamel to cover, uniformly and completely, the surface of the fired ware.

Cowper Stove. See HOT-BLAST STOVE.

Cracking off. The severing of shaped glass-ware from the MOIL (q.v.).

Crackle. (1) A multiply crazed or cracked surface on art pottery or glass. To produce the effect on pottery the glaze is compounded so as to have a higher thermal expansion than the body; the craze pattern is sometimes emphasized by rubbing colouring matter, such as umber, into the fine cracks. With glass, the ware is cracked by quenching in water; it is then reheated and shaped.

(2) A crackled vitreous enamel—the surface appearing to be wrinkled due to its mottled texture—can be produced by the wet process of application.

Crank

Crank. This word is used in the pottery industry in two related senses: (1) A thin refractory bat (Fig. 4, p. 158) used as an item of KILN FURNITURE (q.v.) in the glost-firing of wall tiles. A number of cranks, each supporting one or more tiles, are built up to form a stack; the cranks are kept apart by refractory distance-pieces known as DOTS.

(2) A composite refractory structure for the support of flatware during glost- and decorating-firing; the crank is designed to prevent the glazed surfaces of the ware from coming into contact with other ware or kiln furniture.

Craquelé. See CRACKLE.

Crawler. Local term for an apron-feeder to a pan mill used in brickmaking.

Crawling. (1) A defect that sometimes occurs during the glazing of pottery, irregular areas that are unglazed, or only partially glazed, appearing on the fired ware. The cause is a weak bond between glaze and body; this may result from greasy patches or dust on the surface of the biscuit ware, or from shrinkage of the applied glaze slip during drying.

(2) A similar defect liable to occur in vitreous enamelling when one coating of enamel is fired over another coating that has already been fired. Causes include a too-heavy application of enamel, poorly controlled drying, and the use of enamel that has been too finely ground.

Crazing. The formation of a network of surface cracks. A typical example is the crazing of a glaze; this is caused by tensile stresses greater than the glaze is able to withstand. Such stresses may result from MISMATCH (q.v.) between the thermal expansions of glaze and body, or from MOISTURE EXPANSION (q.v.) of the body; in the special case of glazed tiles fixed to a wall, a third cause is movement of the wall or of the cementing material between the tile and the wall. Crazing of vitreous enamelware may also occur, the system of fine cracks penetrating through the enamel to the base-metal. The crazing of cement and concrete is due partly to natural shrinkage and partly to volume changes following surface reaction with CO_2 present in the atmosphere. (For CRAZING TESTS see under AUTOCLAVE; HARKORT TEST; PUNCH TEST; RING TEST; SINGER'S TEST; STEGER'S TEST; TUNING-FORK TEST.)

Crazing Pot. Popular name in the pottery industry for an AUTOCLAVE (q.v.).

Creep. The slow deformation that many materials undergo

when continuously subjected to a sufficiently high stress. With most ceramics, creep becomes measurable only when the stress is applied at a relatively high temperature.

Cremer Kiln. A German design of tunnel kiln that can be divided into compartments by a series of metal slides to permit better control of temperature and atmosphere. The fired ware is cooled by air currents through permeable refractory brickwork in the kiln roof or by water-cooling coils. (G. Cremer, Brit. Pats. 697 644; 30/9/53; 740 639; 16/11/55; 803 691; 29/10/58.)

Crespi Hearth. A type of open-hearth steel furnace bottom characterized by the fineness of the particles of dolomite used for ramming; after it has been burned-in, the hearth is very dense and resistant to metal penetration. (G. B. Crespi, Brit. Pat. 507 715; 10/8/38.)

Crimping. The production of a rolled or curled edge to the base-metal prior to vitreous enamelling.

Cristobalite. The crystalline form of silica stable at high temperatures; its m.p. is 1723°C. Cristobalite is formed when quartz is heated with a mineralizer at temperatures above about 1200°C and is itself characterized by a crystalline inversion, the temperature of which varies but is generally between 200°C and 250°C; this inversion is accompanied by a change in length of about 1%. Cristobalite is a principal constituent of silica refractories, causing their sensitivity to thermal shock at low temperatures; it is also present in many pottery bodies and is synthesized for use as a refractory powder in the investment casting of metals. A silica deposit stated to consist of about 85% cristobalite and 15% quartz has been recently discovered in S. Africa.

Critical Speed. (1) The maximum safe speed of rotation of an abrasive wheel; at higher speed vibration becomes dangerous.

(2) In ball-milling, the speed of rotation above which the balls remain against the casing, as a result of centrifugal force, without cascading; this speed is given by the equation: $N = 54.18R^{-\frac{1}{2}}$ where N is the rev/min and R is the radius of the interior of the mill, less the radius of the ball, in feet.

Crizzle. See CHECK.

Crockery. A popular term for ceramic tableware.

Crookes Glass. A glass, usually containing cerium, that absorbs ultra-violet light and is used for protective goggles, etc. This glass resulted from the work of Sir Wm. Crookes, in 1914, for the Glass Workers' Cataract Committee of the Royal Society.

Cross-bend Test. (1) Term sometimes used for TRANSVERSE STRENGTH TEST (q.v.).

(2) A test to determine the resistance of vitreous enamelware to cracking when it is distorted.

Cross-fired Furnace. A glass-tank furnace heated by flames that cross the furnace perpendicular to the direction of flow of the glass; the furnace has several pairs of ports along its melting end (cf. END-FIRED FURNACE).

Crouch Ware. Light-coloured Staffordshire salt-glazed stoneware of the early 18th century; it was made from a clay from Crich, Derbyshire, the word 'crouch' being a corruption.

Crowding Barrow. A hand-barrow for bricks; it has a base and front, but no sides.

Crown. A furnace roof, particularly of a glass-tank furnace.

Crown Blast. The procedure of blowing air at roof level into the exit end of a tunnel kiln to counteract the natural flow of gases in this part of the kiln.

Crown Brick. See KEY BRICK and CENTRE BRICK.

Crown Glass. Glass of uneven thickness and slightly convex (thus producing some optical distortion), hand-made by blowing and spinning (cf. OPTICAL CROWN GLASS).

Crush Dressing. Shaping the face of an abrasive wheel to a required contour by means of steel rolls.

Crushing Rolls. A unit frequently used for the size reduction of brick clays, etc. A pair of steel rolls are arranged horizontally and adjacent, with a gap between them of a uniform width equivalent to the maximum size of particle allowable in the crushed product; the rolls are rotated in opposition, and sometimes at different speeds, so that the clay is carried downwards and 'nipped' between the approaching surfaces of the rolls (cf. EXPRESSION ROLLS).

Crushing Strength. The maximum load per unit area, applied at a specified rate, that a material will withstand before it fails. Typical ranges of values for some ceramic materials are:

Fireclay and silica refractories:	2000–5000 p.s.i.
Common building bricks:	2000–6000 p.s.i.
Engineering bricks Class 'A':	>10 000 p.s.i.
Sintered Alumina:	>50 000 p.s.i.

Cryolite. Natural sodium aluminium fluoride, Na_3AlF_6; m.p. 980°C; sp. gr. 2·9. Because of its low m.p. and its fluxing action, it is used in the manufacture of enamels and glass and in the ceramic coatings of welding rods. Opal glass is often made from batches containing about 10% cryolite; a similar preparation is sometimes used in white cover-coat enamels.

Crypto System. An impulse system of oil firing, more particularly for the top-firing of annular kilns. Trade-name: R. Aebi & Cie., Zurich.

Cryptoflorescence. Term proposed by S. A. McIntyre and R. J. Schaffer (*Trans. Brit. Ceram. Soc.*, **28**, 363, 1929) for soluble salts that have crystallized in the interior of a clay building product and are therefore hidden.

Crystal Glass. A popular, but misleading, name for a type of decorative glass that is usually deeply cut so that the brilliance resulting from its high refractive index is fully displayed. English Crystal Glass, or English Full Crystal, has the basic composition 55% SiO_2, 33% PbO, 12% K_2O; a so-called 'Half Crystal' contains only about 15% PbO.

Crystalline Glaze. A glaze containing crystals of visible size to produce a decorative effect. Typical examples are glazes containing zinc silicate crystals and the AVENTURINE (q.v.) glazes.

C.T. Nozzle. Trade-name: a refractory nozzle for steel-pouring designed to give a *Constant Teeming* rate (hence the name). The nozzle consists of an outer fireclay shell and a refractory insert of different composition. Strictly speaking, the term refers to a particular type of insert developed for the teeming of free-cutting steels. (Thos. Marshall & Co. (Loxley) Ltd., Brit. Pats. 832 280, 6/4/60; 904 526, 29/8/62.)

Cubing Rolls. CRUSHING ROLLS (q.v.) having projections and used for breaking down hard 'slabby' clays into a cube-like product that is more suitable for feeding to a secondary grinding unit.

Cuckhold. An iron tool for cutting off lumps of prepared clay, from a pug, ready for the hand-moulding of building bricks.

Cull. US equivalent of the English WASTER (q.v.).

Cullet. Broken glass that can be recharged to the glass furnace. The word is derived from the French *collet*, the little neck left on the blowing iron when bottles were hand blown; these 'collets' were returned to the glass-pot and remelted. FACTORY CULLET or DOMESTIC CULLET is from the same glass-works at which it is to be used; FOREIGN CULLET is from a different glass-works.

Culm. A Carboniferous shale used for brickmaking in the Exeter area. (Dialect word for coal dust or soot.)

Cummings' Sedimentation Method. An approximate method of particle-size analysis having the merit of giving a weight/size distribution directly. (D. E. Cummings, *J. Industr. Hyg. Toxicol.*, **11**, 245, 1929.)

Cup Gun. A spray gun, particularly as used for touching-up vitreous enamelware, with a container for the enamel slip forming an integral part of the gun.

Cup Wheel. An abrasive wheel shaped like a cup; such wheels are used, for example, in the grinding of flat surfaces. Diamond cup-wheels are employed in the grinding of tungsten carbide tools.

Cupel. A small, refractory, tapered cylinder (broad end up) with a shallow depression in the top; cupels are made from bone ash or calcined magnesia and are used for the assay of non-ferrous ores.

Cupola. A shaft furnace used in a foundry for the melting of iron. Cupolas are generally lined with fireclay refractories covered with a ganister-clay mixture. For the production of cast-iron with a low sulphur content, a basic lining is sometimes used; the lining is in this case built of chrome–magnesite or dolomite refractories, or it may be rammed with a monolithic basic refractory composition.

Cupola Brick. See KEY BRICK.

Cupping. A process in which vitreous enamel slip is poured over selected areas of a piece of ware while it is being drained, to ensure that the overall thickness of application shall be uniform.

Curb Bend. A special shape of wall tile (see Fig. 6, p. 307).

Curie Point. The temperature (of importance in special electroceramics) at which a material changes from ferroelectric to non-ferroelectric, or from ferromagnetic to paramagnetic, behaviour. Named after Pierre Curie who discovered the effect in 1895.

Curing. The process of keeping freshly placed concrete moist to ensure complete hydration so that maximum strength is developed. Compounds are available for spraying on concrete to retard loss of moisture during the curing period (see ASTM – C156 and C309). Pre-cast concrete units are often STEAM CURED (q.v.).

Curling. A defect in vitreous enamelling that is similar to CRAWLING (q.v.).

Curtain Arch. An arch of refractory brickwork that supports the wall between the upper part of a gas-producer and the gas uptake.

Curtain Wall. See SHADOW WALL.

Curtains. An enamelling defect, in the form of dark areas having the appearance of drapery, liable to occur in sheet-steel ground-coats. The probable cause is boiling or blistering when the ground-coat is being fired.

Cut Glass. Glass-ware into which a pattern has been ground by means of an abrasive wheel; the grinding is followed by polishing (cf. BRILLIANT CUT).

Cut Glaze. A faulty glaze, spots or patches being bare or only very thinly covered. The common cause is contaminated areas on the biscuit-ware, i.e. patches of oil, grease, dust, or soluble salts. A fault resulting in a similar appearance is KNOCKING (q.v.).

Cut-off Scar. Marks on the base of a glass bottle made by the Owen's suction machine; the 'scar' is largely caused during the final blowing operation, however.

Cutlery-marking. See SILVER-MARKING.

Cutting-off Table or Cutter. A frame carrying a tightly stretched wire, or a system of such frames and wires, that operates automatically at a short distance from the mouthpiece of a pug or auger to cut off clots or finished bricks or pipes from the extruded column.

Cutting-off Wheel or Parting Wheel. A thin abrasive wheel of the type used for cutting-off or for making slots; such wheels generally have an organic bond.

Cutting Tools. See TOOL TIPS.

Cutty Clay. A variety of English ball clay that was formerly used for making tobacco pipes.

CVR. Abbreviation for CONTINUOUS VERTICAL RETORT (q.v.).

Cyanide Neutralizer. See under NEUTRALIZER.

Cylinder Process. An old method of making flat glass by blowing molten glass to form a cylinder, which is then cracked open and flattened in a special furnace known as a flattening kiln. In Belgium, France and Germany, the cylinder was made of a length to correspond to that of the glass sheet and the circumference corresponded to the width. Bohemian practice was the reverse of this and only small sheets (about 3 ft square) could be made.

Cylindrical Screen Feeder. One type of feeder for plastic clay. It consists of a vertical cylindrical screen through which clay is forced by blades fixed to a vertical shaft that rotates within the cylinder. This machine not only feeds, but also mixes and shreds the clay.

Czochralski Technique. A method of growing single crystals of refractory oxides, and of other compounds, by pulling from the pure melt; the compound must melt congruently. (J. Czochralski, *Z. phys. Chem.*, **92,** 219, 1917.)

D

Danielson–Lindemann Deflection Test. A procedure for assessing the ability of vitreous enamelware to suffer a small degree of bending without the enamel cracking (R. R. Danielson and W. C. Lindemann, *J. Amer. Ceram. Soc.*, **8**, 795, 1925). The procedure has been standardized by the American Ceramic Society (*Bull. Amer. Ceram. Soc.*, **7**, 360, 1928; **9**, 269, 1930).

Danner Process. A method for the continuous production of glass tubing invented by Edward Danner in the USA in 1917. Glass flows from a tank furnace on to a mandrel, which is inclined and tapered and slowly rotates. The mandrel is hollow and air is blown through it to maintain a hole through the glass, which is continuously drawn from the lower end of the mandrel as tubing.

Danny. An open crack at the base of the neck of a bottle.

Dapple. External or internal surface irregularity in a glass container.

Darcy's Law; Darcy. Darcy's Law states that the rate of flow of a fluid, subjected to a low pressure difference, through a packing of particles is very nearly proportional to the pressure drop per unit length of the packing. This Law forms the basis of methods for the determination of the permeability of ceramics. The DARCY is the c.g.s. unit of permeability: a material has a permeability of 1 darcy if in a section 1 cm^2 in area perpendicular to the flow, 1 ml of fluid of unit viscosity flows at a rate of 1 cm/s under a pressure differential of 1 atm. (The Law was propounded by a Frenchman, H. P. G. Darcy, when designing the fountains at Dijon in 1856.)

Datolite. A boron mineral approximating in composition to $CaO.B_2O_3.2SiO_2.H_2O$; it occurs in Russia and elsewhere. Trials have shown that it is a suitable flux for use in glazes for structural clay products.

Davis Revergen Kiln. The word 'Revergen' is a trade-mark. A gas-fired tunnel kiln of the open-flame type; the flame does not come in actual contact with the ware. The combustion air is preheated by regenerators (hence the name) below the kiln. The design was introduced by Davis Gas Stove Co. Ltd, Luton, England; this firm has since been absorbed by Gibbons Bros Ltd, Dudley, England.

Day Tank. A periodic glass-tank furnace, usually consisting of a single chamber, from which glass is worked out by hand; the furnace is operated (charging, melting, and working) on a 24-h cycle.

DCL Fusion-cast Refractory. A US fusion-cast refractory, e.g. glass-tank block, made by a process that largely eliminates the cavities liable to occur as a result of shrinkage during cooling; the mould is L-shaped and is tilted while it is being filled so that the shrinkage cavities concentrate in the smaller leg of the L (the 'lug'), which is then sawn off and discarded. (DCL = Diamond Cut Lug.)

Dead-burned. Term applied to a refractory raw material, and especially to magnesite, after it has been heated at a sufficiently high temperature for the crystal size to increase so that the oxide becomes relatively unreactive with water. Magnesite is dead-burned in shaft kilns or rotary kilns at a temperature of 1600–1700°C.

De-airing. The removal of air from plastic clay or body, from the moist powder in dry-pressing, from casting slip, or from plaster during blending. There are various devices for submitting these materials to a partial vacuum during their processing. De-airing is most commonly practised in extrusion, shredded plastic clay being fed to the pug, or auger, via a de-airing chamber.

Debiteuse. A refractory block having a vertical slot; it is used in the FOURCAULT PROCESS (q.v.) of sheet-glass manufacture, being depressed below the molten glass which is drawn upward through the slot. (French word for a feeding device.)

Decalcomania (USA). A particular type of transfer printing, now known in England as LITHOGRAPHY (q.v.). The term is derived, via the French, from two Greek words: *decal* (off the paper), and *mania*, this form of printing having had a short, but extreme, popularity with young ladies in 1860–65.

Deck. The refractory top of a car used in a tunnel kiln or bogie kiln.

Decking. The stacking of vitreous enamelware in several layers ready for firing.

Decolorizer. A material added to glass to counteract the colour imparted by impurities such as iron; the decolorizer may be an oxidizing agent, removing the colour by chemical action, or it may counteract the colour already present by introducing the complementary colour.

Decorating Firing or Enamel Firing. The process of firing pottery-ware after the application of coloured or metallic

79

De-enamelling

decoration; the temperature is usually 700–800°C and this fixes the decoration and makes it durable.

De-enamelling. The removal of vitreous enamel from the base metal; this can be done by sand-blasting or by solution in alkali.

Deep Cut. Alternative name for CUT GLASS (q.v.).

Deflecting Block or Spreader Block. A block of refractory material, triangular in cross-section, that is built into a coke-oven below a charging hole; the sharp edge of the block is uppermost and this deflects or spreads the stream of descending coal so that it comes to rest more uniformly in the oven.

Deflocculation. The dispersion of a clay slip by the addition of a small amount of suitable electrolyte, e.g. sodium silicate and/or sodium carbonate.

Defluorinated Stone. CHINA STONE (q.v.) from Cornwall, England, from which the small amount of fluoride present has been removed by flotation.

Deformation Eutectic. The composition within a ceramic system (e.g. china clay, flint and feldspar) which, when heated under specified conditions, deforms at a temperature lower than that required to produce deformation in any other composition within the system. This term is used more particularly in USA.

Deformation Temperature. The temperature at which, when a ceramic material is heated under specified conditions, the rate of subsidence becomes equal to the rate of thermal expansion. With glass, this temperature corresponds to a viscosity of 10^{11}–10^{12} poises.

Dégourdi. The preliminary low-temperature (800–900°C) firing of feldspathic porcelain, as practised in Europe; the second (glost) firing is at approx. 1400°C. (French word meaning 'warming' as distinct from the high-temperature—*grand feu*—glost firing.)

Delft Ware. An early type of porous earthenware covered with a tin-opacified glaze; named from Delft, Holland, but the process was already in use in England in the 16th century. In USA the term is defined (ASTM – C242) as a calcareous earthenware having an opaque white glaze and monochrome on-glaze decoration.

Demijohn. A glass container for wine or spirits; it has a narrow neck and a capacity of over 2 gallons. The name is derived from the French *Dame Jeanne*, a popular 17th-century name for this type of large bottle.

Dense. When applied to structural clay or refractory products the term generally signifies 'of low porosity'; when applied to a

80

glass it means 'of high refractive index' (in this context the term is sometimes expanded to OPTICALLY DENSE).

Density. See APPARENT SOLID DENSITY; BULK DENSITY; DENSITY FACTORS; PACKING DENSITY; TRUE DENSITY.

Density Factors for Glass. Factors for calculating the density of a glass from its composition; the original set of factors was that of A. Winkelmann and O. Schott (*Ann. Physik. Chem.*, **51**, 730, 1894). Numerous amendments to these have since been put forward; probably the most reliable are those of M. L. Huggins and K. H. Sun, *J. Amer. Ceram. Soc.*, **26**, 4, 1943.

Dental Porcelain. Feldspathic porcelain, shaped, tinted and fired for use as false teeth; the firing is sometimes carried out in a partial vacuum to remove small air bubbles and thus ensure maximum density and strength.

Derby Press. Trade-name; a machine for the re-pressing of wire-cut building bricks. (Bennett & Sayer Ltd., Derby, England.)

Devitrification. The change from the glassy to the crystalline state; it may occur either as a fault or by controlled processing to produce a devitrified ceramic; see DEVITRIFIED GLASS.

Devitrified Glass. A type of ceramic material that, while in the form of a molten glass, is shaped by one of the conventional glass-making processes, and is subsequently devitrified in a controlled manner so that the finished product is crystalline. The crux of the process is the precipitation, during cooling of the shaped ware, of nucleating agents previously added in small amounts to the glass batch; the nucleated article is then heated to a temperature at which the nucleated crystals can grow. Devitrified-glass products can be made in a wide range of compositions; the properties can thus be varied, but typically the ware is impermeable and has high strength and good thermal-shock resistance. Uses include RADOMES (q.v.), high-temperature bearings and domestic oven-ware.

Devitrite. A crystalline product of the devitrification of many commercial glasses; the composition is $Na_2O.3CaO.6SiO_2$; its field of stability in the ternary system is small and far removed from its own composition—when heated to 1045°C devitrite decomposes into wollastonite and a liquid. First named by G. W. Morey and N. L. Bowen (*Glass Industry*, **12**, 133, 1931).

DF Stone. Abbreviation for DEFLUORINATED STONE (q.v.).

Diamantini. See GLASS FROST.

Diamond Pyramid Hardness (DPH). A hardness test based on the measurement of the depth of indentation made by a loaded diamond; for details see B.S. 427. As applied to the testing of glass, this procedure affords a measure of the yield point of the glass structure.

Diamond Wheel. An abrasive wheel consisting of graded industrial diamonds set in a ceramic, metal, or resinoid bond.

Diaspore. One of the monohydrates of alumina. It occurs, mixed with a certain amount of clay (and thus more properly termed DIASPORE CLAY), in Missouri (USA) and in Swaziland (S. Africa). Diaspore, after strong calcination to remove the water and eliminate further firing shrinkage, is used as a raw material for the manufacture of high-alumina refractories.

Diatomite; Diatomaceous Earth. A sedimentary material formed from the siliceous skeletons of diatoms, which are minute vegetable organisms living in water (both fresh and marine). Large deposits occur in USA, Denmark, and France; there are smaller deposits in Ireland and elsewhere. Because of its cellular nature (porosity about 80%) and the fact that it can be used up to a temperature of about 800°C, diatomite is used as a heat- and sound-absorbing material; it is employed either as a powder for loose-fill, as shaped and fired bricks, or as an aggregate for lightweight concrete.

Dice. The small, roughly cubical, fragments produced when toughened glass is shattered.

Dice Blocks. See THROAT.

Dickite. $Al_2O_3 . 2SiO_2 . 2H_2O$. This is the best crystallized of the kaolin minerals, the crystals consisting of regular sequences of two basic kaolin layers. Dickite is comparatively rare; it is occasionally found in sandstones.

Didier–March Kiln. A coal-fired tunnel kiln; typically, there are four fireboxes—two on each side.

Didymium. A mixture of rare earth oxides, chiefly the oxides of LANTHANUM, NEODYMIUM and PRASEODYMIUM (q.v.).

Die. (1) An attachment at the exit of an extruder designed to give the final shape to an extruded clay column. A brick die usually has an internal set of steel plates arranged to permit lubrication of the internal surface presented to the clay column.

(2) In dry-pressing and plastic shaping, this term is often applied to the metal mould into which the moist powder or plastic clay is charged prior to pressing.

Die Pressing. Term used in some sections of the industry for DRY PRESSING (q.v.).

Dielectric. A material that is capable of sustaining an electrical stress, i.e. an electrical insulator. Electroceramics of high dielectric constant include the titanates, stannates, and zirconates; with suitable compositions the dielectric constant can attain 20 000, the power factor varying from 1×10^{-4} to 500×10^{-4}. They are used in high-capacity condensers at radio-frequencies.

Differential Thermal Analysis (DTA). A method for the identification and approximate quantitative determination of minerals; in the ceramic industry, DTA is particularly applied to the study of clays. The basis of this technique is the observation, by means of a thermocouple, of the temperatures of endothermic and/or exothermic reactions that take place when a test sample is heated at a specified rate; in the differential method, one junction of the thermocouple is buried in the test sample and the other junction is buried in an inert material (calcined Al_2O_3) that is heated at the same rate as the test sample. In the DTA of a clay, the major effect is the endotherm resulting from the evolution of the water of constitution. The temperature of the peak of this endotherm varies according to the particular clay mineral present; the area of the endotherm (as measured on DTA curve) affords a means of assessing the quantity of the mineral present.

Differential Thermogravimetry (DTG). A technique for the study of the changes in weight of a material when heated; it has been applied, for example, in following the dehydration process of clay minerals.

Diffusion Sintering. Term used for true solid-state sintering by those who allow a wider meaning than this to the word SINTERING (q.v.) itself, when used without qualification.

Diffusivity. See THERMAL DIFFUSIVITY.

Dilatancy. The behaviour exhibited by some materials of becoming more fluid when allowed to stand and less fluid when stirred; dilatancy is shown by some ceramic bodies that are deficient in fine ($< 2\ \mu$) particles (cf. RHEOPEXY and THIXOTROPY).

Dimming Test. To determine the durability of optical glass the surface is subjected to the action of air saturated with water vapour at a definite temperature (usually 80°C) for a specified period. Any dimming of the surface is then observed.

Dimple. A fault in vitreous enamelware appearing as a small shallow depression. Causes include rusting or over-oxidation of the base metal, and the contamination of one enamel by another.

DIN. Prefix to specifications of the German Standards Association: Deutschen Normenausschusses, Berlin W.15.

Dinas Brick. The original name for a silica refractory, so-called from Dinas, in S. Wales, where the silica rock was quarried. The name is preserved in the German *dinas-stein* and the Russian equivalent.

Diopside. $MgO.CaO.2SiO_2$; m.p. 1392°C; thermal expansion (100–1000°C) $7·5 \times 10^{-6}$. There is a deposit in New York State. Trials have been made with synthetic diopside as a h.f. electro-ceramic. It is formed as a devitrification product of soda–lime glass if the CaO is partially replaced by MgO; it is also formed when siliceous slags attack dolomite refractories.

Dipping. (1) BISCUIT-FIRED POTTERY (q.v.) is dipped into a suspension of the glaze ingredients in water; the dipped ware is then dried and GLOST-FIRED (q.v.).

(2) In vitreous enamelling, the base-metal can be dipped in slip and drained (wet process) or it can be first heated and then dipped in powdered frit (dry process).

Dipping Weight. See PICK-UP.

Direct-arc Furnace. See under ELECTRIC FURNACES FOR MELTING and REFINING METALS.

Direct Firing. The firing of pottery or vitreous enamelware in a fuel-fired kiln or furnace without protection of the ware from the products of combustion.

Direct Teeming or Top Pouring. The transfer of molten steel from a ladle, through one or more refractory nozzles, directly into the ingot moulds.

Dirty Finish. See FINISH.

Dirty Ware. Foreign matter that occasionally disfigures pottery-ware as taken from the kiln; potential sources of the 'dirt' include the atmosphere, both in the factory and in the kiln, the placers' hands, the kiln lining, and the kiln furniture.

Disappearing-filament Pyrometer. An optical pyrometer consisting of a small telescope with an electrically heated filament placed in its focal plane. A hot surface within a kiln or furnace is focused through the telescope and the current supplied to the filament is adjusted until the apparent temperature of the filament and furnace coincide, the filament then disappearing in the general colour of its background. The corresponding temperature is read from a scale on the instrument.

Disappearing-highlight Test. A test to determine the degree of attack of a vitreous-enamelled surface after an acid-resistance test; (see ASTM – C282).

Discharge-end Block. See NOSE-RING BLOCK.

Disintegration Index. A measure of the durability of a hydraulic cement proposed by T. Merriman (*Engng. News Record*, **104**, 62, 1930). The test involves shaking with a lime–sugar solution followed by titration of one aliquot against HCl with phenolphthalein as indicator and another with methyl orange as indicator. The Disintegration Index is the difference between the two titrations. The test was superseded by the test now known as the MERRIMAN TEST (q.v.).

Disintegrator. A machine used for the size reduction of some ceramic materials. A rotor is rapidly revolved within a casing, both rotor and casing having fixed hammers which impact on the material being ground (cf. HAMMER MILL).

Disk Feeder. A type of clay feeder for attachment to the base of a storage bin. There are various types. In one of these there is a short fixed cylinder with a side outlet; below the cylinder is a revolving horizontal disk. In another design the disk is stationary, the clay being discharged by moving arms inside the cylinder.

Disk Wheel. An abrasive wheel of a type that is usually mounted on a plate so that grinding can be done on the side of the wheel.

Dispex. Trade-name; ammonium polyacrylate—sometimes used as a deflocculant in clay slips.

Dissector. A person employed to classify defective pottery-ware according to the nature of the fault.

Disthene. Obsolete name for KYANITE (q.v.).

Diver Method. A technique for the determination of particle size by sedimentation. The specific gravity at a given depth in a sedimenting suspension is determined by means of small loaded glass 'divers' of known specific gravities in a range between the specific gravity of the dispersion medium and that of the homogeneous suspension. If a 'diver' is placed under the surface of a sedimenting suspension it will descend to a level where its weight is equal to the weight of suspension displaced; it will then continue to descend at the same rate as the largest particles at the level of its geometrical centre of gravity and at a greater rate than all the particles in the suspension located above that level. (S. Berg. Ingeniorvidenskabelige Skrifter, No. 2, 1940.)

Division Wall. A wall of refractory bricks between two adjacent settings in a bench of gas retorts.

Dobbin. A type of dryer used in the tableware section of the pottery industry; the ware, while still in the plaster mould, is placed on horizontal turntables within the drying cabinet; the turntables can be rotated about a vertical axis so that the ware moves from the working opening into the interior of the dryer where moisture is removed from the mould and the ware by means of hot air.

Dobie. Term sometimes applied to a hand-shaped clay building brick before it has been fired; from ADOBE (q.v.).

Docking. The immersion of building bricks in water as soon as they are taken from the kiln; this is done only when the bricks are known to contain lime nodules and is a method for the prevention of LIME BLOWING (q.v.).

Doctor Blade. (1) A thin, flexible, piece of steel used for smoothing a surface, e.g. for cleaning excess colour from the engraved copper plate used in printing on pottery.

(2) A blade used for parting thin ceramic sheets or wafers of the type used in miniature condensers.

Document Glass. A glass that absorbs ultraviolet rays and thus protects documents from deterioration.

Dod Box. An old device for extruding rods or strips of a pottery body for use in the making of cup handles or of basket-ware. The term may be a corruption of WAD BOX (q.v.) or it may be from DOD the old name for the Reed Mace or Bulrush.

Dod Handle. A cup- or jug-handle made by the old DOD BOX (q.v.) method.

Dodecacalcium Heptaluminate. $12CaO \cdot 7Al_2O_3$; m.p. 1455°C. A constituent of high-alumina hydraulic cement. This compound was formerly believed to be pentacalcium trialuminate ($5CaO \cdot 3Al_2O_3$).

Dodge-type Jaw Crusher. A jaw-crusher with one jaw fixed, the other jaw being pivoted at the bottom and oscillating at the top; the output is low but of uniform size (cf. BLAKE-TYPE JAW CRUSHER).

DOFP. Direct-On Finish Process of vitreous enamelling (US abbreviation).

Dog-house. In an open-hearth steel furnace, the arched refractory area through which a metallic burner (for oil-firing) is inserted; in a glass-tank furnace, the refractory-lined extension into which batch is fed.

Dog's Teeth or Dragon's Teeth. A fault sometimes found on the edges of a rectangular extruded column of clay, the greater friction at the corners of the die holding the clay back relative to the centre of the extruding column; if this corner friction is too great it results in a regular series of tears along the edges of the column. Methods for curing the fault are increasing the moisture content of the clay, improving the lubrication of the die, or enlarging the corners of the die at the back of the mouthpiece.

Dolly. A gathering iron with a refractory tip used in the making of glass-ware in semi-automatic machines.

Dolly Dimples. A slight defect in cast-iron vitreous enamelware, blisters in a leadless enamel having almost completely healed.

Doloma. Calcined dolomite, i.e. a mixture of the oxides CaO and MgO; the term was introduced by the Basic Furnace Linings Committee of the Iron and Steel Institute (I.S.I. Spec. Rept. 35, 1946).

Dolomite. The double carbonate of calcium and magnesium, $(Ca,Mg)CO_3$. Dolomite occurs abundantly in many countries: in England it extends as a belt of rock from Durham to Nottinghamshire; in Wales the Carboniferous Limestone has been dolomitized locally. It is used as a source of magnesia in glass production but the principal use is as a refractory material, for which purpose it is calcined. Because of the free lime present calcined dolomite rapidly 'perishes' in contact with the air; it may be tar-bonded to give it partial protection, or it can be stabilized by firing it, mixed with steatite or other siliceous material, so that the lime becomes combined as one or more of the calcium silicates. Stabilized dolomite refractory bricks find some use in the lining of electric steel furnaces and rotary cement kilns; tarred dolomite bricks are much used in the newer 'oxygen steelmaking' processes.

Domain. In a ferroelectric crystal, e.g. barium titanate, a 'domain' is a small area within which the polarization is uniform. If the crystal is exposed to a high electric field, those domains in which the polarization is in a favourable direction will grow at the expense of other domains. The presence of domains in a ferroelectric crystal is the cause of hysteresis.

Dome Brick. A brick in which both the large and the side faces are inclined towards each other in such a way that, with a number of these bricks, a dome can be built. (See Fig. 1, p. 37.)

Dome Plug. A refractory shape, usually made of aluminous fireclay or of a refractory material of still higher alumina content, used in the top of the dome of a HOT-BLAST STOVE (q.v.).

Dorfner Test

Dorfner Test. A test for stress in glazed ware proposed by J. Dorfner (*Sprechsaal*, **47**, 523, 1914): a cylinder of the ware is partly glazed and the shrinkage of the glazed portion is noted.

Dorr Mill. A TUBE MILL (q.v.) designed for operation as a closed-circuit wet-grinding unit.

Dorry Machine. Apparatus for testing the abrasion resistance of a ceramic; the flat ends of cylindrical test-pieces are abraded under standardized conditions by movement in contact with a specially graded sand.

Dot. A small refractory distance-piece for separating CRANKS (q.v.) and SETTERS (q.v.).

Dottling. The setting of pottery flatware horizontally on THIMBLES (q.v.).

Double Draining. A further period of flow of slip from dipped vitreous enamelware after the initial draining has finished and the enamel appears to have set. A possible cause of this trouble is excessive alkalinity of the slip caused by the solution of alkalis from the frit; alternatively, the amount of electrolyte added to the slip may have been incorrect.

Double-face Ware. Vitreous enamelware that has a finish coat on both sides.

Double-frit Glaze. A glaze containing two frits of different compositions. As an example, a glaze may contain a lead frit and a leadless frit; the glaze is thus rendered highly insoluble by the inclusion in the second frit of those constituents liable to increase lead solubility.

Double-roll Verge Tile. A single-lap roofing tile having a roll on both edges so that verges on the two sides are similar.

Double-screened Ground Refractory Material. A US term defined as: A refractory material that contains its original gradation of particle sizes resulting from crushing, grinding, or both, and from which particles coarser and finer than two specified sizes have been removed by screening (cf. SINGLE-SCREENED).

Double Standard. A brick (particularly a refractory brick) that is twice as wide as a standard square, e.g. $9 \times 9 \times 3$ in.

Double-thread Method. A procedure for determining the coefficient of thermal expansion of a glass by forming a thread by fusing a fibre of the glass under test to a fibre of a glass of known expansion; from the curvature of the double-thread, when cold, the coefficient of expansion of the glass under test can be calculated. (M. Huebscher, *Glashütte*, **76**, 57, 1949.)

Down-draught Kiln. A kiln in which the hot gases from the fireboxes first rise to the roof, then descend through the setting and are finally withdrawn through flues in the kiln-floor.

Down-draw Process. The production of glass tubing by continuously drawing molten glass downward from an orifice.

Downtake or Uptake. One of the two vertical passages, built of refractory bricks, leading from the ports to the slag-pockets of an open-hearth furnace. As such a furnace operates on the regenerative principle, the direction of gas-flow being periodically reversed, the identical passages at the two ends of the furnace alternately serve as Downtake for the waste gases leaving the furnace and Uptake for the hot air for combustion and (in gas-fired furnaces) the fuel gas.

Dozzle. See under CORE.

DPH. Abbreviation for DIAMOND PYRAMID HARDNESS (q.v.).

Drag-ladle or Dragade. To make quenched CULLET (q.v.) by ladling molten glass into water.

Drag-line. A type of mechanical excavator often used in the winning of brick-clays; a 'bucket', suspended from a boom, is lowered on to the clay and is then dragged towards the excavator by a wire rope—thus filling the 'bucket' with clay. An advantageous feature of a drag-line is its ability to work clay below the level of the excavator itself.

Dragged. A surface texture on clay facing bricks produced by a tightly stretched wire contacting the column of clay as it is extruded from the pug in the wire-cut process; this texture is also known as RIPPLED.

Dragon's Teeth. See DOG'S TEETH.

Drain Casting or Hollow Casting. Terms used (more particularly in USA) for the slip-casting process for making hollow-ware, the excess slip being drained by inversion of the mould.

Drain Lines. Lines or streaks liable to appear in badly drained wet-process vitreous enamelware after it has been fired.

Drain Tile. US term for an unglazed field-drain pipe. The properties of clay drain tiles are specified in ASTM – C4; of concrete drain tiles in ASTM – C412.

Draw. See LOAD.

Draw Bar. In the PITTSBURGH PROCESS (q.v.) of drawing sheet glass, the position of the sheet is defined by a refractory block (the draw-bar) submerged in the molten glass.

Drawing Chamber. The part of a tank furnace for flat glass from which the sheet of glass is drawn.

Drawn Stem. See STEMWARE.

Dredging. A dry process of vitreous enamelling in which powdered frit is sifted on to the surface of the hot base-metal.

Dressing. The process of removing, from the face of an abrasive wheel, those grains that have become dulled during use.

Dressler Kiln. The first successful muffle-type tunnel kiln was that built by Conrad Dressler in 1912. The name is now applied to a variety of kilns designed and built by the Swindell–Dressler Corp., Pittsburgh, USA.

Drop Arch. An auxiliary brick arch projecting below the general inner surface of the arched roof of a furnace, brick conduit or like structure.

Dropping. See SAGGING.

Drum. (1) Term sometimes applied to the mouth of a port in a glass-tank furnace.

(2) A wooden former of the type that was used in making the side of a saggar by hand.

Drum-head Process. A process used in Europe for the shaping of flatware; it was developed on account of the 'shortness' of the feldspathic porcelain body. A slice of the pugged body is placed on a detachable 'drum-head' which fits on the BATTING-OUT (q.v.) machine. The 'drum-head', with the shaped disk lying on it, is then removed and inverted over the jigger-head, the bat then being allowed to fall on the mould for its final jiggering.

Dry Body. An unglazed stoneware type of body. The term has been applied, for example, to CANE WARE, JASPER WARE and BASALT WARE (q.v.).

Dry Edging. A fault sometimes occurring in pottery manufacture as a result of insufficient glaze application; it is shown by rough edges and corners.

Dry Mix. See DRY PROCESS.

Dry Pan. An EDGE-RUNNER MILL (q.v.) used for grinding relatively dry material in the refractories and structural clayware industries. The bottom has a solid inner track on which the mullers rotate and outer perforated grids through which the ground material is screened, oversize being ploughed back to the grinding track.

Dry Pressing. The shaping of ceramic ware under high pressure (up to 14 000 p.s.i.), the moisture addition being kept to a minimum (5–6%) or, with some materials, eliminated by the use of a plasticizer, e.g. a stearate. Dry pressing is used in the shaping of wall and floor tiles (when it is often referred to as Dust Pressing), most high-grade refractories, abrasive wheels, the Fletton type of building brick (the moisture content for pressing is in this case 19–20%), and many articles in the electroceramic industry. The process is also sometimes referred to as Semi-dry Pressing.

Dry Process or Dry Mix. (1) Term used in the US whiteware industry and defined (ASTM – C242) as the method of preparation of a ceramic body by which the constituents are blended dry; liquid may then be added as required for subsequent processing.

(2) The process of cement manufacture in which the batch is fed to the kiln dry.

Dry Process Enamelling. In this method of vitreous enamelling, the base metal is preheated to a temperature above that at which the enamel to be used will mature (usually 850–950°C); the finely powdered enamel is then applied to the hot ware which is then fired to complete the maturing process.

Dry-rubbing Test. A test to determine the degree of attack of a vitreous-enamelled surface after an acid-resistance test; (see ASTM – C282).

Dry Spray. A fault sometimes occurring in vitreous enamelware when the enamel has been applied by spraying; the fault appears as a rough texture.

Dry Strength. The mechanical strength of a ceramic material that has been shaped and dried but not fired; it is commonly measured by a transverse strength test.

Dryer. See Chamber Dryer; Dobbin; Hot-floor; Mangle; Tunnel Dryer.

Dryer Scum. See Scum.

Drying Shrinkage. Ceramic ware (and particularly clayware) that is shaped from a moist batch shrinks during drying; the drying shrinkage is usually expressed as a linear percentage, e.g. the drying shrinkage of china clay is usually 6–10%, that of a plastic ball clay is 9–12%. To produce ware (e.g. electroceramics or refractory bricks) of high dimensional accuracy, the drying and firing shrinkages must be low; this is achieved by reducing the proportion of raw clay and increasing the proportion of non-plastic material in the batch, which is then shaped by dry-pressing, for example.

DTA. Abbreviation for Differential Thermal Analysis (q.v.).

DTG. Abbreviation for Differential Thermogravimetry (q.v.).

Dulling. A glaze fault characterized by the ware having poor gloss when drawn from the kiln; the cause is surface devitrification, which may result from factors such as SULPHURING (q.v.) or too-slow cooling.

Dumont's Blue. Alternative name for SMALT (q.v.).

Dumortierite. A high-alumina mineral, $8Al_2O_3 \cdot 6SiO_2 \cdot B_2O_3$; it occurs sufficiently abundantly in Nevada, USA, for use in the ceramic cores of sparking plugs although bodies of still higher alumina content are now more generally used for this purpose.

Dump. An item of KILN FURNITURE (q.v.) designed for use in a RING—a bottomless saggar—for the support of large hollow-ware, e.g. basins. Dumps may also be used as spacers in a CRANK (q.v.).

Dunite. A rock consisting essentially of olivine but sometimes also containing chromite; it occurs in many parts of the world and is used in the manufacture of forsterite refractories.

Dunnachie Kiln. A gas-fired chamber kiln designed by J. Dunnachie (Brit. Pat. 3862; 1881). The first such kiln was built at Glenboig, Scotland, in 1881 for the firing of firebricks. Important features are the solid floor and the space between the two lines of chambers.

Dunt, Dunting. A crack, or the formation of cracks (which may be invisible), in ware cooled too rapidly after it has been fired.

Dust. A fault, in electric lamp bulbs or valves, resulting from local concentrations of seed or finely-divided foreign matter; also known as SPEW. See also DUSTING.

Dust Coat. A coating of vitreous enamel that has been sprayed thinly and relatively dry.

Dust Pressing. See DRY PRESSING.

Dusting. (See also DUST) (1) Spontaneous falling to a powder, particularly of material containing a large amount of CALCIUM ORTHOSILICATE (q.v.), which suddenly expands when it is cooled from red heat.

(2) In dry-process vitreous enamelling, a synonym of DREDGING (q.v.).

(3) In wet-process vitreous enamelling, a defect during spraying resulting in localized concentrations of almost dry slip.

(4) The cleaning of an applied coating of vitreous enamel slip after it has dried, preparatory to firing.

Dutch Kiln. An early type of up-draught intermittent kiln for the firing of bricks; it had a number of small chimneys in the roof.

Dutch Oven. A simple furnace of small size and usually fired with solid fuel; it can be constructed outside a newly-built furnace, for example, and used as an air-heater for drying-out and warming-up (cf. DUTCH KILN).

Dye Absorption or Dye Penetration. A test for porosity in ceramic products that are nominally non-porous. It is applied, for example, to porcelain insulators for which BS. 137 stipulates that there shall be no sign of dye penetration after a fractured specimen has been immersed for 24 h in a 0·5% solution of fuchsine in alcohol under a pressure of 2000 p.s.i.

Dyer Method. A procedure for shaping the socket of a clay sewer-pipe proposed by J. J. Dyer (*Brick Clay Record*, **105**, No. 3, 27, 1944).

E

'E' Glass. A fibre glass of low alkali content ($\leq 1\% Na_2O$).

Earthenware. Non-vitreous, opaque, ceramic whiteware; it is normally glazed and its most common use is as tableware. The general body composition is (per cent): china clay, 25; ball clay, 25; calcined flint, 35; china stone, 15. The biscuit firing temperature is 1100–1150°C.

Easing Air. The air that is admitted through the feed-holes of an annular kiln at one stage in the firing of FLETTON (q.v.) bricks; the purpose is to check the rapid rise of temperature consequent on the ignition of the organic matter present in such bricks.

Easy Fired. Clayware, particularly earthenware, is said to be easy fired if it has been fired at too low a temperature and/or for too short a time.

Eaves Course; Eaves Tile. A course of special-size roofing tiles —EAVES TILES—for use at the eaves of a roof to obtain the correct lap.

Edge Bowl. A hollow bowl about 7 in. deep and containing the slot through which glass is drawn in the PITTSBURGH PROCESS (q.v.).

Edge Lining. The painting, by hand or machine, of a coloured line round the edge of pottery.

Edge-runner Mill. A crushing and grinding unit depending for its action on heavy mullers, usually two in number, that rotate relative to a shallow pan which forms the base; the pan bottom may be solid or perforated (cf. END-RUNNER MILL).

Edging. (1) The removal of dried vitreous enamel cover-coat from the edge of ware, to reveal the underlying coating of enamel.

(2) The application of differently coloured slip around the rim of enamel-ware.

Efflorescence. A deposit of soluble salts that sometimes appears on the surface of building bricks after they have been built into a wall. If the salts are derived from the bricks themselves they consist chiefly of $CaSO_4$, $MgSO_4$, K_2SO_4 and Na_2SO_4; soluble sulphates present in the raw clay can be rendered insoluble by the addition of $BaCO_3$ to the clay while it is being mixed; this precipitates the sulphate as insoluble $BaSO_4$. Efflorescence may arise, however, from soluble salts in the mortar or, if a wall has no damp course or is backed by soil, from the soil itself.

Efflorwick Test. A test for the likelihood of the formation of EFFLORESCENCE (q.v.) on a clay building brick. A cylinder, made by shaping and firing a red clay known to be free from soluble salts, is allowed to absorb any soluble salts dissolved by distilled water from the crushed sample to be tested; the clay cylinder is then dried and examined for efflorescence. The conditions of the test have been standardized by the New York State College of Ceramics (*Brick & Clay Record*, **104**, No. 5, 25, 1944).

Eggshell Porcelain. A very thin, and hence highly translucent, porcelain originally made by the Chinese and Japanese for the European market. Bone China has also for long been available of 'eggshell' thinness.

Eggshelling. (1) A glaze fault resulting in potteryware coming from the glost-kiln with an egg-shell appearance. The fault is caused by gas bubbles that have burst on the surface of the glaze, which has subsequently failed to heal; the glaze is too viscous at the firing temperature used.

(2) A similar surface fault in vitreous enamelware (unless this surface appears on the ground-coat, when it can be an advantage as providing a good base for the cover-coat).

Eirich Mixer. An under-driven wet-pan mixer. The original design was that of two Germans L. Eirich and J. Eirich (Brit. Pat. 379 265; 25/8/32).

Elastic After-effect. When glass and certain ceramic materials are subjected to stress for a long period they remain partly deformed when the stress is removed; the elastic after-effect is the ratio of the deformation remaining after a given time to the deformation immediately after removal of the stress.

Electret. An electrical analogue of a permanent magnet: a material that is 'permanently' electrified and exhibits electrical charges of opposite sign at its extremities. In order to retain their charge for a long period (days or weeks) ceramic electrets must be polarized at high temperature; materials that have been treated in this way include the titanate dielectrics.

Electric Furnaces for Melting and Refining Metals. Several types of electric furnace are used in the metallurgical industries— both ferrous and non-ferrous; all these furnaces are lined with refractory materials, the larger furnaces generally being bricked, the smaller furnaces usually having a monolithic refractory lining which is rammed into place. The chief types of such furnaces are: DIRECT ARC, in which the electric current passes through the charge; INDIRECT ARC, in which the arc is struck between the electrodes only; INDUCTION FURNACE, in which the metal charge is heated by eddy-currents induced in it. Induction furnaces may be operated at high frequency (h.f. induction furnaces) or at low frequency (l.f. induction furnaces).

Electrical Porcelain. Porcelain made for use as an electrical insulating material. A typical batch composition is 18% ball clay, 22% china clay, 30% quartz, 15% china stone, 15% feldspar. Low-tension porcelain for the insulators used on normal supply lines and high-tension porcelain for the high-voltage grid are of essentially the same composition, but the latter is generally made to a lower porosity. Large insulators may be jolleyed or, where necessary, thrown and turned; some types of insulator for suspension lines are warm-pressed. Relevant British Standards include: B.S. 16, 137, 223, 1540, 1598 and 2133.

Electrocast Refractory. A refractory material that has been made by FUSION-CASTING (q.v.).

Electroceramics. A group of ceramic materials of various compositions having electrical and other properties that render them suitable for use as insulators for power-lines and in many electrical components. In terms of tonnage made, ELECTRICAL PORCELAIN (q.v.) is the most important; more specialized types include CORDIERITE, STEATITE, TITANATE CERAMICS and ZIRCON PORCELAIN (q.v.).

Electrofusion. The process of fusion in an electric furnace. See FUSION-CASTING.

Electrode Ring or Bull's Eye. Special refractory shapes, in the roof of an electric arc steel furnace, forming an opening through which an electrode is inserted.

Electrolyte. A compound which, when dissolved in water, partially dissociates into ions, i.e. into electrically charged atoms, molecules, or radicals. Electrolytes are added to clay slips and to vitreous-enamel slips to control their flow properties.

Electrolytic Pickling. A method (not much used) for the preparation of the base-metal for vitreous enamelling; chemical PICKLING (q.v.) is assisted (ANODE PICKLING) or replaced (CATHODE PICKLING) by electrolysis.

Electro-osmosis. The de-watering and partial purification of clay by a process of ELECTROPHORESIS (q.v.) first proposed by Elektro-Osmose A. G. (Brit. Pats., 135 815–20; 25/6/18). The process has had only limited application because of its high cost; it has been used at Karlovy Vary (formerly Karlsbad) in Czechoslovakia, and at Grossalmerode and Westerwald (Germany).

Electrophoresis. The movement of fine particles in a suspension as a result of the application of an electric field; use is made of this effect in the electrical lubrication of the dies in some clayworking machinery, the migration of the clay particles leaving a concentration of water between the clay and the metal die.

Electrostatic Spraying. A process in which particles that are to be sprayed are given an electrostatic charge opposite to that on the ware to be sprayed; this attracts the sprayed particles to the ware. Although technically applicable to vitreous enamelling, this method of spraying has so far been little used in the ceramic industry.

Elevator Kiln. A kiln into which a setting of ware is raised from below; the ware is set (outside the kiln) on a refractory base which is subsequently elevated by jacks into the firing position. Kilns of this type have been used, for example, in the firing of abrasive wheels.

Elutriation. The process of separation of particles, according to their size and/or density, by submitting them to an upward current of water, air or other fluid (cf. SEDIMENTATION).

Embossing. The decoration of pottery by means of a raised pattern (flowers, figures, etc.); the effect is usually obtained by depressions in the plaster mould in which such ware is made (cf. SPRIGGING).

Emery. A naturally occurring, impure, CORUNDUM (q.v.); used as an abrasive.

Emissivity. A surface property, being the ratio of its emissive power for heat to that of a BLACK-BODY (q.v.) for a given wavelength and at the same temperature. Some reported values for refractory materials are:

Type of Refractory	1000°C	1500°C
Fireclay	0·8	0·7
Silica	0·85	0·7
Sillimanite	0·55	0·6
Sintered alumina	0·4	0·4
Chrome–magnesite	0·85	0·7

Emley Plastometer. An instrument designed primarily for assessment of the plasticity of building plaster; it has also been used for the testing of clay. The material to be tested is placed on a porous disk which is mounted on a vertical shaft; as the shaft revolves it rises, pressing the sample against a conical metal disk, the motion of which is resisted by a lever. Equilibrium is reached when the force of the sample under test against the metal disk is equal to the stress acting through the lever; the average relative tangential force for the first 5-min. period is taken as an index of plasticity. (W. E. Emley, *Trans. Amer. Ceram. Soc.*, **19**, 523, 1917.)

Emperor Press. Trade-name; a dry-press brickmaking machine of the rotary-table type. (Sutcliffe Speakman Ltd., Leigh, Lancs., England.)

Enamel. See VITREOUS ENAMEL.

Enamel-back Tubing. Glass tubing, the back half of which (the tube being held vertically) is seen to consist of white or coloured PLY GLASS (q.v.).

Enamel Colour. A ceramic colour for the on-glaze decoration of pottery.

Enamel Firing. In the British pottery industry this term is synonymous with DECORATING FIRING (q.v.).

Enamelling Iron. Cold-rolled sheet specially made from steel of very low carbon content for the vitreous-enamel industry.

Encapsulation. The sealing of an electronic component, particularly of a semi-conductor, generally with a ceramic sealing compound (cf. POTTING MATERIAL).

Encaustic Tiles

Encaustic Tiles. Ceramic tiles in which a pattern is inlaid with coloured clays, the whole tile then being fired.

End Arch. A brick shape used for the construction of arches and sprung roofs; the large faces are inclined towards each other in such a way that one of the end faces is smaller than the other. (See Fig. 1, p. 37.)

End Feather. See FEATHER BRICK.

End-fired Furnace. A type of glass-tank furnace in which the ports are in the back wall (cf. CROSS-FIRED FURNACE).

End Runner. See RISER BRICK.

End-runner Mill. A small grinding unit, primarily for laboratory use, operating on the principle of the pestle and mortar; the runner is set eccentrically in the mortar, which is mechanically driven (cf. EDGE-RUNNER MILL).

End Skew. A brick (particularly a refractory brick) with one end completely bevelled at an angle of 60°. This bevel can be towards a large face (END SKEW ON FLAT) or towards a side face (END SKEW ON EDGE). Both types of brick are used in the springing of an arch. (See Fig. 1, p. 37.)

End Wall. (1) The vertical refractory wall, furthest from the furnace chamber, of the downtake of an open-hearth steel furnace.

(2) One of the two vertical walls terminating a battery of coke-ovens or a bench of gas retorts; it is generally constructed of refractory bricks and heat-insulating bricks with an exterior facing of building bricks.

(3) cf. GABLE WALL.

Endell Plastometer. See GAREIS–ENDELL PLASTOMETER.

Endellite. Obsolete name for the clay mineral HALLOYSITE (q.v.); some authorities, however, would preserve the name for those halloysites containing an excess of water.

Endothermic Reaction. A chemical reaction that takes place with absorption of heat. The dehydration of kaolinite is a reaction of this type.

Enforced-order Mixer. See FRENKEL MIXER.

Engine-turning Lathe. A lathe having an eccentric motion and used to incise decorations on pottery-ware before it is fired.

Engineering Bricks. Building bricks that have been shaped in a mechanical press from a clay such as an Etruria Marl which, when fired at a high temperature, will vitrify to produce a brick of great strength and low water absorption. If firing is under reducing conditions, blue bricks are produced; with oxidizing firing the product is a red engineering brick. In either case the bricks must conform, in the UK, to B.S. 1301 which sub-divides them into two classes: Class A—min. crushing strength 10 000 p.s.i., max. water absorption 4·5 wt %; Class B—min. crushing strength 7 000 p.s.i., max. water absorption 7 wt %.

English Bond. A brick wall built with alternate header and stretcher courses.

English Kiln. A transverse-arch chamber kiln with a system of flues and dampers above the chambers permitting any two chambers to be connected. It was designed by A. Adams in 1899 for the firing of building bricks made from highly bituminous clays.

English Pink. See CHROME-TIN PINK.

English and Turner Factors. See THERMAL EXPANSION FACTORS FOR GLASS.

English Translucent China. Ceramic tableware, etc., introduced in 1959 by Doulton Fine China Ltd.; in contrast to English bone china it is feldspathic, but differs from Continental porcelain in that it is biscuit fired at a higher temperature than the glost fire.

Engobe. A coating of slip, white or coloured, applied to a porous ceramic body to improve its appearance; a glaze is sometimes applied over the engobe, as in sanitary fireclay. A typical engobe for sanitary fireclay consists of 10% ball clay, 40% china clay, 20% flint and 30% china stone; some of the china stone may be replaced by feldspar and considerable variation is possible in the proportions of the other constituents. (French word.)

Engraving. A method of decoration. For application to pottery, the pattern is engraved on a copper plate or roller; the incised pattern is then filled with specially prepared colour which is transferred to the ware by transfer paper. As a method of decorating glass-ware, the pattern is cut directly into the glass surface by means of copper wheels; the depth of cut is shallower than in intaglio work.

Enslin Apparatus. This apparatus for the determination of the water-absorption capacity of clays was originally designed by O. Enslin (*Chem. Fabrik.*, **6**, 147, 1933) for testing soils. It consists of a U-tube, one arm of which is connected via a 3-way tap to a calibrated horizontal capillary tube; the other arm ends in a funnel with a sintered-glass base on which is placed a weighed sample of clay. Water is allowed to contact the sample and the amount absorbed is read from the capillary tube. The result is expressed as a percentage of the weight of the dry clay.

Envelope Kiln. Alternative name, particularly in USA, for TOP-HAT KILN (q.v.) or SHUTTLE KILN (q.v.).

Equilibrium Diagram or Phase Diagram. Equilibrium diagrams of interest to the ceramist represent crystallographic changes and the melting (or conversely the solidification) behaviour of compounds, mixtures, and solid solutions under conditions of chemical equilibrium. As an example, the Al_2O_3–SiO_2 equilibrium diagram,

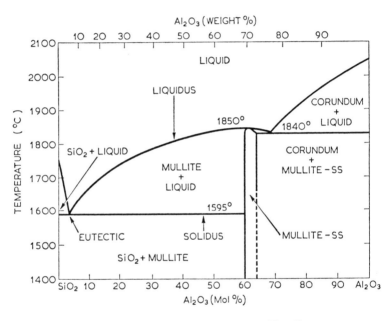

FIG. 2. EQUILIBRIUM DIAGRAM: Al_2O_3—SiO_2 SYSTEM.

(After S. Aramaki and R. Roy, *J. Amer. Ceram. Soc.*, **45**, No. 5, 229, 1962).

one of the most important for the ceramist, is shown in Fig. 2. Items of interest are:

(1) The melting points of three compounds are shown, namely, that of silica (1723°C), mullite (1850°C) and alumina (2050°C).

(2) There are two EUTECTICS (q.v.), that between silica and mullite melting at 1595°C, and that between alumina and mullite melting at 1840°C.

(3) The ability of mullite to take a small amount of alumina into SOLID SOLUTION (q.v.).

For a useful discussion of equilibrium diagrams from the standpoint of the ceramist see E. M. Levin, H. F. McMurdie and F. B. Hall, 'Phase Diagrams for Ceramists', Part 1. (*Amer. Ceram. Soc.*, 1956); see also E. M. Levin and H. F. McMurdie, 'Phase Diagrams for Ceramists,' Part 2 (*Amer. Ceram. Soc.*, 1959).

Equivalent Particle Diameter or Equivalent Free-falling Diameter. A concept used in evaluating the size of fine particles by a sedimentation process; it is defined as the diameter of a sphere that has the same density and the same free-falling velocity in any given fluid as the particle in question (cf. PARTICLE SIZE).

Erosion. Wear caused by the mechanical action of a fluid, e.g. of molten steel flowing through the refractory nozzle in a ladle, or of waste gases flowing through the downtake of an open-hearth furnace (cf. ABRASION and CORROSION).

Erz Cement. A ferruginous hydraulic cement formerly made in Germany; it has now given place to FERRARI CEMENT (q.v.).

Estuarine Clays. Clays that were deposited, during the course of geological time, in estuaries and deltas. Estuarine clays of the Middle Jurassic occur in Yorkshire and Lincolnshire and are used as raw material for making building bricks.

E.T.C. ENGLISH TRANSLUCENT CHINA (q.v.).

Ethyl Silicate. See SILICON ESTER.

Etruria Marl. A brick-clay occurring in the Carboniferous System and used for the manufacture of bricks and roofing tiles, particularly in the Midlands and North Wales. These clays have a high iron content; they fire to a red colour under oxidizing conditions but under reducing conditions they fire to the blue colour of the well-known Staffordshire engineering brick.

Etruscan Ware. BASALT WARE (q.v.) having an encaustic decoration, mainly in red or white in imitation of early Italian Etruscan pottery.

Eucryptite. A lithium mineral, $Li_2O \cdot Al_2O_3 \cdot 2SiO_2$. When heated, the β form expands in one direction and contracts in another direction; it is a constituent of special ceramic bodies having zero thermal expansion.

Eurite. A feldspathic mineral occurring on the island of Elba and used locally as a ceramic raw material.

Eutectic. An invariant point on an equilibrium diagram. In

101

Excavator

the Al_2O_3–SiO_2 system (see Fig. 2, p. 100) there is a eutectic between silica and mullite and a eutectic between alumina and mullite. The eutectic temperature is that at which a eutectic composition solidifies when cooled from the liquid state. The eutectic composition has been defined as 'that combination of components of a minimum system having the lowest melting point of any ratio of the components; in a binary system it is the intersection of the two solubility curves.'

Excavator. See DRAG-LINE; MECHANICAL SHOVEL; MULTI-BUCKET EXCAVATOR; SHALE CUTTER or SHALE PLANER; SINGLE-BUCKET EXCAVATOR.

Exchangeable Bases. See IONIC EXCHANGE.

Exfoliation. The property of some hydrous silicates, notably VERMICULITE (q.v.), of permanently expanding concertina-wise when rapidly heated to a temperature above that at which water is evolved; cf. BLOATING and INTUMESCENCE.

Exothermic Reaction. A reaction that takes place with evolution of heat. Such a reaction occurs at 900–1000°C when most clays are fired.

Expanded Clay. See LIGHTWEIGHT EXPANDED CLAY AGGREGATE.

Explosive Pressing. A process for compaction by the blast of an explosion within the mould containing the powder to be compacted; the rate of energy release has been stated to be approx. 25 000 ft lbf/s and densities as high as 97% of the theoretical have been attained.

Exposed Finish Tile. US term (ASTM – C43) for a hollow clay building block the surfaces of which are intended to be left exposed or painted; the surface may be smooth, combed, or roughened.

Expression Rolls. A pair of steel rolls which, when rotated, will force a clay column through a die or along a cutting-off table (as in the shaping of bats for roofing-tile manufacture); cf. CRUSHING ROLLS.

Extra-duty Glazed Tile. US term for a ceramic tile with a glaze that is sufficiently durable for light-duty floors and all other surfaces inside buildings provided that there is no serious abrasion or impact.

Extrusion. The forcing of clay or other material through a die; this is done by a continuous screw on a shaft rotating centrally in a steel cylinder (an 'auger'), or by a series of knives obliquely mounted on such a shaft (a 'pug'), or, occasionally, by means of a piston (a 'stupid'). Ceramic material may be extruded either to mix and consolidate it (as in the extrusion of pottery body prior to jiggering) or to give the body its final shape (as in the extrusion or bricks and pipes).

Eye. An opening in the SIEGE (q.v.) of a pot furnace for glass melting; gas and air for combustion enter the furnace through the eye.

Eykometer. An instrument for the determination of the yield point of clay suspensions; details have been given by A. G. Stern (*U.S. Bur. Mines*, *Rept. Invest.* 3495, 1940).

F

°F. Degrees Fahrenheit; for conversion table to Centigrade see Appendix 3.

Face. When referring to abrasive wheels this term signifies that part of the wheel against which is applied the work to be ground.

Faced Wall. A wall in which the facing and backing are bonded in such a way that the wall acts as a unit when loaded (cf. VENEERED WALL).

Facing Brick. A clay building brick having an appearance, and weather-resistance, fitting for use in the outside leaf of a wall or in an external brick panel. Such bricks are made in a variety of textures and colours; physical properties range from a crushing strength of about 1000 to 15 000 p.s.i. and a water absorption of about 5 to 30 wt %.

Faggot. Defined in the S. African specification for building bricks as 'A facing brick of the normal face dimensions but with the smallest bed dimension less than the height'.

Fahrenheit. For conversion table to Centigrade, see Appendix 3.

Faience. Originally the French name for the earthenware made at Faenza, Italy, in the 16th century; the ware had a tin-opacified glaze and in this resembled Maiolica and Delft ware. The meaning has now changed. In France, *faience* is any glazed porous ceramic ware; *faience fine* is equivalent to English EARTHENWARE (q.v.). In England, the term 'faience' now refers to glazed architectural ware, e.g. large glazed blocks and slabs (*not* dust-pressed glazed tiles). In USA, *faience ware* signifies a decorated earthenware having a transparent glaze.

Faience Tiles and Mosaic

Faience Tiles and Mosaic. US terms defined (ASTM – C242) as follows: FAIENCE TILE—glazed or unglazed tiles, generally made by the plastic process, showing characteristic variations in the face, edges, and glaze that give an 'art' effect; FAIENCE MOSAIC —Faience tiles that are less than 6 in^2 in facial area and usually $\frac{5}{16}$ in. to $\frac{3}{8}$ in. thick.

Fairlight Clay. A cretaceous clay of the Hastings (England) area that finds use in making building bricks.

Falling Slag. Blast-furnace slag that contains sufficient CALCIUM ORTHOSILICATE (q.v.) to render it liable to fall to a powder when cold. Such a slag is precluded from use as a concrete aggregate by the limits of composition specified in B.S. 1047.

False Set. Premature, but temporary, stiffening of portland cement resulting from overheating of the added gypsum during grinding of the clinker (cf. FLASH SET).

Falter Apparatus. Apparatus designed by A. H. Falter (*J. Amer. Ceram, Soc.*, **28**, 5, 1945) for the determination of the SOFTENING POINT (q.v.) of a glass by the fibre-elongation method as defined by J. T. Littleton (*J. Soc. Glass Tech.*, **24**, 176, 1940).

Fantail. The opening in the refractory brickwork that separates the slag pocket from the regenerator of an open-hearth steel furnace.

Faraday Effect. The rotation of the plane of polarization of a beam of polarized light as it passes through glass that is located in a magnetic field; first observed by Michael Faraday in 1830.

Farren Wall. A cavity wall (4-in. cavity) for house construction introduced in USA; (*Brick & Clay Record*, **95**, No. 4, 38, 1939).

Fat Clay. Term sometimes used for a clay that is highly plastic.

Faugeron Kiln. A coal-fired tunnel kiln of a design proposed in 1910 by E. G. Faugeron (French Pat. 421 765; 24/10/1910); the distinctive feature is the division of the tunnel into a series of chambers by division-walls on the cars and drop-arches in the roof. Such kilns have been used for the firing of feldspathic porcelain.

Fayalite. Ferrous orthosilicate, $2FeO.SiO_2$; m.p. 1205°C. This low-melting mineral is formed when ferruginous slags attack, under reducing conditions, aluminosilicate refractories.

Feather. (1) A fault, in glass, of feather-like appearance and caused by SEED (q.v.) produced by foreign matter picked up by the glass during its shaping.

(2) A fault in wired glass resulting from bending of the transverse wires.

104

Feather Brick. A specially moulded brick of the shape that would be produced by cutting a standard square diagonally from one edge. This could be done in three ways to produce a FEATHER END (or END FEATHER) a FEATHER-END-ON-EDGE, or a FEATHER SIDE (or SIDE FEATHER) depending on whether the diagonal terminates, for a $9 \times 4\frac{1}{2} \times 3$ in. brick, at a $4\frac{1}{2}$-in. edge, a 3-in. edge or a 9-in. edge. (See Fig. 1, p. 37).

Feather Combing. A method of decoration sometimes used by the studio potter: the ware is first covered with layers of variously coloured slips and a sharp tool is then drawn across the surface while it is still moist.

Feather End. See under FEATHER BRICK.

Feather Side. See under FEATHER BRICK.

Feathering. A glaze fault caused by devitrification. It is particularly liable to occur in glazes rich in lime. To prevent the fault the initial rate of cooling in the kiln, after the glaze has matured, should be rapid.

Feed Shaft. See FIRE PILLAR.

Feeder. (1) A device for supplying raw material, e.g. clay, to a preparation machine.

(2) A mechanical system for the production of gobs of glass for a forming machine.

Feeder Channel. The part of the forehearth of a tank furnace, producing container-glass or pressed glass-ware, that carries the molten glass from the working end to the feeder mechanism.

Feeder Gate or Feeder Plug. A shaped refractory used to adjust the rate of flow of molten glass in the feeder channel.

Feeder Sleeve or Feeder Tube. A cylindrical tube that surrounds the feeder plunger in a glass-forming machine.

Feeder Spout or Feeder Nose. The part of the feeder in a glass-tank furnace containing an opening in which the orifice ring is inserted; it forms the end of the forehearth.

Feldspar. The internationally agreed spelling is now *feldspar*; until recently the usual spelling in Britain was *felspar*. The feldspars are a group of minerals consisting of aluminosilicates of potassium, sodium, calcium, and, less frequently, barium. The most common feldspar minerals are:

Orthoclase and Microcline: $K_2O \cdot Al_2O_3 \cdot 6SiO_2$
Albite: $Na_2O \cdot Al_2O_3 \cdot 6SiO_2$
Anorthite: $CaO \cdot Al_2O_3 \cdot 2SiO_2$

These minerals rarely occur pure, isomorphous substitution

being common; most potash feldspars contain sodium, and soda feldspars usually contain both potash and calcium. There is a continuous series of solid solutions, known as plagioclase feldspars, between albite and anorthite.

Feldspars are the chief components of igneous rocks and occur in economic quantity, and in adequate purity for ceramic use, in USA (N. Carolina, S. Dakota, New England), Canada (Ontario, Quebec, Manitoba), Scandinavia, France, and many other countries.

Over half the world's output is consumed by the glass industry; in USA and Europe it is a component (20–30%) of practically all pottery bodies, but in England its use is largely confined to electrical porcelain and vitreous sanitaryware. Feldspar constitutes 25–50% of most vitreous enamel batches.

Feldspar Convention. See under RATIONAL ANALYSIS.

Felite. The name given to one of the crystalline constituents of portland cement clinker by A. E. Törnebohm (*Tonindustr. Ztg.*, **21**, 1148, 1897). Felite is now known to be one form of $2CaO . SiO_2$.

Felspar. See FELDSPAR.

Feret's Law. States that the strength (S) of cement or concrete is related to its mixing ratio by the equation $S = K [c/(c + w + a)]^2$ where c, w and a are the absolute volumes of the cement, water and air in the mix. This relationship was proposed by R. Feret at the beginning of the century.

Ferrari Cement. A sulphate-resistant cement consisting principally of $3CaO . SiO_2$, $2CaO . SiO_2$ and $4CaO . Al_2O_3 . Fe_2O_3$. The sulphate resistance results from the formation of a protective film of calcium ferrite around the calcium aluminate crystals formed by hydrolysis of the brownmillerite. (F. Ferrari, *Cemento Armato*, **38**, 147, 1941.)

Ferric Oxide. Fe_2O_3, the mineral hematite; m.p. 1565°C, but when heated in air it loses oxygen below this temperature to form magnetite, Fe_3O_4; sp. gr. 5·24. Ferric oxide is present in the surface layers of blue engineering bricks; it gives the characteristic sparkle to aventurine glazes.

Ferrites. Specifically, this term refers to a group of special ceramics having FERROMAGNETIC (q.v.) properties, combining high magnetic permeability (up to 4300) and low coercive force (0·1–4 oersteds). The ferrites are synthesized from Fe_2O_3 and the oxide, hydroxide, or carbonate of one or more divalent metals, i.e. Zn, Mn, Ni, Co, Mg or Cu. The firing process must be carefully controlled to maintain the required state of oxidation. Mg–Mn ferrites have a square hysteresis loop; $BaO.6Fe_2O_3$ is a 'hard', or permanent, magnetic material.

Ferroelectric. A material is ferroelectric if it is inherently polarizable and if the direction of its polarization can be controlled by the application of an electric field. Important features of ceramic ferroelectrics are the CURIE POINT (q.v.), and the hysteresis and non-linearity in the relationship between the polarization and the applied electric field; a further important property of these ceramics is their PIEZOELECTRIC (q.v.) nature.

Ferromagnetic. A material exhibiting high magnetic susceptibility and permeability, and magnetic hysteresis; when heated, the magnetization becomes zero at a temperature known as the CURIE POINT (q.v.). For some properties of Ceramic Ferromagnetics see FERRITES.

Ferrous Oxide. FeO; m.p. 1420°C; sp. gr. 5·7. This lower oxide tends to be formed under reducing conditions; it will react with SiO_2 to produce a material melting at about 1200°C, hence the fluxing action of ferruginous impurities present in some clays if the latter are fired under reducing conditions.

Fettling. (1) The removal, in the 'clay state', and usually by hand, of excess body left in the shaping of pottery-ware at such places as seams and edges (cf. SCRAPPING).

(2) The process of repairing a steel-furnace hearth, with dead-burned magnesite or burned dolomite, between tapping and re-charging the furnace.

Feying. A local term in the English brick industry for the cleaning up of a clay-pit floor after an excavator has been at work.

Fiberglas. Trade name: Owens-Corning Fiberglas Corp., USA.

Fibre (Ceramic). Typically, these fibres are made from a batch consisting of alumina and silica (separate or already combined as kaolin or kyanite) together with a borosilicate flux; zirconia may also be present. Other types of ceramic fibre are made from fused silica and from potassium titanate. These fibres are used in the producton of lightweight units for thermal, electrical, and sound insulation; they have also been used for high-temperature filtration, for packing, and for the reinforcement of other ceramic materials (see *Inorganic Fibres* by C. Z. Carroll-Porczynski, 1958).

Fibreglass. Trade name: Fibreglass Ltd., St. Helens, England.

Fidler–Maxwell Kiln. A straight tunnel kiln designed to be fired with gas, coal, or oil; a distinctive feature was the use of

Field-drain Pipe

cast-iron recuperators in the cooling zone. (F. Fidler and J. G. Maxwell, Brit. Pat. 141 124 and 141 125; 6/1/19.)

Field-drain Pipe. An unglazed, fired clay pipe, generally 3 in. or 4 in. dia. and about 1 ft long, for the drainage of fields; occasionally these pipes have a flattened base, or longitudinal ribs, to facilitate alignment during laying. For specification (which includes pipes from 2½ to 12 in. dia.) see B.S. 1196.

Figuier's Gold Purple. A tin-gold colour, produced by a dry method; it has been used for porcelain decoration.

Figured Glass. Flat glass with a pattern on one or both surfaces.

Filler. A general term for a solid material used (in a particular set of conditions) primarily to occupy space, e.g. china clay in paper. In vitreous enamelling a filler (or plugging compound) consisting of a mixture of inorganic compounds is used to fill any holes in iron castings, thus providing a uniform surface on which to apply the enamel.

Film Strength. The mechanical strength, determining its resistance to damage by impact or abrasion, of the dried coating of vitreous enamel slip prior to firing.

Filter Block. The properties required of a clay filter-block for use in trickling filters in the chemical industry are specified in ASTM – C159 (cf. CERAMIC FILTER).

Filter Candle. A porous ceramic tube, which may be rounded and closed at one end, made with a high porosity and of substantially uniform pore size.

Filter Cloth. Cloths form the actual filtering medium in the FILTER PRESSES (q.v.) used to de-water clay slips. Originally, these cloths were made of cotton, subsequently of twill with a backing cloth of jute; cotton and jute cloths need to be proofed against mildew, usually by cuprammonium treatment. Nylon or Terylene filter cloths have the advantage of being rot-proof and are stronger than cotton; they are now much used in the ceramic industry.

Filter-press. Equipment for the dewatering of a slurry by compression between two filter-cloths, which are supported by metal frames; a number of such filtering compartments are set side by side, the compression being applied through the whole series of units. Filter-presses are much used in the pottery industry, in which it is customary to blend the constituents of the body in slip form.

Fin. A fault, sometimes occurring in pressed or blown glass-ware,

108

in the form of a thin projection following the line between the parts of the mould. (Also called FLASH).

Final Set. The time required for a hydraulic cement to develop sufficient strength to resist a prescribed pressure. In the usual VICAT'S NEEDLE (q.v.) test this stage of the setting process is defined as that at which the needle point will, but its circular attachment will not, make a depression on the surface of the cement. In B.S. 12, the conditions of the test are 14·4–17·8°C (58–64°F) and $\not< 90\%$ r.h.; for normal cements, the Final Set must not exceed 10 h (the usual time is 3–4 h).

Findlings Quartzite. A compact, cemented quartzite of a type occurring in Germany as erratic blocks—hence the name, which is the German word for 'foundling'. This type of quartzite, which is used as a raw material for silica-brick manufacture, is composed of about 60% of quartz grains set in 40% of a chalcedonic matrix. In Germany the term is being displaced by the more informative term 'cemented quartzite'.

Fine Grinder or Pulverizer. A machine for the final stage of size reduction, i.e. to -100 mesh. Such machines include ball mills, tube mills, and ring-roll mills.

Fineness Modulus. Defined in USA as one-hundredth of the sum of the cumulative values for the amount of material retained on the series of Tyler or US sieves (excluding half sizes) up to 100 mesh. Example:

On   4 mesh 2%
On   8 mesh 13
On 16 mesh 35
On 30 mesh 57
On 50 mesh 76
On 100 mesh 93
———————
Sum = 276 ÷ 100 = 2·76 Fineness Modulus
———————

Finger-car. A small four-wheeled bogie having two uprights from which project pairs (usually 10 in number) of 'fingers'; these can be raised or lowered by a lever and cam. Finger-cars are used in the KELLER SYSTEM (q.v.) of handling bricks.

Finial. An ornamental piece of fired clayware formerly much used to finish the ends of roof ridges.

Fining. See REFINING.

Finish. The top section of a glass bottle. The finish may be

Finish Mould

DIRTY (rough or spotted) or SHARP; other faults include: BULGED FINISH—distended so that the bung will not fit; FLANGED FINISH—having a protruding fin of glass; OFFSET FINISH—unsymmetrical relative to the axis of the bottle.

Finish Mould. See NECK MOULD.

Fireback. The refractory shape, sometimes made as two or more sections to improve thermal-shock resistance, that forms the back of an open domestic fire. A typical batch composition is 60–70% fireclay and 30–40% grog. B.S. 1251 specifies three sizes and a minimum P.C.E. of Cone 26.

Firebox. One of the small refractory-lined chambers, built wholly or partly in the wall of a kiln, for combustion of the fuel.

Firebrick. Popular term for a FIRECLAY REFRACTORY (q.v.).

Firebridge. See BRIDGE.

Fire-Chek Keys. Trade-name: pyrometric cones made by Bell Research Inc., E. Liverpool, Ohio, USA.

Fireclay; Fireclay Refractory. A clay, commonly associated with the Coal Measures, that is resistant to high temperatures; it normally consists of kaolinite together with some free silica, other impurities rarely exceeding a total of 5%. The most important deposits in the UK are in the Central Valley of Scotland and Ayrshire, in the Stourbridge district, S. W. Yorkshire, and the Swadlincote area. In these districts fireclays of various grades are worked for the manufacture of refractories; in Scotland, S.W. Yorkshire, and Swadlincote the fireclays are also used for making sanitary-ware and glazed sewer-pipes; in a few other areas fireclays are also worked for less exacting purposes. The P.C.E. of fireclay, and of refractories made from it, generally increases with the Al_2O_3 content from about 1550°C for the poorer qualities to 1750°C for grades containing 42% Al_2O_3 when fired; B.S. 1902 defines a fireclay refractory as containing less than 78% SiO_2 and less than 38% Al_2O_3, those grades containing 38–45% Al_2O_3 being termed Aluminous Fireclay Refractories. The principal uses of fireclay refractories are in blast-furnace linings, casting-pit refractories, boiler-furnace linings, and many general purposes where the conditions are not too severe.

Fireclay Plastic Refractory. A US term defined as: A fireclay material tempered with water and suitable for ramming into place to form a monolithic furnace lining that will attain satisfactory physical properties when subjected to the heat of furnace operation.

Fire Crack; Firing Crack. A fire crack is a crack in glass-ware caused by local thermal shock; a firing crack is a crack caused in clay-ware, or in a non-clay refractory, by too-rapid firing.

Fire Finished. Glass-ware that has received its final surface gloss by heating the ware, usually in a flame.

Fire Mark. A defect, in the form of a minute indentation, that may occur in vitreous enamelware if it is over-fired.

Fire Pillar. One of the vertical shafts, beneath each firehole, left in a setting of bricks in a top-fired kiln. (Also known as a FEED SHAFT.)

Fire Polished. See FIRE FINISHED.

Fireproofing Tile. US term for a hollow fired-clay building block for use as a protection for structural members against fire.

Fire Resistance. This term has at times been used indiscriminately to denote the resistance of a material to ignition or to the spread of flame. In the relevant British Standard (B.S. 476, Pt. 1) the meaning is restricted to the performance of complete elements of a building structure without regard to the performance of the materials of which they are composed. In USA, fire tests for building construction and materials are the subject of ASTM – E119.

Firesand. An intermediate product in the manufacture of silicon carbide.

Firestone. A siliceous rock (80–90% SiO_2) suitable for use, when trimmed, as a furnace lining material in its natural state. Firestone has been worked in England at Mexborough and Wickersley, and in Co. Durham, but is now little used.

Fire Travel. The movement of the zone of highest temperature around the gallery of an annular kiln. A typical rate of fire travel is one chamber per day, often a little faster.

Fired-on. Decoration fused into the surface of glazed pottery or glass-ware.

Firing. The process of heat treatment of ceramic ware in a kiln to develop a vitreous or crystalline bond, thus giving the ware properties associated with a ceramic material.

Firing Expansion. The increase in size that sometimes occurs when a refractory raw material or product is fired; it is usually expressed as a linear percentage expansion from the dry to the fired state. Firing expansion can be caused by a crystalline conversion (e.g. of quartz into cristobalite, or of kyanite into mullite plus cristobalite), or by BLOATING (q.v.) cf. AFTER EXPANSION.

Firing Range. See FIRING TEMPERATURE.

Firing Shrinkage. The decrease in size that usually occurs when ceramic ware is fired; it is usually expressed as a linear percentage contraction from the dry to the fired state. Firing shrinkage always occurs with shaped products containing plastic clay and often amounts to 5–6% (cf. AFTER CONTRACTION).

Firing Temperature. Typical ranges of firing temperature of ceramic wares are as follows:

	°C
Aluminium enamels	430– 540
Wet-process cast-iron enamels	620– 760
Decorating firing of pottery	700– 800
Sheet-iron cover-coat enamels	730– 840
Sheet-iron ground-coat enamels	760– 870
Dry-process cast-iron enamels	820– 920
Building bricks	900–1100
Glost firing of pottery	1050–1100
Engineering bricks	1100–1150
Earthenware biscuit	1100–1150
Salt-glazed pipes	1100–1200
China biscuit	1200–1250
Fireclay refractories	1200–1400
Hard porcelain	1300–1400
Silica refractories	1400–1450
Basic refractories	1550–1650

Firsts. Pottery ware that has been selected as virtually free from blemishs (cf. SECONDS, LUMP).

Fish-scaling. A fault sometimes occurring in vitreous enamelware, small semicircles ('fish scales') of enamel becoming detached —often some days after the ware has been fired. The source of this defect is always the ground-coat and the ultimate cause is release of hydrogen followed by a build-up of pressure; the hydrogen may originate in the base metal, in the water used in the preparation and application of the enamel, or in the furnace atmosphere during the process of firing.

Fittings (UK); Trimmers (USA). Special sizes and shapes of wall and floor tiles (see Fig. 6, p. 307). For another meaning of Fittings see ODDMENTS.

Fitz Mill. A type of fine-grinding unit used, for example, in the preparation of the body for sparking plugs; trade name—W. J. Fitzpatrick Co., USA.

Fixing Block. A building unit that may be made of clay, light-weight concrete or breeze, and that is sufficiently soft to permit nails to be driven in but sufficiently non-friable to hold the nails firmly in position. Clay fixing blocks are made to have a high porosity, the pores being of controlled size.

Flabby Cast. A fault sometimes encountered in the casting of pottery-ware. The article appears satisfactory as cast but subsequently deforms, either as a result of a thixotropic effect or because the interior of the cast is still fluid; if the cause is thixotropy, the amount of Na silicate and Na_2CO_3 used as deflocculant should be increased; the second cause is most common in the casting of thick ware and is prevented by increasing the casting rate.

Flaking. The detachment of thin patches of refractory from the silica lining of a gas retort. It is probable that flaking is initiated during the working period of the retort and that the removal of SCURF (q.v.) may be a contributory, but not the principal, factor.

Flambé. See ROUGE FLAMBÉ.

Flame Plating. This term has been proposed for the application of a thin coating of refractory material to a surface by the introduction of the plating powder, oxygen, and acetylene into a chamber where the explosive gas mixture is detonated, the plating powder thus being melted and projected on the inner surface of of the chamber and on any object within it. The process was developed in USA by Linde Air Products Co., the original patent being in the name of R. M. Poorman (US Pat. 2 714 563; 2/8/55).

Flame-spraying. The process of coating a surface (of metal or of a refractory) by spraying it with particles of oxides, carbides, silicides or nitrides that have been made molten by passage through an oxy-acetylene or oxy-hydrogen flame; the coating material can be fed into the flame either as a powder or as a continuous rod. The object is to provide a thin protective coating, usually to prevent oxidation—as in the flame-spraying of alumina on to steel.

Flameless Combustion. Term sometimes used for SURFACE COMBUSTION (q.v.).

Flanged Finish. See FINISH.

Flare Bed. The refractory-lined duct that conveys gas from the producers to the combustion chambers in a setting of horizontal gas retorts.

Flaring Cup. An abrasive wheel shaped like a cup but with a larger diameter at the outer (open) edge than at the bottom of the cup, the bottom being the disk that fits on the spindle.

Flash, Flashing. (1) The formation, by surface fusion or vitrification, of a film of different texture and/or colour on clay products or on glass-ware. In the firing of clay products flashing may occur unintentionally; it is then a defect because of its uncontrolled nature. Bricks that are intentionally flashed make possible pleasing architectural effects. Flashed glass-ware is made by fusing a thin film of a different glass (usually opaque or coloured) on the surface of the ware.

(2) In structural brickwork, a sheet of impervious material secured over a joint through which water might otherwise penetrate.

(3) A fault in glass-ware (see FIN).

(4) The fin of excess body formed during the plastic-pressing of ceramic ware, e.g. electrical porcelain; it is removed by an auxiliary process.

(5) Alternative name for CASTING SPOT (q.v.).

Flash Set. Premature and permanent stiffening of portland cement that has not been adequately retarded; cf. FALSE SET.

Flash Wall. A continuous wall of refractory brickwork built inside a downdraught kiln in front of the fireboxes; its purpose is to direct the hot gases towards the roof of the kiln and to prevent the flames from impinging directly on the setting.

Flask. Local term for the mould, made of wood or cast iron, used in the shaping of crucibles for steel melting.

Flat Arch. See JACK ARCH.

Flat Drawn. Sheet glass made by the vertical drawing process.

Flatter. See STONER.

Flatting. A process for trueing-up hand-made fireclay refractories while they are still only partially dried. Handmaking is now little used except for some special shapes.

Flat-ware. Plates, saucers, dishes, etc. (cf. HOLLOW-WARE).

Flemish Bond. A brick wall built with alternate headers and stretchers in each course. Single Flemish Bond is so bonded on one face of the wall only; Double Flemish Bond is bonded in this way on both faces of the wall.

Fletton. An English building brick made in the Fletton district, near Peterborough, by the semi-dry process from Oxford Clay; this clay is shaly and contains much organic matter, which assists in the firing process. The crushing strength varies from about

2000 to 4500 p.s.i. and the water absorption from about 17 to 25 wt %.

Flexural Strength. See MODULUS OF RUPTURE.

Flint. Nodular chalcedonic silica from the chalk deposits of W. Europe and elsewhere. These flint pebbles are calcined and ground for use in earthenware and tile bodies. In USA the term 'flint' is often applied to other finely ground high-silica rocks used in whiteware manufacture. See also CAOLAD FLINT.

Flint Clay. A non-plastic refractory clay of a type found in Missouri (USA), Southern France, and a few other localities. Flint clays are kaolinitic, but the Missouri deposit may also contain diaspore whereas the French may contain boehmite. Flint clays contain (fired) 42–45% Al_2O_3; P.C.E. 34–35.

Flint Glass. 'White Flint' is a term applied to many colourless glasses; 'Optical Flint' is any glass of high dispersion for optical equipment (cf. OPTICAL CROWN). The introduction of lead oxide into the batch to produce the original flint glass, in the 17th century was an English contribution to glass-making; silica was introduced into the batch in the form of crushed flints.

Flint Shot. US term for the sharp sand sometimes used in sand-blasting the base metal as part of the preparation process for vitreous enamelling.

Flintless Stoneware. Defined in the Pottery (Health and Welfare) Special Regulations of 1950 as: Stoneware the body of which consists of natural clay to which no flint or quartz or other form of free silica has been added.

Flip-flop. A glass decanter with a thin base.

Float Glass Process. A process for making sheet glass introduced in 1959 by Pilkington Bros. Ltd, at St. Helen's, England. A ribbon of glass is floated on molten tin, the product being sheet glass with truly parallel surfaces, both fire-polished.

Floater. A refractory shape that is allowed to float on the surface of molten glass in a tank furnace in order to hold back any scum that may be present (cf. RING).

Floating. (1) Slight displacement of a pattern applied to pottery-ware; it is caused by the presence of grease or moisture on the glaze at the time when the decoration was applied.

(2) The smoothing of freshly-placed concrete to give a good surface finish.

Floating Agent. As used in the vitreous-enamel industry this term is the US equivalent of the English term SUSPENDING AGENT (q.v.).

115

Floc Test or Water Test. A test for the durability of hydraulic cement: 1 g of the cement is shaken with 100 ml of water in a test-tube which is then placed on its side and allowed to stand for 7 days; if the amount of floc formed is very small the cement is considered to be durable. The test was proposed by I. Paul (*Proc. Assoc. Highway Officials, N. Atlantic States*, **12**, 140, 1936).

Flocculation. The addition of a suitable electrolyte to a colloidal suspension to agglomerate the particles and so hasten their settling; the sediment thus formed is usually more readily removed from the containing vessel than is the sediment from a deflocculated suspension. An example is the addition of up to 0.05% $CaCl_2$ to flocculate slop glaze.

Flood Casting. Term used in the British sanitary-ware industry for the process of slip casting in which excess slip is removed from the mould by draining. In other sections of the pottery industry the process is referred to merely as 'casting'; in USA the process is known as 'Drain Casting'.

Floor Brick. A brick having mechanical, thermal, and chemical resistance to the conditions to which it is likely to be exposed when used in an industrial floor. Properties are specified in ASTM – C410: Industrial Floor Brick.

Floor Quarry. (UK) or Quarry Tile (USA). A heavy ceramic flooring material. Floor quarries (as distinct from FLOOR TILES q.v.) are usually made by a plastic process. They are relatively thick (1 in.) and generally not less than 8×8 in. in size. They are hard fired to produce a body resistant to heavy abrasion and to attack by most industrial liquids—hence their wide use for factory floors.

Floor Tile. (1) Ceramic tiles, normally unglazed, for flooring. It is difficult to draw a sharp distinction between floor tiles and FLOOR QUARRIES (q.v.) but the former are always dry-pressed, and they are relatively thin and do not normally exceed 6×6 in. in size.

(2) One of the refractory shapes used in the construction of a gas retort; a group of these tiles is laid horizontally to brace the retorts of a vertical setting and to limit the combustion flues.

(3) In the USA the term is used for a hollow fired-clay block for use in the construction of floors and roofs; properties are specified in ASTM – C57.

Floridin. An older name for ATTAPULGITE (q.v.); now a trade-name of the Floridin Co., USA.

Flotation. A process for the concentration of a mineral by crushing the rock in which it is dispersed, suspending the crushed material in water and causing a froth to form, one of the constituents of the mineral adhering to the air bubbles so that it can be floated off, leaving the remainder to settle out. Three types of reagent are required: frothers, collectors, and controllers. Flotation has been applied industrially to the preparation of the following ceramic raw materials; feldspar, talc, magnesite, kyanite, glass-sands, lithium ores.

Flow Blue. A deep cobalt blue which was used for under-glaze printing on pottery. As the name indicates, the colour tended to flow into the glaze, giving a blurred effect; this result was obtained by placing FLOW POWDER (q.v.) in the saggar containing the ware, chlorine evolved from the powder and combined with some of the cobalt, thus rendering it slightly soluble in the glaze.

Flow-button Test. See FUSION-FLOW TEST.

Flow Coating. In vitreous enamelling, the application of enamel slip by allowing it to flow over the metal and then to drain.

Flow Line. A surface defect on glass; see CHILL MARK.

Flow Machine. A machine used in glass-making; molten glass flows into it from a feeder under the action of gravity.

Flow Powder. A mixture formulated to evolve chlorine at the temperature of the glost firing of pottery and used in the production of FLOW BLUE (q.v.). For ware covered with a lead glaze, a suitable composition is (per cent): NaCl, 22; white lead, 40; $CaCO_3$, 30; borax, 8. For use with a leadless glaze a suitable mixture is (per cent): NaCl, 15; $MgCl_2$, 55; KNO_3, 15; $CaCO_3$, 15.

Flow Process. A method of producing flat glass by allowing molten glass to flow continuously from a tank furnace between rollers. The term is sometimes also applied to a process for making hollow-ware, but this is more correctly referred to as the GOB PROCESS; (q.v.).

Flow Test. (1) A test for the consistency of concrete in terms of its tendency to spread when placed on a metal table and jolted under specified conditions; (ASTM – C124).

Flowers. See MOTTLING (of silica refractories).

Flue Liner. A fireclay shape for use in the flues and chimneys of domestic heating appliances; for specifications see B.S. 1181 and ASTM – C315.

Fluid-energy Mill. A size-reduction unit depending for its action on collisions between the particles being ground, the energy

Fluidized Bed

being supplied by a compressed fluid (e.g. air or steam) that enters the grinding chamber at high speed. Such mills will give a product of 5μ or less; they have been used for the fine grinding of frits, kaolin, zircon, titania, and calcined alumina, but the energy consumed per ton of milled product is high.

Fluidized Bed. A cushion of gas between a powder and a porous ceramic support, which is generally in the form of slabs; a current of air or hot gases is forced through the porous ceramic under pressure. The principle is used as a method of conveying powders along a slightly inclined porous ceramic trough; the powder can be simultaneously dried and/or calcined.

Flume. Local term for the alkali vapour volatilized from the glass in a glass-tank furnace; it causes corrosion of the furnace roof, downtakes, and regenerator refractories. (A corruption of 'Fume'.)

Fluorspar. CaF_2; m.p. 1360°C; sp. gr. 3·1. The largest deposits of this mineral are in Mexico, China, USA and USSR, but it is also worked in England and most other European countries. It is used as an opacifier in glass and vitreous enamel. Fluorspar crucibles have recently been used in the melting of uranium for nuclear engineering.

Flux. A substance that, even in small quantities, lowers the fusion point of material in which it is naturally present (e.g. alkalis in clays) or of material to which it has been added (e.g. borax added to glazes).

Flux Factor. A factor for assessing the quality of steelworks-grade silica refractories. It is defined in ASTM – C416 as the percentage of Al_2O_3 in the brick plus twice the total percentage of alkalis; for first-quality (Type A) bricks the flux factor must not exceed 0·50.

Flux Line. The level of the molten glass in a tank furnace; this level is generally marked by a horizontal line of maximum attack on the refractories. The term is also sometimes applied to the boundary between unmelted batch and molten glass in a tank furnace.

Flux-line Block. A refractory block for use in the upper course of the walls of a glass-tank furnace. The flux line is the surface level of the molten glass and attack on the refractories is more severe at this level than beneath the molten glass.

Fly-ash. The ash carried by the waste gases in a furnace or kiln (cf. PULVERIZED FUEL ASH).

118

Flying Arch. In a modern glass-tank furnace the double-walled bridge built across the furnace to separate the working end from the melting and refining end; the flying arch is independent of the general furnace structure.

Foam Glass. Cellular glass, in the form of blocks, usually made by mixing powdered glass with a gasifying agent (e.g. carbon or a carbon compound) the mixture then being heated for a short time to fuse the glass and trap the evolving gas bubbles. Foam glass is used as a structural heat-insulating material (cf. BUBBLE GLASS).

Foamed Clay. Lightweight cellular clayware for heat and sound insulation. Foam is generated in a clay slip, either mechanically or by a chemical reaction that evolves gas bubbles, and the slip is then caused to set. Some insulating refractories are made by this process.

Foamed Concrete. See AERATED CONCRETE.

Fold. See LAP.

Folded Foot. The foot of a wine glass is said to be folded if the outer edge has been raised by folding it back on itself.

Font. A reservoir above the mould for fusion-casting refractories; molten material from the font helps to fill the PIPE (q.v.).

Foot-boards. Wooden boards, hinged together, for hand-shaping the foot of glass stem-ware.

Ford Cup. An orifice-type viscometer. It has been used to a limited extent in the testing of the flow properties of ceramic suspensions. For Ford Cup No. 4 (the commonest size) the following conversion applies:

$$\text{Absolute viscosity in poises} = \frac{\text{time in seconds}}{27} \times \text{sp. gr.}$$

Forehearth. (1) An extension to the bottom of a CUPOLA (q.v.) serving as a reservoir for molten iron.

(2) An extension of a glass-tank furnace from which glass is taken for forming.

Fork. A metal device for placing vitreous enamelware in, and subsequently removing it from, a box furnace.

Forsterite. Magnesium orthosilicate, $2MgO \cdot SiO_2$; m.p. 1890°C. In economic amounts forsterite occurs naturally only in association with FAYALITE (q.v.) in the mineral OLIVINE (q.v.). Forsterite refractories are made from olivine or SERPENTINE (q.v.), dead-burned magnesite being added to the batch to combine with the fayalite and produce the more refractory mineral MAGNESIO-FER-RITE (q.v.). Pure synthetic forsterite is used as a low-loss dielectric ceramic; it is particularly suitable (on account of its relatively high thermal expansion) for making vacuum-tight seals with metallic Ti.

Founder. See TEASER.

Founding (of glass). See REFINING.

Fourcault Process. A method of making sheet glass invented in 1902 by E. Fourcault of Lodelinsard (Belgium). The glass is drawn vertically and continuously from the glass-tank through a DEBITEUSE (q.v.).

Free-blown. Alternative term for OFF-HAND (q.v.).

Free Crushing. The method of running a pair of crushing rolls so that virtually all the crushing occurs as the particles pass between the rolls, only a small proportion of the particles being crushed by bearing on one another. When crushing rolls are operated in this way, the product contains less fines than with CHOKE CRUSHING (q.v.).

Freehand Grinding or Off-hand Grinding. The method of grinding in which the object to be ground is held by hand against an abrasive wheel.

Freeze-casting. A process for making intricate shapes of special ceramic material, e.g. turbo-supercharger blades. Refractory powder, with a small proportion of binder, is made into a thick slip, which is cast into a mould and then frozen; the cast is then dried and sintered.

French Chalk. See STEATITE.

Frenkel Mixer. A screw-type, 'enforced order', mixer of much smaller pitch than the usual shaft mixer; it operates on the convergence–divergence (C–D) principle. (M. S. Frenkel, Brit. Pat. 888 864; 7/2/62.)

Frey Automatic Cutter. Trade name: a machine for cutting an extruding column of clay into bricks by one or more horizontal wires that cut downwards while the clay is moving forward. (G. Willy A.G., Chur, Switzerland.)

Friction Element. Some clutches and brakes for use in severe conditions are now lined with CERMETS (q.v.) Amongst the materials used are corundum and sillimanite as the ceramic component, and Mo, Cr, Fe, and Cu alloys as the metallic bond.

Friction Press or Friction-screw Press. A machine for dry-pressing: a plunger is forced into the mould by a vertical screw the screw shaft being driven by friction disks or rollers—

downwards for pressing by one disk and upwards for release of pressure by a second disk on the opposite side of the driving wheel. This type of press is used for special shapes of tiles and sometimes for making silica refractories, etc.

Frigger. A small whimsically shaped piece of hand-blown glass ware—a glass-blower's bagatelle.

Frischer Ring. A type of ceramic ring for the packing of towers in the chemical industry.

Frit. A ceramic composition that has been fused, quenched to form a glass, and granulated. Frits form an important part of the batches used in compounding enamels and glazes; the purpose of this pre-fusion is to render any soluble and/or toxic components insoluble by causing them to combine with silica and other added oxides.

Fritted Glass. See SINTERED GLASS.

Fritted Glaze. A glaze in which some of the constituents have been previously fused together to form a frit; the constituents so pretreated include lead oxide, which fritting converts into an insoluble silicate, and other constituents that would otherwise be soluble.

Fritted Porcelain. Alternative name for SOFT-PASTE (q.v.).

Fritting Zone. See SOAKING AREA.

Frizzling. A fault liable to develop during the firing of pottery-ware that has been decorated with lithographic transfers; if the varnishes are burned away too rapidly in the early stages of the enamel fire, the colour is liable to crack and curl up. To prevent this fault, the layer of size should be thin and the rate of firing between 200° and 400°C should not greatly exceed 1°C/min.

Frog. A depression in the bedding face of some pressed building bricks to decrease the weight and improve the keying-in of the mortar. The term is probably derived from the same word as applied to the similar depression in the centre of a horse's hoof.

Frost Glass. See GLASS FROST.

Frost's Cement. An early form of hydraulic cement patented in England in 1811 by James Frost; it was made from two parts chalk to one part clay.

Frosting. See ACID FROSTING.

Froth Flotation. See FLOTATION.

Fuch's Gold Purple. A tin-gold colour, produced by a wet method; it has been used in the decoration of porcelain.

Fuller's Earth. A type of clay formerly used in the cleansing of

121

cloth. The clay minerals present are of the montmorillonite and/or beidellite types. It is used for decolorizing oils and has also been used as a bond for foundry sands.

Fuller's Grading Curve. A method of graphical representation of particle-size analysis; the grain size (in fractions of an inch) is shown on the abscissa and the cumulative percentage on the ordinate. In the original paper (W. B. Fuller and S. E. Thompson, *Trans. Amer. Soc. Civil Engrs*, **59**, 67, 1907) the concept of 'ideal' grading curves was introduced, these being selected to be ellipses with straight lines tangent to them; more strictly, the 'ideal' curves are parabolas having the form $d = P^2 D/10\,000$, where d is any selected particle diameter, D is the diameter of the largest particles and P is the percentage finer than d.

Furring Tile. US term for a hollow fired-clay building block for lining the inside of walls and carrying no superimposed load.

Fused Silica. See VITREOUS SILICA.

Fusion-cast Basalt. An abrasion-resistant material made by fusing natural basalt and casting the molten material into moulds to form blocks. The hardness is 8–9 Moh; crushing strength, 70 000 p.s.i. These blocks can be used for industrial flooring and for the lining of bunkers, chutes, and other equipment where abrasion is severe.

Fusion Casting. A process for the manufacture of refractory blocks and shapes of low porosity and a high degree of crystallinity; the refractory batch is electrically fused and, while molten, is cast into a mould and carefully cooled. The usual types of fusion-cast refractory are those consisting of mullite, corundum, and zirconia in various proportions; such refractories find considerable use as tank blocks for glass-melting furnaces.

Fusion-flow Test. A method for the evaluation of the fusion-flow properties of a vitreous enamel or of a glaze. As used for testing enamel frits, the method is standardized in ASTM – C374.

Fuzzy Texture. A fault sometimes occurring in vitreous enamelware, the 'fuzziness' resulting from minute craters and bubbles in and near the enamel surface. Also known as GASSY SURFACE.

G

G Stone. A name that has been used for PYROPHYLLITE (q.v.).

G-Value. The basis of a method of calculation for compounding slips and glazes, the 'G-value' being the grammes of suspended solids per cm^3 of suspension: $G = SP/100$, where S is the sp. gr. of the suspension and P is the percentage of solids in the suspension. (K. M. Kautz, *J. Amer. Ceram. Soc.*, **15**, 644, 1932).

Gable Tile. A roofing tile that is half as wide again as the standard tile. Gable tiles are used to complete alternate courses at the VERGE (q.v.) of a tiled roof.

Gable Wall. The refractory wall above the tank-blocks at the end of a glass-making furnace; it is also sometimes referred to as the END WALL or BACK WALL.

Gadget. A tool, used in the hand-made glass industry, for holding the foot of a wine glass while the bowl is being finished.

Gadolinium Oxide. Gd_2O_3; m.p. 2330°C; sp. gr. 7·41; thermal expansion (20–1000°C), $10·5 \times 10^{-6}$.

Gaffer. The foreman of a SHOP (q.v.) making glass-ware by hand; also called a CHAIRMAN.

Gahnite. See ZINC ALUMINATE.

Gaize. A siliceous rock, containing some clay, found in the Ardennes and Meuse Valley (France). It has been used as a POZZOLANA (q.v.).

Galena. Natural lead sulphide, PbS; formerly used in the glazing of pottery.

Gall. Molten sulphates sometimes formed on the surface of glass in a furnace; the cause is inadequate reduction of the salt cake to the more reactive sodium sulphite. Other aggregation of unfused material floating on molten glass are also sometimes referred to as gall, e.g. batch gall.

Galleting. Small pieces of roofing tile bedded in the top course of single-lap tiles to give a level bedding for the ridge tiles.

Galleyware or Gallyware. Term for the early (16th century) tin-glazed earthenware; the name derives either from the importation of the ware in Mediterranean galleys or from the use of the tin-glazed tiles in ships' galleys.

Gallium Nitride. GaN; m.p. 800°C. A special ceramic of high electrical resistivity (4×10^8 ohm.cm at 20°C).

Gallium Oxide. Ga_2O_3; m.p. $1795° \pm 15°C$.

Ganister. Strictly, this term should be reserved for a silica rock formed by deposition of dissolved silica in the siliceous seat-earth of a coal seam; the true ganister was typically found as the seat-earth of some of the coal seams of the Lower Coal Measures in the Sheffield district of England and was the raw material on which the manufacture of silica refractories in that area was based.

Gap Grading

The term 'ground ganister' is used to denote the highly siliceous patching and ramming material used in cupolas and foundry ladles.

Gap Grading. See GRADING.

Gardner Mobilometer. An instrument for the evaluation of the flow properties of vitreous-enamel slips. It consists of a plunger ending in a disk, which may be solid or may have a standardized system of perforations; the plunger is inserted in a tall cylinder containing slip and is loaded so that it descends through the slip; the time taken to fall through a specified distance is a measure of the mobility of the slip. (H. A. Gardner Laboratory, Bethesda, Maryland, USA.)

Gareis–Endell Plastometer. Consists of two disks between which a cylinder of clay is squeezed; the upper disk is rotated while the lower disk is slowly raised by a revolving drum; a stress/deformation curve is recorded. (F. Gareis and K. Endell, *Ber. Deut. Keram. Ges.*, **15**, 613, 1934.)

Garnet. A group of minerals crystallizing in the isometric system and of general composition $3RO.R_2O_3.3SiO_2$, e.g. $3CaO.Al_2O_3.3SiO_2$ (grossularite), $3MgO.Al_2O_3.3SiO_2$ (pyrope), $3CaO.Fe_2O_3.3SiO_2$ (andradite). Natural garnet occurs in economic quantity in USA, Canada, India, S. Africa, and elsewhere; it is used as an abrasive, particularly in the woodworking industry and for the lapping of bronze worm wheels. Iron garnets are synthesized for use as FERROMAGNETICS (q.v.); the most interesting is YTTRIUM–IRON GARNET (q.v.).

Gas Adsorption Method. A technique for the determination of SPECIFIC SURFACE (q.v.); variants of the method include the BRUNAUER, EMMETT AND TELLER METHOD (q.v.) and the HARKINS AND JURA METHOD (q.v.).

Gas Concrete. See AERATED CONCRETE.

Gas-fire Radiant. The radiants for gas fires are made of refractory material having good resistance to thermal shock. The usual composition is a mixture of clay and crushed fused silica; this is shaped and then fired at about 1000°C.

Gas Pickling. A method of preparing sheet steel for vitreous enamelling by treatment, while hot, with gaseous HCl.

Gas Retort. A refractory structure used for the conversion of coal into coke with the simultaneous distillation of town gas. There are two types: CONTINUOUS VERTICAL RETORT (q.v.) and HORIZONTAL RETORT (q.v.). In the UK the refractories used must meet the specifications issued by the Gas Council (1 Grosvenor Place, London, S.W.1) in collaboration with the Society of British Gas Industries and the British Coking Industry Association.

Gas Turbine. A device for the conversion of the energy of hot gases, derived from internal combustion, into rotary motion of a machine element. The efficiency increases with operating temperature and is at present limited by the safe temperature at which heat-resisting alloys can be used. There has been much research on the possible use of cermets and other special ceramics in these turbines, particularly in the blades.

Gassy Surface. See FUZZY TEXTURE.

Gate. See STOPPER.

Gater Hall Device. See BARRATT–HALSALL FIREMOUTH.

Gather. To take molten glass from a furnace for shaping; the amount of glass so taken (gathered) is called a 'gather' also.

Gathering Hole. An opening in the working-end of a glass-tank furnace, or in the wall of a pot furnace, to permit the gathering of molten glass.

Gaudin's Equation. An equation for the particle-size distribution that can be expected when a material is crushed in a ball mill or rod mill; it is of the form $P = 100(x/D)^m$, where P is the percentage passing a sieve of aperture x, D is the maximum size of particle, and m is a constant which is a measure of dispersion. The equation holds good only if the ratio of size of feed to size of balls is below a critical value which, for quartz, is 1:12. (A. M. Gaudin, *Trans. Amer. Inst. Min. Met. Engrs.*, **123**, 253, 1926.) cf. SCHUHMANN EQUATION.

Gauge. The exposed length of a roofing tile as laid on a roof; also sometimes known as a MARGIN.

Gauging of Cement. The process of mixing cement with water. For the preparation of a cement paste of standard consistency prior to testing, B.S. 12 stipulates that the time of gauging shall be 3–5 min.

Gault Clay. A calcareous clay with a short vitrification range used for making building bricks in S. E. England. Bricks made from this clay are generally porous and cream coloured, but in a few localities red bricks are made from it.

Gehlenite. $2CaO \cdot Al_2O_3 \cdot SiO_2$; m.p. 1593°C. This mineral is sometimes formed by the action of lime on firebricks; it may occur in the slagged parts of blast-furnace linings.

Gelatine-pad Printing. See MURRAY–CURVEX MACHINE.

Generator

Generator. In a water-gas plant, the refractory-lined chamber in which solid fuel is gasified by blowing in steam and air alternately.

Georgian Glass. Wired Glass (q.v.) in which the wire mesh is square.

Germanium Nitride. Ge_3N_4; decomposes at 800°C. A special electroceramic of high resistivity.

Getting. The actual process of digging clay, by hand or by excavator; getting and transporting form the successive stages of winning.

Gibbsite. Aluminium trihydrate, $Al_2O_3.3H_2O$. Occurs in Dutch Guiana, the Congo, and some other areas where lateralization has occurred. A typical analysis is 65% Al_2O_3, 33% H_2O, and 2% impurities, but many samples contain clay. Gibbsite requires calcination at a very high temperature to eliminate all the shrinkage that results from the loss of water.

Gilard and Dubrul Factors. See Thermal Expansion Factors for Glass.

Gild. The painting of pottery with liquid gold; this is subsequently fired on at about 700°C.

Gillmore Needle. Apparatus for the determination of the initial and final set of portland cement. It consists of two loaded rods which slide vertically in a frame: the rod ('needle') for the determination of initial set is $\frac{1}{12}$ in. dia. and weighs $\frac{1}{4}$ lb; the needle for the final set is $\frac{1}{24}$ in. dia. and weighs 1 lb. Details are given in ASTM – C266 (cf. Vicat Needle).

Ginneter. Term in the N. Staffordshire potteries for a woman whose job it is to grind from china-ware, after it has been taken from the glost kiln, any adhering particles of refractory material from the kiln furniture; cf. Sorting. (From *Ginnet*, an old term for a tool used by carpenters to remove excrescences from wood.)

Glaceramic. A term that has been used in USA for devitrified glass products of the type most commonly known as Pyroceram (q.v.).

Glarimeter. See Ingersoll Glarimeter.

Glass. An inorganic material that has been produced by fusion and subsequent cooling, the fused mass becoming rigid without crystallizing. The names of many types of glass indicate their main constituents, e.g. soda-lime glass (the presence of silica being understood); for the system of designating types of optical glass see Optical Glass Classification. A major advance in the understanding of the fundamental nature of a glass was made when

126

W. H. Zachariasen (*J. Amer. Chem. Soc.*, **54**, 3841, 1932) deduced that the characteristic properties of glasses are explicable if the interatomic forces are essentially the same as in a crystal, but if the three-dimensional atomic network in a glass lacks the symmetry and periodicity of the crystalline network. Zachariasen's rules for the formation of an oxide glass are: (1) the sample must contain a high proportion of cations that are surrounded by oxygen tetrahedra or by oxygen triangles; (2) these tetrahedra or triangles must share only corners with each other; (3) some oxygen atoms must be linked to only two such cations and must not form further bonds with any other cations. The ions in a glass are thus divided into NETWORK-FORMING (q.v.) and NETWORK-MODIFYING (q.v.). See also STRUCTON; VITRON.

Glass-bonded Mica. See MICA (GLASS-BONDED).

Glass Eye. A large unbroken blister on vitreous enamelware.

Glass Fibre. Filamentous glass made by mechanical drawing or centrifugal spinning; or by the action of a blast of air or steam to produce Staple Fibre. The unprocessed filaments are known as Basic Fibre; a number of filaments bonded together form a Strand. Long glass fibres are known as 'Silk'; a fleece-like mass of fibres is 'Wool'; felty material is 'Mat'.

Glass Frost. Very thin glass that has been crushed for use as a decorative material (cf. TINSEL).

Glass-to-Metal Seal. Metal components varying in size from fine wires to heavy flanges are sealed to glass for many purposes, e.g. electric lamp bulbs and radio valves. Metals that have been used for this purpose include Pt, Cu, W, Mo and alloys such as Fe–Cr, Ni–Fe and Ni–Fe–Co. (See *Glass-to-Metal Seals* by J. H. Partridge, Sheffield, 1949.)

Glaze. A thin glassy layer formed on the surface of a ceramic product by firing-on an applied coating, by firing in the presence of alkali vapour (as in SALT GLAZING (q.v.)), or as a result of slag attack on a refractory material. A glaze may, however, be partially crystalline (see CRYSTALLINE GLAZE and MATT GLAZE). The term 'glaze' is also applied to a prepared mixture, which may be either a powder or a suspension in water (SLOP GLAZE (q.v.)) ready for application to ceramic ware by dipping or spraying.

Glaze Fit. The matching of the thermal expansion of a glaze to that of the body on which it is held. To prevent CRAZING (q.v.), the glaze must be in compression when the ware has been cooled

Glazing

from the kiln to room temperature; to achieve this, the thermal expansion of the glaze must be less than that of the body.

Glazing. (1) The process of applying a glaze to ceramic ware; the latter may be unfired or in the biscuit state.

(2) The formation of a glazed surface on a refractory material as a result of exposure to high temperature and/or slagging agents. Occasionally, a glaze is deliberately applied to a refractory to seal the surface pores.

(3) The dulling of the cutting grains in the face of an abrasive wheel, generally as a result of operating at an incorrect speed.

Glenboig Fireclay. A fireclay occurring in the Millstone Grit in the region of Glenboig, Lanarkshire, Scotland. A typical per cent analysis (raw) is: SiO_2, 50–51; Al_2O_3, 33; Fe_2O_3, 2·5; alkalis 0·5. The P.C.E. is 32–33.

Glory Hole. An opening in a furnace used in the glass industry for reheating glass-ware preparatory to shaping.

Gloss. A surface property, for example of a glazed ceramic or of vitreous enamelware, related to the reflectivity. For a quantitative definition see SPECULAR GLOSS.

Gloss Point. When a layer of glaze powder is heated, a temperature is reached at which the surface changes its appearance from dull to bright; this temperature has been termed the 'gloss point'. (R. W. P. de Vries, *Philips Tech. Rev.*, **17**, 153, 1955.)

Glost. This word, meaning 'glazed', is used in compound terms such as 'Glost ware', 'Glost firing' and 'Glost placer'.

Glut Arch. A brick arch below the firemouth of a pottery BOTTLE OVEN (q.v.) for the admission of primary air and the removal of clinker.

Gob. A lump of hot glass delivered, or gathered, for shaping.

Gob Process. A method of making hollow glass-ware, gobs of glass being delivered automatically to a forming machine.

Gold Decoration. See BURNISH GOLD and BRIGHT GOLD.

Gold Ruby Glass. See RUBY GLASS.

Gold Scouring. Alternative term for burnishing; see under BURNISH GOLD.

Goldstone Glaze. An AVENTURINE (q.v.) glaze; a quoted composition is (parts): white lead, 198·7; feldspar, 83·4; whiting, 8; ferric oxide, 11·2; flint, 41·4. This glaze matures at cone 04.

Gonell Air Elutriator. A down-blast type of ELUTRIATOR (q.v.) designed by H. W. Gonell (*Zeit. Ver. Deutsch. Ing.*, **72**, 945, 1928);

128

it has found considerable use in Europe for assessing the fineness of portland cement.

Gooseneck. See under BUSTLE PIPE.

Goskar Dryer. A chamber dryer for bricks and tiles designed by T. A. Goskar (*Trans. Brit. Ceram. Soc.*, **37**, 62, 1938). Each chamber has a false floor and a false roof; air enters the chamber via the space above the false roof and is withdrawn via the corresponding space below the false floor.

Gothic Pitch. See under PITCH.

Gottignies Kiln. The original electric MULTI-PASSAGE KILN (q.v.); it was introduced in 1938 by two Belgians, R. Gottignies and L. Gottignies (Belgian Pat. 430 018).

Gouging Test. A procedure for the evaluation of the resistance of a vitreous-enamelled surface to mechanical wear. In the procedure laid down in a Special Bulletin issued by the Porcelain Enamel Institute (Washington, D.C., USA), a small steel ball is rolled on the enamel surface under various loads.

Grade. As applied to an abrasive wheel this term refers to the tenacity of the bond for the abrasive grains, i.e. the resistance to the tearing-out of the abrasive grains during use.

Grading. (1) The relative proportions of the variously sized particles in a batch, or the process of screening and mixing to produce a batch with particle sizes correctly proportioned. A batch with a grading for low porosity will contain high proportions of coarse and fine particles and a low proportion of intermediate size; if a particular particle size, e.g. the medium size, is excluded from the batch this is said to be a GAP GRADING.

(2) In the abrasives industry, the process of testing to determine the GRADE (q.v.) of a wheel; testing machines are available for this purpose.

Grain. Abrasive material, dead-burned magnesite, or similar materials that have been size-graded.

Graining; Graining Paste. The production on vitreous enamelware of a patterned surface resembling grained wood; this is done by transferring, by means of a metal roller, GRAINING PASTE from an etched plate to the enamel surface; this paste consists of colouring oxides and a flux; it is made to the required consistency with an oil, e.g. clove oil.

Grain-size Analysis. See PARTICLE-SIZE ANALYSIS.

Grain Spacing. A term in the US abrasives industry equivalent to STRUCTURE (q.v.).

Granite Ware. This term has had various meanings within the vitreous-enamel industry but the definition given in ASTM – C286 would restrict the meaning to: a one-coat vitreous-enamelled article with a mottled pattern produced by controlled corrosion of the metal base prior to firing.

Granny Bonnet. Term sometimes used for a Bonnet Hip Tile—see under HIP TILE.

Graphite. A crystalline form of carbon occurring naturally in Korea, Austria, Mexico, Ceylon, Madagascar, and elsewhere; it is produced when amorphous carbon is heated, out of contact with air, at a high temperature. The m.p. exceeds 3500°C; the electrical and thermal conductivities are high. Graphite is used (under reducing conditions) as a special refractory and as an electrical heating element; mixed with fireclay it is used in the manufacture of plumbago crucibles and stoppers for use in the metal industry. See also CARBON REFRACTORIES.

Gravity Process. See GOB PROCESS.

Green. Ceramic ware in the condition after it has been shaped but before it has been dried and fired.

Green House. A heated room in which pottery-ware is inspected and stored between the shaping process and the firing process.

Green Spot. A fault that occasionally becomes serious in the manufacture of sanitary fireclay and glazed bricks. The green spots are comparatively large and frequently of an intense colour. The usual causes are the presence of chalcopyrite ($CuFeS_2$) in the raw clay or accidental contamination by a particle of copper or copper alloy, e.g. a chip off a bronze bearing.

Green Strength. The mechanical strength (usually measured by a transverse test) of ceramic ware in the GREEN (q.v.) state.

Grey Stock. A clamp-fired STOCK BRICK (q.v.) that is off-colour.

Grinding. (1) Particle-size reduction by attrition and/or high-speed impact; in the size reduction of ceramic raw materials, coarse crushing usually precedes fine grinding.

(2) Final shaping to close tolerance, e.g. of an electroceramic component, by means of abrasive wheels.

Grinding Aid. An additive to the charge in a ball mill or rod mill to accelerate the grinding process; the additive has surface-active or lubricating properties. Grinding aids find particular use in the grinding of portland cement clinker, but in the UK their use is precluded by the conditions laid down in B.S. 12.

Grinding Wheel. A disk, or comparable symmetrical shape, of bonded abrasive material. The abrasive is either alumina or silicon carbide; the bond may be of the vitrified ceramic type, or it may consist of sodium silicate (here called a SILICATE BOND), resin, rubber, or shellac. A standard marking system for grinding wheels was adopted many years ago by the Grinding Wheel Manufacturers' Association of America; in 1952 this system was also adopted in the UK as British Standard 1814. This marking system is reproduced in Fig. 3.

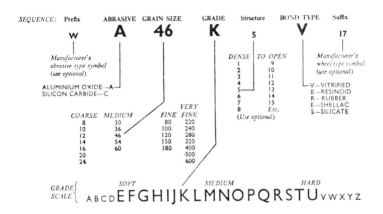

FIG. 3. GRINDING WHEEL MARKING SYSTEM.

(From B.S. 1814, by permission of the British Standards Institution, 2 Park Street, London, W.1.)

Grinstead Clay. A Cretaceous clay for brickmaking in parts of Sussex.

Grisaille. A method of decorating, at one time used on pottery vases, etc., in which different shades of grey were used to produce the effect of low relief; from French word meaning grey shading.

Gritting. The process of forming a smooth surface on blocks of marble, or other natural stones, by means of abrasive blocks known as 'rubbing blocks'.

Grizzle. A clamp-fired STOCK BRICK (q.v.) that is soft and suitable only for internal walls.

Grizzly. The name used in the English structural-clay products industry for a simple, stationary, screen consisting merely of a frame and perforated metal plate, or wire mesh; the screen is set at an angle, the inclination depending on the angle of repose of the material being screened.

Grog. Firebrick, which may already have been used in a furnace lining or elsewhere, crushed to a size suitable for incorporation in a batch, either for remaking into bricks or for use in a refractory cement, ramming, or patching material. The word is probably derived from the French *gros grain* (coarse grain) via the English 'grogram' (coarse cloth) (cf. CHAMOTTE).

Grog Fireclay Mortar. A US term defined as: raw fireclay mixed with calcined fireclay, or with broken fireclay brick, or both, all ground to suitable fineness.

Grossalmerode Clay. A refractory clay from Grossalmerode, about 10 miles E.S.E. of Kassel, West Germany. These clays are of Tertiary origin. They have been used for making glass-pots for five centuries. Aluminous clays are worked in the district but, typically, the Grossalmerode clay contains (raw) 70% SiO_2 and 18% Al_2O_3; P.C.E. 28–29.

Ground-coat. The coating on vitreous enamelware that is applied to the metal. The principal constituents of a sheet-iron ground-coat are feldspar, borax, and silica which, together, constitute 75–90% of the batch; in addition, there is about 10% $Na_2CO_3 + NaNO_3$, 5% fluorspar, 0·5% cobalt oxide and 1·5% MnO_2. The chief constituents of ground coats for cast iron are also feldspar, borax, and silica, but the batches are more variable than for ground coats on sheet iron.

Ground-hog Kiln. US term for an art-potter's kiln (usually fired with solid fuel) partly buried in a convenient hillside to support the roof and conserve heat.

Ground Laying. A process for the application of a uniform coating of colour to pottery-ware by painting with oil the area to be coloured and then dusting powdered colour over the ware; the colour sticks only where oil has been first applied. The process is now used only for the decoration of some expensive types of china and porcelain.

Ground Mass. See MATRIX.

Grout. Mortar or cement mixed to a fluid consistency for use in filling the joints between brickwork or tiles.

Growan. A term in the Cornish china clay industry for incompletely and unevenly decomposed granite.

Guard Ring. An arrangement in thermal-conductivity apparatus designed to ensure that heat shall flow, through the sample actually under test, in a direction perpendicular to the hot and cold faces, i.e. no heat flows through the sides of the test-piece.

Guide Tube. A fireclay tube having a spigot and socket, for use in the TRUMPET (q.v.) assembly in the bottom-pouring of molten steel. Two sizes are specified in B.S. 2496: $4\frac{5}{8}$ in. o.d. and $2\frac{7}{8}$ in. i.d. for tubes up to 12 in. long; 6 in. o.d. and $3\frac{15}{16}$ in. i.d. for tubes 12–15$\frac{1}{2}$ in. long.

Gunning. The placing of a refractory powder or slurry by means of a cement gun; most commonly gunning is practised as a method of repairing furnace linings without the need to cool them to room temperature.

Guthrie Kiln. A variant of the BELGIAN KILN (q.v.), a trough replacing the transverse grate; the design was patented by H. Guthrie in 1877.

GVC. Abbreviation for GLAZED VITRIFIED CLAY: term applied to glazed clay pipes.

Gypsum. Natural hydrated calcium sulphate, $CaSO_4 . 2H_2O$, from which PLASTER OF PARIS (q.v.). is produced. In England gypsum occurs in the Newark and Tutbury zones of the Keuper Marl, and in the Purbeck Beds of Sussex. It occurs abundantly in USA, Canada, France, USSR and elsewhere.

Gyratory Crusher. A primary crusher for hard rocks such as the quartzite used in silica brick manufacture. The material to be crushed is fed into the space between a vertical steel cone and a similarly shaped steel casing; the cone rotates eccentrically to the casing (cf. CONE CRUSHER).

H

Habla Kiln. A ZIG-ZAG KILN (q.v.) that may be archless or with a permanent flat roof; top-fired with fine slack; output 25 000—50 000 bricks/week. It was designed by a Czech, A. Habla (Brit. Pats. 242 051 and 311 884).

Hack. A double row of building bricks, with a 1-ft space between them and 8–9 bricks high, set on boards in the open to dry; the top of a completed hack is protected from rain by a 'cap'. Side protection, when necessary, is provided by LEE BOARDS usually pronounced LOO or LEW BOARDS. Hack drying has now largely been replaced by artificial drying under more controlled conditions.

Hackle Marks

Hackle Marks. See RIB MARKS.

Hafnium Boride. HfB_2; m.p., 3250°C; sp. gr. (theoretical), 11·2; thermal expansion, $5·3 \times 10^{-6}$; resistivity (20°C), 10–12 μohm.cm.

Hafnium Carbide. HfC; m.p. 3890°C; theoretical density, 12·2 g/ml.; hardness, 2533 (K100); modulus of rupture, 34 000 p.s.i. (25°C), 12 500 p.s.i., (2000°C); modulus of elasticity 45×10^6 p.s.i. (20°C); thermal expansion, $8·8 \times 10^{-6}$ (25–2500°C).

Hafnium Nitride. HfN; m.p. 3310°C; theoretical density, 14·0 g/ml. This special ceramic can be prepared by sintering Hf in dry N_2 at 1000°C.

Hafnium Oxide. HfO_2; m.p. 2790°C; sp. gr. 9·68; thermal expansion (100–1300°C) 5·8. The low-temperature (monoclinic) form changes to the high-temperature (tetragonal) form at 1700°C.

Hafnium Silicate. A compound analogous to zircon, hence the suggested name HAFNON. It can be synthesized from the oxides at 1550°C. Thermal expansion (150–1300°C), $3·6 \times 10^{-6}$.

Hafnium Titanate. Special refractory compositions have been made by sintering mixtures of HfO_2 and TiO_2 in various proportions. The m.p. of these sintered bodies was approx. 2200°C; there appeared to be a phase change at about 1850°C. Some of the compositions had negative thermal expansions.

Hafnon. See HAFNIUM SILICATE.

Haigh Kiln. A US type of CHAMBER KILN (q.v.) designed by H. Haigh; it is U-shaped with an open space between the two lines of chambers, permitting the chambers to be fired from both sides. In one such kiln, as used at a US brickworks, the chambers are 12–15 ft wide, 16 ft long and 10 ft high; this particular kiln is gas-fired at 1150°C.

Hair Line. (1) Fine CORD (q.v.) on the surface of glass-ware; see also AIR LINE.

(2) A fault sometimes seen in vitreous enamelware. The hair lines are a series of fine cracks that have healed in the later firings (thus differing from crazing). The cause is excessive stress in the enamel.

(3) A line of separation sometimes found near the centre of thick ceramic ware that has been shaped by SOLID CASTING (q.v.).

Hair-pin Furnace. A type of furnace used in vitreous enamelling; see U-TYPE FURNACE.

Half-bat. A building brick of half the normal length; also called a SNAP-HEADER.

134

Halifax Hard Bed. A siliceous fireclay of the Lower Coal Measures extensively worked in Yorkshire, England. It contains (fired) 68–74% SiO_2, 20–25% Al_2O_3, 2·5–3·0% Fe_2O_3 and 1·0–1·5% alkalis.

Hall's Factors. For calculating the thermal expansion of a glass; see THERMAL EXPANSION FACTORS FOR GLASS.

Halls Kiln. An annular kiln with permanent walls dividing the chambers; there are openings at the ends of these partition walls to give a zig-zag fire travel, but there are also large trace holes through these walls. The kiln was designed by G. Zehner of Wiesbaden, Germany. The name 'Halls Kiln' derives from the hall-like appearance of the long narrow chambers.

Halloysite. A clay mineral of the kaolinite group approximating in composition to $Al_2O_3 . 2SiO_2 . 4H_2O$. Halloysite is a constituent of the clays of New Zealand and Brazil; it also occurs in Missouri and Utah (USA), in N. Africa, Japan, and elsewhere. The special feature of halloysite is the tubular shape of the crystals resulting from the rolling up of the fundamental sheet structure; this causes difficulty when halloysitic clays are used in ceramic manufacture because of the consequent high and non-uniform shrinkage.

Hammer Mill. An impact mill consisting of a rotor, fitted with movable hammers, that is revolved rapidly in a vertical plane within a closely fitting steel casing. Hammer mills are sometimes used for the size reduction of clay shales, glass cullet, and some of the minerals used in the ceramic industry (cf. DISINTEGRATOR; IMPACT MILL).

Hard. As applied to a glass, glaze, or enamel, this word means that the softening temperature is high; such a product, when cold, is also likely to be hard in the normal sense.

Hard-paste. The old term (still used by collectors) for true feldspathic porcelain, as distinct from fritted porcelain, i.e. SOFT-PASTE (q.v.).

Hardening-on. Biscuit ware that has been decorated prior to glazing is heated at a comparatively low temperature (c. 700°C) to volatilize the oils used in decorating and to fuse the flux in the applied colour so that the decoration remains fixed during the subsequent dipping in the liquid suspension.

Hardinge Mill. A ball mill for continuous operation in which the casing is in the form of two conical sections joined by a short, central, cylindrical section; the material to be ground is fed into the

135

Harkins and Jura Method

mill through one of the hollow trunnions, being discharged through the opposite trunnion. Named after H. W. Hardinge, USA, who invented this design of ball mill in 1906.

Harkins and Jura Method. A gas-adsorption method for the determination of the specific surface of a powder. The sample is first evacuated and then exposed to a vapour near to its saturation pressure; the wetted powder is then immersed in the liquid itself and the rise in temperature is measured. From this, the surface energy change, and hence the surface area, is calculated. (W. D. Harkins and G. Jura, *J. Chem. Phys.*, **11**, 430, 1943.)

Harkort Test. Although H. Harkort was a German, details of his so-called crazing test for pottery-ware were first published in USA (*Trans. Amer. Ceram. Soc.*, **15**, 368, 1913). The test-piece is heated to 120°C and then plunged into cold water; the cycle is repeated, with successive increases of 10°C in temperature, until crazing can be detected after the quenching. The criteria are: no cracks after quenching from 150°C—crazing at room temperature likely after 3–4 months; 160°C—15 months; 170–180°C—no crazing after $2\frac{1}{2}$ years (with a few exceptions); 190°C—no crazing after $2\frac{1}{2}$ years without exception. It is now known that crazing may result from moisture expansion as well as from thermal shock as in the Harkort Test.

Harrop Kiln. One of the variety of tunnel kilns built to the designs of Harrop Ceramic Service Co., Columbus, Ohio. The early examples of Harrop Kilns in the N. Staffordshire potteries were of the large open-flame type.

Harrow. See under WASH-MILL.

Hartman Formula. Relates the refractive index (n) of a glass to the wavelength (λ) of the incident light: $n = n_0 + c(\lambda - \lambda_0)^{-\alpha}$ where λ_0, c and α are constant and α may have a value from 1 to 2. This is an improvement on the CAUCHY FORMULA (q.v.).

Hassall Joint. A type of joint for glazed pipes designed by Wm. Hassall (of John Knowles & Co., Woodville, Burton-on-Trent) in the late 19th century. Bitumen rings are attached to the outside of the spigot and the inside of the socket of the fired pipe; a thin smear of cement is rubbed around these rings just before the pipes are laid, the spigot of one pipe then being pushed into the socket of the next. Liquid cement is finally poured through holes in the socket to complete the joint.

Hastings Beds. A series of clay and sand deposits in the Lower Cretaceous of S.E. England; the Fairlight Clays at the base of these deposits have been used for brickmaking near Hastings and Bexhill.

Hawk Pug. Trade name: a de-airing pug having two barrels in tandem, the first rotating around a fixed bladed shaft, whereas the second is fixed while the bladed internal shaft rotates. Each shaft is set slightly off-centre. (Service Engineers, Stoke-on-Trent, England.)

Hawse Clay. A local English term for a clay that is crumbly but becomes plastic when worked up with water.

Haydite. Trade name: a lightweight expanded clay aggregate, made in USA, named after inventor—S. J. Hayde (US Pat. 1 255 878 and 1 707 395).

Header. A brick laid with its length across the width of a wall, thus leaving its smallest face exposed. A course of such bricks is known as a Header Course (cf. STRETCHER).

Healing. The ability of a glaze to cover any areas of the ware that may have been damaged before the glaze has been fired on; this depends on a correct combination of surface tension and viscosity at the glost-firing temperature.

Healy–Sullivan Process. See HYDROGEN-TREATING PROCESS.

Hearth. The lower part of a furnace, particularly the part of a metallurgical furnace on which the metal rests during its extraction or refining. In a blast furnace the hearth is constructed of aluminous fireclay or carbon blocks; in a basic open-hearth furnace or basic electric-arc steel furnace, the hearth is constructed of basic refractory bricks which are covered with dead-burned magnesite or burned dolomite rammed in place.

Heat-work. An imprecise term denoting the combined effect of time and temperature on a ceramic process. For example, prolonged heating at a lower temperature may result in the same degree of vitrification in a ceramic body as a shorter period at a higher temperature; a similar amount of heat-work is said to have been expended in the two cases.

Hecht's Porcelain. A German refractory porcelain similar to MARQUARDT PORCELAIN (q.v.).

Hectorite. Originally known as HECTOR CLAY from its source near Hector, California, USA. It is a hydrous magnesium silicate related to montmorillonite and forming an end-member of the saponite series.

Heel Tap. See SLUGGED BOTTOM.

Hercules Press. Trade name: a semi-dry brickmaking machine

Hercynite

of the rotary-table type. (H. Alexander & Co. Ltd., Leeds, England.)

Hercynite. Ferrous aluminium spinel, $FeO \cdot Al_2O_3$; m.p. 1780°C. Hercynite is sometimes present, in solid solution with other spinels, in aluminous chrome ores; it can also be formed when ferruginous slag attacks an aluminous refractory under reducing conditions. When heated under oxidizing conditions, hercynite changes into a solid solution of Fe_2O_3 and Al_2O_3.

Hermansen Furnace. A recuperative pot-furnace for melting glass; the first furnace of this type was built in Sweden in 1907 and was soon afterwards introduced into England and elsewhere.

Herreshoff Furnace. A multiple-hearth roasting furnace; such a furnace has been used in USA for the calcination of fireclay.

Hessian Crucible. A type of refractory crucible for use in the small-scale melting of non-ferrous metals. These crucibles are made from the GROSSALMERODE CLAYS (q.v.).

Hexite–Pentite Theory. A theory (long since discarded) for the atomic structure of clays proposed by W. Asch and D. Asch (see *Trans. Brit. Ceram. Soc.*, **13**, 90, 1914). The theory was based on the hypothesis that silicates are built up of hexagonal and pentagonal rings of hydrated silica and hydrated alumina.

High-alumina Cement. (1) A special hydraulic cement made by fusing (or occasionally by sintering) a mixture of limestone and bauxite. Although the setting time is similar to that of portland cement, the final strength is virtually attained in 24 h. The composition generally lies within the following limits (per cent): CaO, 36–42; Al_2O_3, 36–42; Fe oxides, 10–18; SiO_2, 4–7; B.S. 915 stipulates $\nless 32\%$ Al_2O_3 and Al_2O_3:CaO (by wt) equal to 0·85–1·3. High-alumina hydraulic cement, if mixed with crushed refractory material, can be used as a refractory concrete.

(2) A refractory cement (non-hydraulic) of high alumina content.

High-alumina Refractory. A refractory of the alumino-silicate type. In the UK it is defined in B.S. 3446 as containing over 45% Al_2O_3.

In the USA four classes of high-alumina refractory are defined (ASTM – C27): 50%-Al_2O_3 (P.C.E. $\nless 34$; Al_2O_3 content $50 \pm 2 \cdot 5\%$); 60%-Al_2O_3 (P.C.E. $\nless 35$; Al_2O_3 content $60 \pm 2 \cdot 5\%$); 70%-Al_2O_3 (P.C.E. $\nless 36$; Al_2O_3 content $70 \pm 2 \cdot 5\%$); 80%-Al_2O_3 (P.C.E. $\nless 37$; Al_2O_3 content $80 \pm 2 \cdot 5\%$). The ASTM propose to add three more classes, namely 85%-Al_2O_3, 90%-Al_2O_3 and 99%-Al_2O_3.

138

In Western Europe, P.R.E. divide high-alumina refractories into two types: Group I— $\geqslant 58\%$ ($Al_2O_3 + TiO_2$); Group II— 46–58% ($Al_2O_3 + TiO_2$).

High-duty Fireclay Brick. Defined in ASTM – C27 as a fireclay refractory with a P.C.E. $\ll 31\frac{1}{2}$. There are three types: Regular, Spall Resistant, Slag Resistant. Further properties are specified for the last two types.

High Early-strength Cement. US term equivalent to the UK Rapid-hardening Portland Cement (q.v.). It is Type III Cement of ASTM Specification C150, which stipulates a crushing strength $\geqslant 1700$ p.s.i. after 1 day in moist air and $\geqslant 3000$ p.s.i. after 1 day in moist air followed by 2 days in water.

High-frequency Heating. See Induction Heating.

High-frequency Induction Furnace. See under Electric Furnaces for Melting and Refining Metals.

High-velocity Thermocouple. See Suction Pyrometer.

Hind Effect. The porosity of a dried cast (deflocculated) pottery body is less than that of the same body, dried, if made by the plastic (flocculated) process; the observation was first made by S. R. Hind (*J. Inst. Fuel.* **24,** 116, 1951).

Hip Tile. A specially shaped roofing tile for use on a descending ridge that forms the junction of two faces of a roof. Hip tiles are available in a number of patterns varying from the close-fitting hip of angular section to the broadly curved types known as Bonnet Hips and Cone Hips.

Hob-mouthed Oven. A pottery Bottle Oven (q.v.) in which the firemouths projected 18–24 in. outwards, coal being fed to the fire from the top.

Hoffmann Kiln. A top-fired Longitudinal-arch Kiln (q.v.) of a type introduced by F. Hoffmann in 1858; variants of this type of kiln have remained the basis of the firing process in the building brick industry up to the present day.

Hofmeister Series. Cations arranged in order of decreasing flocculating power for clay slips: H^+, Al^{3+}, Ba^{2+}, Sr^{2+}, Ca^{2+}, Mg^{2+}, NH_4^+, K^+, Na^+, Li^+. (F. Hofmeister, *Arch. Exptl. Path. Pharmakol.*, **24,** 247, 1888.)

Hog's Back. A type of Ridge Tile (q.v.).

Hoimester. Trade name: a device for controlling the softness of an extruded column of clay; two small sensing rollers run on the clay column, their axial positions changing with any change in softness of the clay; this change can be made to adjust the rate of water addition to the mixer and so bring the consistency of the clay back to normal. (Buhler Brothers, Uzwil, Switzerland.)

Holdcroft Bars. Pressed and prefired ceramic bars ($2\frac{1}{4} \times 1\frac{5}{16} \times \frac{1}{4}$ in.) made of blended materials so proportioned that, when placed horizontally with only their ends supported, and when heated under suitable conditions, they will sag at a stated temperature. The bars are numbered from 1 (bending at 600°C) to 40 (1550°C). These bars were introduced in 1898 by Holdcroft & Co., and are now available from Harrison & Sons Ltd., Stoke-on-Trent, England.

Holding Ring. A BEAD (q.v.) in the finish of a glass bottle for use when the parison is transferred to the blow mould; sometimes known as the BACK RING or TRANSFER RING.

Holey Boy. A type of perforated floor in intermittent down-draught kilns firing salt-glazed pipes.

Hollow Blocks. See HOLLOW CLAY BLOCKS.

Hollow Casting. See DRAIN CASTING.

Hollow Clay Blocks. Fired clay building blocks, usually of comparatively large size, with cells (air spaces) within the block; they are used for wall construction and, with metal reinforcement, for floors and prefabricated panels. For specifications see B.S. 1190; ASTM – C34, C56, C112 and C212 (cf. PERFORATED BRICK).

Hollow Neck. A fault in the neck of a bottle resulting from an insufficiency of glass in the gob from which the bottle was made; if the bottle is of the screw-necked type, the faulty neck is said to be BLOWN AWAY.

Hollow-ware. Cups, basins, etc. (cf. FLAT-WARE).

Hommelaya Process. Trade name: a procedure in which cobalt is deposited on sheet steel before it is enamelled, the purpose being to eliminate the need for a ground coat; (Hommel Co., Pittsburgh, USA).

Hone. A block of fine abrasive, generally SiC; the block is appreciably longer than it is broad or wide. Hones are used for fine grinding, particularly of internal bores.

Honeycomb. Ceramic honeycombs have been developed to provide a light sheathing material for the protection of the outer surface of supersonic aircraft, missiles, and space-craft from the heat generated by air friction. A high-alumina ceramic is generally used; experiments have also been made with zirconia, thoria, and various carbides. The refractory powder is mixed with a binder and rolled into a thin flexible sheet; the core is shaped by corrugated rolls. The sheet and core are bonded with the same mixture and fired together to produce the honeycomb.

Hood. (1) A protective casing, with an exhaust system, for use in the carrying out of dusty operations (e.g. TOWING (q.v.)) in pottery manufacture.

(2) Alternative name for POTETTE (q.v.).

Hooded Pot. See under POT.

Horizontal Retort. (1) An intermittent unit for the production of town gas from coal; it is constructed of segments of silica or siliceous refractory material (cf. CONTINUOUS VERTICAL RETORT).

(2) An intermittent unit formerly used for the production of zinc. Horizontal zinc retorts were generally made from a siliceous fireclay, suitably grogged.

Horse; Horsing. (1) A 'horse' is a heavy stool with a convex rectangular top on which roofing tiles, after they have been partially dried, can be given the slight curvature (about 10-ft radius) necessary to ensure ventilation in a tiled roof; the process itself is termed 'horsing'.

(2) In the old bottle-ovens for firing pottery, a 'horse' was used by the placers to enable them to reach the top parts of the bungs of saggars.

Hospital. In a vitreous-enamel factory, the department in which ware that has faults, not too serious for remedy, is repaired.

Hot-blast Circulating Duct. See BUSTLE PIPE.

Hot-blast Main. A duct, lined with refractory material, through which hot air passes from a HOT-BLAST STOVE (q.v.) to the BUSTLE PIPE (q.v.) of a blast furnace.

Hot-blast Stove or Cowper Stove. A unit for heating the air delivered to the tuyeres of a blast furnace. It is a cylindrical furnace, about 80 ft high and 20 ft dia., lined with fireclay refractories. There is a combustion chamber up one side; fire-clay checker bricks (usually special shapes known as STOVE FILLINGS (q.v.)) fill the remainder of the space. The checker bricks are heated by the combustion of blast-furnace gas in the combustion chamber and then, on reversal of the direction of gas flow, they deliver heat to incoming air which then passes to the blast-furnace tuyeres. These stoves were first proposed by E. A. Cowper (hence their alternative name) in 1857 and were first used at Middlesbrough, England, in 1860. The idea of heating the blast, however, had been put forward by J. B. Neilson, a Scotsman, in 1824.

Hot Floor

Hot Floor. A drying floor, heated from below by hot gases or by steam pipes, for special shapes of heavy clayware, refractories, etc.

Hot-metal Ladle. A ladle for the transfer of molten iron from a blast furnace to a mixer furnace and from there to a steel furnace; alternatively, the ladle may transfer molten pig-iron direct from blast furnace to steel furnace. Such ladles are generally lined with fireclay refractories but for severe conditions high-alumina and basic refractories have been tried with some success.

Hot-metal Mixer. A large holding furnace for molten pig-iron. The capacity of these furnaces, which are of the tilting type, is up to 1400 tons. Hot-metal mixers may be ACTIVE (i.e. the pig-iron is partially refined while in the furnace) or INACTIVE (i.e. the pig-iron is merely kept molten until it is required for transfer to a steelmaking furnace). In either case the bottom and walls of the furnace are made of magnesite refractories and the roof of silica refractories.

Hot Patching. The repair of the refractory lining of a furnace while it is still hot; this is most commonly done by spraying a refractory slurry through a cement gun. See also AIR-BORNE SEALING and SPRAY WELDING.

Hot-plate Spalling Test. A spalling test designed specifically for the testing of silica refractories; it has been standardized in ASTM – C439.

Hot Preparation. See STEAM TEMPERING.

Hot Pressing. (1) A shaping process for special ceramics, pressure being applied to powder in a mould at a high temperature; by this method it is often possible to produce small articles of zero porosity.

(2) A JIGGERING (q.v.) process in which a heated profile tool or plunger is used; some electrical insulators are made in this way.

Hot-top. A container lined with refractory or heat-insulating tiles and used at the top of an ingot mould in the casting of steel; the molten steel is allowed to fill the mould and the hot-top, the extra metal remaining molten so that it sinks into the ingot as the latter cools and shrinks, thus preventing the formation of a 'pipe'.

Hotel China. See AMERICAN HOTEL CHINA.

Hourdis. French term for a large hollow clay building block (from the name of the inventor).

Hovel. The large conical brick structure that enclosed the old type of pottery oven or glass-pot furnace.

142

HREX. Symbol for a special shape of wall tile—Round Edge External Corner, Left Hand (see Fig. 6, p. 307).

HTI. Abbreviation for High-Temperature Insulating refractory; see INSULATING REFRACTORY.

Hulo System. A system for the handling of building bricks from setting in the kiln to delivery at the building site. (Trade name: Van-Huet, Pannerden, Holland.)

Humboldt Rotary Kiln. A kiln designed for burning cement; the batch is fed to the kiln as a suspension in hot gases with consequent economy in fuel consumption.

Humidity Dryer. A dryer for clay-ware (particularly bricks) in which the relative humidity (r.h.) of the atmosphere in the dryer is controlled so that the initial stages of drying take place at high r.h., this being progressively reduced as drying of the clay-ware proceeds.

Huntington Dresser. See STAR DRESSER.

Hutch. Scottish term for a waggon of about 1 yd^3 capacity used for hauling clay.

Hydrargillite. An obsolescent name for GIBBSITE (q.v.).

Hydraulic Press. A machine used for dry pressing, or semi-dry pressing; the pressure (up to 14 000 p.s.i.) is applied hydraulically to the top, and sometimes also to the bottom, mould-plate. Such presses are used, for example, in dry pressing basic refractories.

Hydraulic Refractory Cement. A REFRACTORY CEMENT (q.v.) containing aluminous hydraulic cement, e.g. Ciment Fondu, so that it sets at room temperature.

Hydraulicking. The method used to win china clay; a high-pressure (150 p.s.i.) jet of water is aimed at the exposed vertical clay face and the clay is washed down. The suspension is then pumped to settling tanks where the impurities (chiefly mica) settle out; the purified suspension of china clay is then concentrated, filtered, and dried.

Hydrogen-treating Process. A method for the preparation of sheet steel for vitreous enamelling by first driving hydrogen into the surface of the steel (this is effected electrolytically) and then removing the hydrogen by immersion of the steel in boiling water. The process was introduced by J. H. Healy and J. D. Sullivan (US Pat. 2 754 222; 10/7/56).

Hydrometer. A glass instrument, shaped rather like a fisherman's float, for the determination of the specific gravity of a liquid, e.g. of clay slip or of vitreous-enamel slip.

143

Hygrometer. An instrument for measuring the relative humidity of air, e.g. in a dryer for clayware.

Hysil. A borosilicate glass of high thermal endurance and chemical resistance used for chemical ware; made by Chance Bros., England.

Hyslop Plasticity Diagram. A diagram relating the extensibility (E) of a clay, as determined by a penetration method, to its softness (S); the relationship is of the form $E = KS^n$, where K and n are constants. (J. F. Hyslop, *Trans. Brit. Ceram. Soc.*, **35**, 247, 1936.)

Hysteresis. The degree of lag in the reaction of a material to a change in the conditions (of mechanical, electrical, or magnetic stress) to which it is exposed. Ceramic ferro-electrics, for example, exhibit hysteresis when subjected to a changing external electric field; hysteresis is also shown by plastic clay when it is stressed cyclically.

I

IC Silicon Carbide. Abbreviation for Impregnated-Carbon Silicon Carbide; it contains free carbon and silicon and the bulk density is comparatively low (2·60).

Ignition Arch. A refractory arch in a boiler furnace fitted with a mechanical grate; its purpose is to assist the ignition of the fuel as the latter moves under the hot brickwork.

Illite. A group name for micaceous clay minerals of variable composition. The name derives from the fact that it was from Illinois (USA) clays and shales that samples were isolated by R. E. Grim, R. H. Bray, and W. F. Bradley (*Amer. Mineral.*, **22**, No. 7, 813, 1937) who first proposed this name. As there has been some confusion as to the exact meaning of the term, the following statement in the original paper should be noted: '. . . the term *illite* . . . is not proposed as a specific mineral name, but as a general term for the clay mineral constituent of argillaceous sediments belonging to the mica group.'

Ilmenite. Ferrous titanate, $FeO \cdot TiO_2$; m.p. 1365°C. This mineral is the principal constituent of the heavy minerals in the beach sands of Australia and elsewhere; it is thus the chief source of titania, which is used in the ceramic industry as an opacifier and as a constituent of some ceramic dielectrics.

Image Furnace. Apparatus for the production of a very high temperature in a small space by focusing the radiation from the sun (SOLAR FURNACE) or from an electric arc (ARC-IMAGE FURNACE) by means of mirrors and/or lenses. Such furnaces have been used for the preparation and study of some special ceramics.

Imbibition. The particular case of absorption or adsorption of a liquid by a solid in which the solid increases in volume. A typical example is the swelling of Na-bentonite when it takes up water.

Imbrex. See ITALIAN TILES.

IMM Sieve. See under SIEVE; for mesh sizes see Appendix 1.

Impact Mill. A crushing unit in which a rapidly moving rotor projects the charged material against steel plates; impact mills find use in the size-reduction of such materials as feldspar, perlite, etc. (cf. DISINTEGRATOR; HAMMER MILL).

Impact Resistance. The resistance of a material to impact. A test for this property (which is clearly of importance in the assessment of tableware, for example) must proceed by a series of blows—usually delivered by a swinging pendulum—of increasing severity; the major difficulties in the impact testing of ceramics are to produce homogeneous test-pieces and to determine the proportion of the impact force causing fracture that is actually used in fracturing the test-piece. In USA a standard impact test for ceramic tableware is described in ASTM – C368.

Impervious. In the USA this word has a defined meaning (ASTM – C242) as applied to ceramic ware, namely, that degree of vitrification shown by complete resistance to dye penetration; the term generally implies that the water absorption is zero, except for floor tiles and wall tiles which are considered to be impervious provided that the water absorption does not exceed 0·5%.

Inclusion. See NON-METALLIC INCLUSION; STONE.

Incongruent Melting. A solid melts incongruently if, at its melting point, it dissociates into a liquid and a solid of different composition; for example ORTHOCLASE melts incongruently at about 1170°C to form LEUCITE and a liquid that is richer in silica.

Indenting. In structural brickwork, the omission of a suitable series of bricks so that recesses are left into which any future work can be bonded.

Indiana Method. A technique for the determination of the quantity of entrained air in concrete on the basis of the difference in the unit weights of a concrete sample with and without air. The name derives from the fact that air-entrained cement was first used to any extent in Indiana, USA. (*Proc. A.S.T.M.*, **47**, 865, 1947.)

Indirect-arc Furnace. See under ELECTRIC FURNACES FOR MELTING AND REFINING METALS.

Induction Furnace. See under ELECTRIC FURNACES FOR MELTING AND REFINING METALS.

Induction Heating. The heating of an electrically conducting material by the effect of induced electric currents, which may be set up by a high-frequency field in a small object or a low-frequency field in a large object (as in a large induction furnace for steel melting). High-frequency heating has been proposed (Brit. Pat. 898 647; 14/6/62) as a method for the firing of glaze on ceramic tiles.

Industrial Floor Brick. See FLOOR BRICK.

Infra-red Drying. Drying by exposure to infra-red radiation from specially designed electric lamps or gas burners. The process has found some application in the vitreous-enamel industry and in the pottery industry.

Infrasizer. An air elutriator for the fractionation of powders into seven grades within the size range 100 mesh to 7μ. It was designed by H. E. T. Haultain (*Trans. Canadian Inst. Min. Met.*, **40**, 229, 1937; *Mine Quarry Engng.*, **13**, 316, 1947).

Infusorial Earth. An obsolete name for DIATOMITE (q.v.).

Ingersoll Glarimeter. An instrument designed primarily to measure the gloss of paper; it has been used to evaluate the abrasion resistance of a glaze in terms of loss of gloss after a specified degree of abrasion. The apparatus was designed by L. R. Ingersoll (*J. Opt. Soc. Amer.*, **5**, 213, 1921).

In-glaze Decoration. A method of decorating pottery; the decoration is applied on the surface of the glaze, before the glost fire, so that it matures simultaneously with the glaze. Such decoration can be applied only by hand-painting, spraying, or silk screen.

Initial Set. The time interval between the GAUGING (q.v.) of a hydraulic cement and its partial loss of plasticity. In the VICAT'S NEEDLE (q.v.) test, the end of this period is defined as the time when the needle will no longer descend through the cement test block to within 5 mm of the bottom. In B.S. 12, the test conditions are 14·4–17·8°C (58–64°F) and $\ngtr 90\%$ r.h.; for normal

146

cements the Initial Set must not be less than 30 min (the usual period is 1–2 h).

Injection Moulding. A process sometimes adopted for the shaping of non-plastic ceramics, e.g. alumina. A plasticizer such as polystyrene or phenol formaldehyde composition is mixed with the ceramic powder and the batch is then warmed and injected into the die.

Insulating Refractory. A refractory material having a low thermal conductivity and used for hot-face insulation in a furnace. Such products have a porosity of 60–75%. If they are made from refractory clay this high porosity is produced by the incorporation of a combustible in the batch or, less commonly, by foaming or by chemical means. In the USA, ASTM – C155 classifies insulating refractories as follows:

Group Identification	Reheat Change not more than 2% after 24h at	Bulk Density not Greater than
Group 16	1550°F (845°C)	34 lb/ft³
Group 20	1950°F (1065°C)	40 lb/ft³
Group 23	2250°F (1230°C)	48 lb/ft³
Group 26	2550°F (1400°C)	54 lb/ft³
Group 28	2750°F (1510°C)	60 lb/ft³
Group 30	2950°F (1620°C)	68 lb/ft³

Insweep. Term applied to the lower part of a glass container if the sides curve inward or taper towards the base.

Intaglio. The decoration of glass-ware by cutting a pattern to a depth intermediate between that of deep cutting and engraving. (The term Intaglio Printing has been applied to the decoration of pottery-ware by transfer-printing from a copper plate.)

Interlocking Tile. A large type of roofing tile having one or more longitudinal ridges and depressions that interlock with the complementary contour of neighbouring tiles on the roof. Such tiles are commonly made in a REVOLVER PRESS (q.v.).

Intermediate Crusher. A machine of a type suitable for size reduction from about 8 to 20 mesh, e.g. a pan mill or ball mill; (note, however, that a ball mill can more properly be used as a FINE GRINDER (q.v.)).

Intermediate Piece. See MATCHING PIECE.

Intermittent Kiln or Periodic Kiln. A batch-type kiln in which

147

Intumescence

goods are set, fired, cooled, and then drawn. The principal types in the ceramic industry are ROUND KILNS (or BEEHIVE KILNS), RECTANGULAR KILNS, BOTTLE OVENS, BOGIE KILNS and TOP-HAT KILNS. (See separate entries under these headings.)

Intumescence. The property of some silicates, notably of PERLITE (q.v.), of expanding permanently, when heated, to form a completely vesicular structure; cf. BLOATING and EXFOLIATION.

Inversion; Inversion Point. An instantaneous change in the crystalline form of a material when it is heated to a temperature above the INVERSION POINT; the change is reversed when the material is cooled below this temperature. An example of importance in ceramics is the $\alpha \rightleftharpoons \beta$ quartz inversion at 573°C.

Investment Casting. A process for the casting of small metal components to a close tolerance. In the usual process a wax replica of the part to be cast is coated ('invested') with refractory powder, suitably bonded, and the whole is then warmed (150°C) to melt out the wax—hence the alternative name Lost-wax Process; the refractory mould is then fired at 1000–1100°C. The refractory used may be powdered sillimanite or alumina, or specially prepared cristobalite; the latter is particularly used in the application of the process in dentistry.

Inwall. US term for the refractory lining of the STACK (q.v.) of a blast furnace.

Iodoeosin Test or Mylius Test. For determining the durability of optical glass. The amount of free alkali in a freshly fractured surface is determined by means of iodoeosin; the surface is then exposed to moist air at 18°C for 7 days and the free alkali is again determined. Any increase in free alkali is taken as a measure of lack of durability; a decrease indicates stability. The test is not valid for many modern optical glasses. (F. Mylius, *Z. Anorg. Chem.*, **67**, 200, 1910; *Silikat Z.*, **1**, 2, 25, 45, 1913.)

Ionic Exchange. The replacement of ions on the surface, or sometimes within the lattice, of materials such as clay. The ions become adsorbed to balance a deficiency of charge in the clay structure, e.g. in a montmorillonite in which some Mg^{2+} has been replaced by Al^{3+}, or to satisfy broken bonds at the edges of the clay crystals. Ionic exchange capacity is generally expressed in milli-equivalents per 100 g: typical values are—kaolinite, 1–3; ball clay, 10–20; bentonite, 80–100.

Ipro Brick. An I-shaped clay paving brick designed for use in

148

roadmaking, particularly on soils of poor bearing capacity—as in Holland.

Irising. A surface fault, in the form of stained patches, sometimes found on flat glass that has been stacked with surfaces in contact. The term originates from the interference colours that often accompany the fault. It is caused by moisture. If the glass is annealed in an acid atmosphere and adequately washed irising is unlikely to occur; separation of the stacked sheets by paper also prevents this trouble.

Iron Modulus. The $Al_2O_3:Fe_2O_3$ ratio in a hydraulic cement. In portland cement this modulus usually lies between 2 and 3.

Iron Notch. Alternative name for the TAPHOLE (q.v.) of a blast furnace.

Iron Ore Cement. See ERZ CEMENT.

Iron Oxide. See FERROUS OXIDE; FERRIC OXIDE; MAGNETITE.

Iron Spot. A dark, sometimes slaggy, spot on or in a refractory brick, resulting from a localized concentration of ferruginous impurities; such spots can cause carbon deposition, or even disintegration of the brick, if the latter is exposed to CO attack at 400–500°C or to hydrocarbons at 800–900°C.

Iron-Zirconium Pink. See ZIRCONIUM-IRON PINK.

Ironing. A fault that may arise during the firing of decorated ware having cobalt blue bands, etc.; the band appears dull and may have a reddish scum caused by the crystallization of cobalt silicate.

Ironstone China. See MASON'S IRONSTONE CHINA.

Irwin Consistometer. A simple capillary-flow viscometer designed by J. T. Irwin (*J. Amer. Ceram. Soc.*, **21**, 66, 1938) for testing vitreous enamel slips. Two capillaries are placed in the lower end of a long wide-bore tube; both capillaries have the same bore but one is twice as long as the other. The wide outer tube is filled with the slip to be tested and is then allowed to discharge through the two capillaries into separate measuring cylinders. The rate of flow through each capillary can thus be calculated and a curve can be drawn from which both the yield value and mobility can be read.

Irwin Slump Test. A work's test for assessing the setting-up of vitreous enamel slips for spraying. It was first described by J. T. Irwin (*Finish*, p. 28, Sept. 1946) in the following words; 'A ground-coated plate is placed on a table and a steel cylinder, $1\frac{13}{16}$ in. internal diameter and $2\frac{1}{2}$ in. high, is placed on the plate.

149

ISCC-NBS Colour System

The cylinder is filled with enamel to be tested. The cylinder is then lifted vertically, with a rapid motion, by means of a hook and cord attached to the top of the cylinder and passing over a pulley to a weight. When the weight is released, the cylinder is raised vertically. This action results in a "pancake" of enamel on the test plate. The diameter of this "pancake" is a function of the set or stiffness of the enamel slip.'

ISCC–NBS Colour System. A system for the designation of colours drawn up in USA by the Inter-Society Color Council and the National Bureau of Standards. For full details see NBS Circular 553, 1955.

ISO. International Standards Organisation, 1 Rue de Varembé, Geneva.

Isostatic Pressing. A process sometimes used for the shaping of ceramic components, e.g. vacuum tube enclosures and envelopes, sparking plugs, and similar items, from Al_2O_3, BeO, or other special ceramics. The powder to be pressed is put in a rubber or plastic bag (often called the TOOLING) which is then placed in a container and subjected to hydrostatic pressure, generally 10 000–20 000 p.s.i. The merit of this process is the uniform manner in which pressure is applied over the whole surface, resulting in uniform density in the shaped component.

Istra. Trade-name: a HIGH-ALUMINA CEMENT (q.v.) made in Yugoslavia.

Italian Tiles. Roofing tiles, of a type common in Italy, designed for use in pairs: a flat under-tile (the TEGULA) being laid adjacent to a rounded over-tile (the IMBREX). cf. SPANISH TILES.

J

Jack Arch. A sprung arch of bricks specially shaped so that the outer surface of the arch is horizontal, the inner surface being either horizontal or arched with a large radius. Also known as a FLAT ARCH.

Jack Bricks. Refractory bricks for the glass industry; they are perforated to accommodate the prongs of a fork truck and used beneath a newly-set pot.

Jam-socket Machine. A machine for shaping the sockets in clay sewer-pipes. The pipes are extruded plain, cut to length and fed to the jam-socket machine in which a ram, having the internal profile of the socket, is forced into the end of the pipe. The machine was introduced by Pacific Clay Products Co., Los Angeles, USA (*Brick Clay Record*, **124**, No. 4, 55, 1954).

Jamb. The brickwork, or other material, forming one of the vertical sides of an opening in a wall of a building or furnace, e.g. a door-jamb.

Jamb Brick. See BULLNOSE.

Jamb Wall. See BREAST WALL.

Jar Mill. A small BALL MILL (q.v.), the revolving cylinder being a vitreous ceramic jar; such mills are used in the grinding of small batches of ceramic colours and vitreous enamels.

Jardiniere Glaze. A former type of unfritted lead glaze containing PbO, K_2O, CaO, ZnO, Al_2O_3 and SiO_2. There were soft (cone 02) and hard (cone 4) types.

Jasper Ware. A fine coloured stoneware first made in 1774 by Josiah Wedgwood who referred to it as being 'peculiarly fit for cameos, portraits and all subjects in bas relief; as the ground may be of any colour throughout, and the raised figures of a pure white'. The body contains barytes and the colours include blue, lavender, and sage green (typically blue with white bas relief). A quoted body composition is: 26% ball clay, 18% china clay, 11% flint, 45% barytes.

Jaw Crusher. A primary crusher for hard rocks, e.g. the quartzite used in making silica refractories. The crusher has two inclined jaws, one or both being actuated by a reciprocating motion so that the charge is repeatedly 'nipped' between the jaws. For different types see under BLAKE, DODGE and SINGLE-TOGGLE.

Jena Glass. Various types of glass (the Jena chemically resistant and optical glasses are particularly well known) made by Jenaer Glaswerk Schott and Gen., Jena, Thuringia, Germany.

Jeroboam. A 4-quart wine bottle.

Jet Dryer. A type of dryer in which the ware is dried chiefly by the action of jets of warm air directed over the surfaces of the ware; the principle has been successfully applied in the drying of pottery.

Jet-enamelled Ware. A type of 18th century porcelain decorated with black on-glaze transfers; (cf. JET WARE).

Jet Impact Mill. See FLUID-ENERGY MILL.

Jet Ware. Pottery-ware, chiefly tea-pots, having a red clay body and a black, manganese-type, glaze; (cf. JET-ENAMELLED WARE).

Jeweller's Enamel. A vitreous enamel fusing at a low temperature and suitable for application to copper, silver, and other non-ferrous metals in the production of jewelry or badges. Because the principal European centre for enamelled jewelry was Limoges, France, the chief types have French names: BASSE-TAILLE, CHAMPLEVE, CLOISONNE, PLIQUE-A-JOUR (for details see under each name).

Jigger; Jiggering. A jigger is a machine for the shaping (jiggering) of pottery-ware by means of a tool fixed at a short distance from the surface of a plaster mould that is mechanically rotated on the head of a vertical spindle. In the making of flatware a suitable quantity of prepared body from the pug is first shaped into a disk by a batting-out machine placed alongside the jigger; in semi-automatic jiggers the movements of the two machines are co-ordinated. The disk of body is placed by hand on a plaster mould which has been fixed on the head of the vertical spindle of the jigger; the upper surface of the mould has the contour required in the upper surface of the finished ware. The mould is then set in motion by a clutch mechanism and a tool having the profile of the bottom of the ware is brought down on the plastic body which is then forced to take the required shape between mould and tool. The essentially similar process applied to the shaping of hollow-ware, e.g. cups, is known as JOLLEYING. In both processes water is necessary as a lubricant.

Jiggerman. A workman in a glassworks whose job is to return SCULLS (q.v.) to the charging end of a glass-tank furnace. (cf. JIGGERER, the man who operates a JIGGER (q.v.) in a pottery.)

Jockey Pot. A POT (q.v.) for glassmaking of such size and shape that it can be supported in the furnace by two other pots.

Joggle. See NATCH.

Joint Line. A visible line on imperfect glass-ware reproducing the line between separate parts of the mould in which the glass was made. Also known as a PARTING LINE, MATCH MARK, MITRE SEAM, MOULD MARK or SEAM.

Jolley; Jolleying. Terms applied to the shaping of hollow-ware in the same senses as JIGGER (q.v.) and JIGGERING are applied to the shaping of flat-ware.

Jolt Moulding. A process sometimes used for the shaping of refractory blocks. A mould is charged with prepared batch which is then consolidated by jolting the mould mechanically; top pressure may simultaneously be applied via a mould plate. cf. TAMPING.

Journey. The period of emptying of a glass-pot before it is again filled with batch; the term is also applied to a shift in which an agreed number of pieces of glass-ware are made.

Jumbo. (1) A type of TRANSFER LADLE (q.v.) for the conveyance of molten iron.

(2) A hollow clay building block of a type made in USA, its size is $11\frac{1}{2} \times 7\frac{1}{2} \times 3\frac{1}{2}$ in. and it has two large CELLS (q.v.) and a $1\frac{1}{2}$-in. SHELL (q.v.); the weight is 15–16 lb.

Jumpers. See POPPERS.

K

Kady Mill. Trade-name: a high-speed dispersion unit consisting principally of a bottom-feeding propeller, a main dispersion head containing a rotor and a stator, and an upper shroud and propeller. (Steele & Cowlishaw Ltd., Stoke-on-Trent, England.)

Kaldo Process. A process for the production of steel by the oxygen-blowing of molten iron in a rotating, slightly inclined, vessel; the latter is usually lined with tarred-dolomite refractories. The name is derived from the first letters of the name of the inventor, Professor Kalling, and of the Domnarvet Steelworks, Sweden, where the process was first used.

Kaliophilite. $K_2O.Al_2O_3.2SiO_2$. Transforms at about 1400°C into an orthorhombic phase that melts at approx. 1800°C. Crystals of this mineral have been found in fireclay refractories that have been attacked by potash vapour.

Kalsilite. $K_2O.Al_2O_3.2SiO_2$ together with a small amount of Na_2O. This mineral is sometimes formed when alkali vapour attacks fireclay refractories.

Kandite. Group name for the kaolinite minerals, i.e. kaolinite, nacrite, dickite and halloysite. (*Clay Minerals Bull.*, **2**, 294, 1955.)

Kaolin. Name derived from Chinese *Kao-Lin*, a high ridge where this white-firing clay was first discovered. See CHINA CLAY.

Kaolinite. The typical clay mineral of china clay and most fireclays; its composition is $Al_2O_3.2SiO_2.2H_2O$. The unit lattice consists of one layer of tetrahedral SiO_4 groups and one layer of octahedral $Al(OH)_6$ groups; a kaolinite crystal (which is hexagonal) consists of a number of these alternate layers. Disorder can, and normally does, occur in the stacking of these layers so

Kaolinization

that the ceramic properties of a kaolinitic clay may vary. When heated, kaolinite loses water over the range 450–600°C; the D.T.A. endothermic peak is at 580°C. When fired at temperatures above 1100°C mullite is formed.

Kaolinization. The geological breakdown of alumino-silicate rocks, usually but not necessarily of the feldspathic type, to produce kaolinite.

Kassel Kiln. An old type of intermittent, rectangular, fuel-fired kiln which diminished in cross-section towards the end leading to the chimney; it originated in the Kassel district of Germany.

Kavalier Glass. An early type of chemically resistant glass characterized by its high potash content; it was first made by F. Kavalier at Sazava, Czechoslovakia, in 1837, and the Kavalier Glassworks still operates on the same site.

Keatite. A form of silica resulting from the crystallization of amorphous precipitated silica at 380–585°C and water pressures of 350–1250 bars; sp. gr. 2·50. Named from its discoverer, P. P. Keat (*Science*, **120**, 328, 1954).

Keene's Cement or Keene's Plaster. A mixture of gypsum, that has been calcined at a dull red heat, and 0·5–1% of an accelerator usually potash alum or K_2SO_4. Also known as PARIAN CEMENT OR PARIAN PLASTER.

Kek Mill. Trade-name: (1) A pin-disk mill depending for its action on high-speed centrifugal force; (2) a beater mill in which a four-armed 'beater' revolves horizontally at high speed between upper and lower serrated disks. (Kek Ltd., Manchester.)

Keller System. A method of handling bricks to and from a chamber dryer; the bricks are placed on STILLAGES (q.v.) which are lifted by a FINGER-CAR (q.v.), carried into the dryer and set down on ledges projecting from the walls of the drying chamber; when dry the bricks are carried in the same manner to the kiln. The system was patented by Carl Keller of Laggenbeck, Westphalia, Germany, in 1894. Since that date the firm has introduced additional equipment to advance the degree of mechanization in the heavy-clay industry.

Kelly Ball Test. An on-site method for assessing the consistency of freshly mixed concrete in terms of the depth of penetration, under its own weight of 30 lb, of a metal hemisphere, 6 in. dia. (J. W. Kelly and M. Polivka, *Proc. Amer. Concrete Inst.*, **51**, 881, 1955.)

Kelly Sedimentation Tube. A device for measuring the rate of

154

settling of particles from a suspension, and hence for particle-size analysis. To the lower part of a sedimentation vessel, a capillary tube is joined and is bent through 90° so that it is vertical to a level a little above that of the suspension in the sedimentation vessel; above this level the capillary tube is inclined at a small angle to the horizontal. As particles settle in the main vessel the position of the meniscus in the capillary tube moves downward, affording a means of assessing the rate of settling of the particles. This apparatus, designed by W. J. Kelly (*Industr. Engng. Chem.*, **16**, 928, 1924), has been used for the particle-size determination of clays.

Kelvin Temperature Scale (°K). The Centigrade scale displaced downwards so that Absolute Zero (-273°C) is 0°K; 0°C $= 273$°K; etc. Also known as the ABSOLUTE TEMPERATURE, °Abs.

Keramzit. Trade-name: a Russian expanded-clay aggregate.

Keratin. The protein in hair. An extract obtained by the treatment of hair with caustic soda is used, under the name keratin, as a retarder to control the rate of setting of plaster—in making pottery moulds for example.

Kerb. In the UK wall-tile industry the accepted spelling is 'curb'; (see CURB BEND).

Kerr Effect and Kerr Constant. The birefringence produced in glass, or other isotropic material, by an electric field; the effect was discovered in 1875 by J. Kerr, a British physicist. The Absolute Kerr Constant has been defined as the birefringence produced by unit potential difference.

Kervit Tiles. Trade-name derived from the words 'Keramik' and 'Vitrum', denoting the mixed nature of the material which is made by casting a ceramic slip containing about 30% of ground glass. The slip is poured on to a refractory former that is coated with a separating material, e.g. a mixture of bentonite and limestone; the tiles, while on the formers, are fired at 950–1000°C then trimmed. The process was first used in Italy; it was introduced into England in 1960. (M. Korach and Dal Borgo, Brit. Pat., 468 010; 10/6/36.)

Kessler Abrasion Tester. Apparatus designed by the National Bureau of Standards, USA, for the determination of the abrasion resistance of floor tiles and quarries. A notched steel wheel is mounted on an overhanging frame so that a definite and constant weight bears on the test-piece as the wheel revolves; No. 60

Ketteler-Helmholz Formula

artificial corundum is fed at a specified rate between the wheel and the test-piece, which is mounted in an inclined position. (D. W. Kessler, *Nat. Bur. Stand. Tech. News Bull.*, **34**, 159, 1950.)

Ketteler–Helmholz Formula. A formula for the optical dispersion of a glass:

$$n^2 = n_\alpha{}^2 + \Sigma M_m (\lambda^2 - \lambda^2{}_m)^{-1}$$

where n is the refractive index of the glass for a wavelength λ, n_α is the index for an infinitely long wavelength, and $\Sigma \lambda_m$ are the wavelengths of the absorption bands for each of which there is an empirical constant M_m.

Kettle. (1) A heated iron or steel vessel in which gypsum is partially dehydrated to form plaster-of-paris.

(2) A vessel for containing molten glass.

Keuper Marl. A Triassic clay much used for brickmaking in the E. and W. Midlands, the W. of England, and the Cardiff area. This type of brick-clay usually contains a considerable amount of lime and iron oxide; magnesium carbonate and gypsum may also be present in significant quantities. Keuper Marl is of variegated colour, hence the name—from German *Köper* (spotted).

Key Brick. A brick with opposite side faces inclined towards each other so that it fits the apex of an arch, see Fig. 1, p. 37. In furnace construction such bricks are also sometimes referred to as BULLHEADS, CUPOLA BRICKS or CROWN BRICKS.

Kibbler Rolls. Toothed steel rolls of a type frequently used in the crushing and grinding of brick clays; from an old word *Kibble*—to grind.

Kick's Law. A law relating to the energy consumed in the crushing and grinding of materials. The energy required to produce identical changes of configuration in geometrically similar particles of the same composition varies as the reduction in the volume or weight of the particles. (F. Kick, *Das Gesetz der Proportionalen Widerstand und Seine Anwendung*, Leipzig, 1885.)

Kidney. See POTTER'S HORN.

Kieselguhr. German name, formerly used also in Britain but now obsolete, for DIATOMITE (q.v.).

Killas. Term used in the Cornish china-clay mines for the altered schistose or hornfelsic rocks in contact with the granite and often considerably modified by emanations from the latter.

Kiln. A high-temperature installation used for firing ceramic ware or for calcining or sintering. Kilns for firing ceramic ware

are of three main types: INTERMITTENT, ANNULAR and TUNNEL; for calcining and sintering, SHAFT KILNS, ROTARY KILNS and MULTIPLE-HEARTH FURNACES are used (see entries under each of these headings).

Kiln Car. A car for the support of ware in a tunnel kiln or bogie kiln; the car has four wheels and runs on rails. In a tunnel kiln, through which a number of these cars are moved end-to-end, the metal undercarriages are protected from the hot kiln gases by continuous sand-seals. The car deck is constructed of refractory and heat-insulating material.

Kiln Furniture. General term for the pieces of refractory material used for the support of pottery-ware during kiln firing; since the use of clean fuels and electricity has made possible the open-setting of ware, a multiplicity of refractory shapes have been introduced for this purpose. For individual items see COVER, CRANK, DOT, DUMP, PILLAR, PIN, PIP, POST (or PROP), PRINTER'S BIT, RING, SADDLE, SAGGAR, SETTER, SPUR, STILT, THIMBLE; see also Fig. 4, p. 158.

Kiln Scum. See SCUM.

Kimmeridge Clay. A clay of the Upper Jurassic system used as a raw material for building bricks in Dorset and Oxford; the lower beds often contain gypsum and lime, the upper beds are shaly and bituminous.

Klebe Hammer. A device for the compaction of test-pieces of cement or mortar prior to the determination of mechanical strength; a weight falling from a fixed height on the test-piece mould ensures compaction under standardized conditions.

Klein Turbidimeter. Apparatus designed by A. Klein (*Proc. A.S.T.M.*, **34**, Pt. 2, 303, 1934) for the determination of the specific surface of portland cement. The sample is suspended in castor oil in a dish and the turbidity is measured photoelectrically. Because of the high viscosity of the suspending liquid the particles do not settle significantly. The specific surface is deduced from the turbidity by means of a calibration curve.

Kling. A type of TRANSFER LADLE (q.v.) for the conveyance of molten iron.

Klingenberg Clay. A refractory clay containing (unfired) 32–37% Al_2O_3; P.C.E. 32–34. From Klingenberg-am-Main, S.W. Germany.

Klinker Brick. A type of building brick, which may be either of engineering-brick or facing-brick quality, made in Germany and

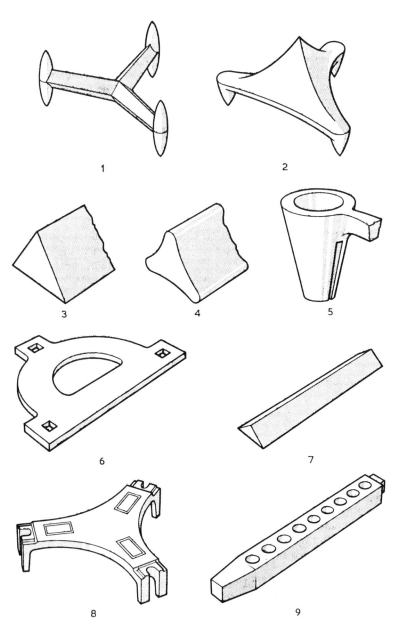

FIG. 4. KILN FURNITURE.

(The relative sizes are *not* to scale: 1. Stilt. 2. Spur. 3 & 4. Saddles.
5. Thimble. 6. Crank. 7. Pin. 8. Crank. 9. Post or Upright.)

Holland from clays that generally have a long vitrification range; the bricks have low water absorption and high crushing strength. The colour may be yellow, red or variegated.

Klompje Brick. A miniature facing brick of a type used in S. Africa, for decorative construction.

Knocking. The accidental removal, during the period between glaze application and the glost firing, of a patch of glaze from the surface of ceramic ware.

Knockings. US term for the oversize material remaining after ceramic slip has been screened.

Knock-out. See BUTTON.

Knoop Hardness. A pyramidal-diamond indentation test is used; it has been applied to glasses and glazes, which have Knoop hardness numbers of 300–500. The test was introduced by F. Knoop, C. G. Peters and W. B. Emerson (*J. Res. Nat. Bur. Stand.*, **23**, 39, 1939).

Knot. A fault, in glass, consisting of a small inclusion of glass having a different composition (and so revealed by its different refractive index) from that of the surrounding glass.

Knotts. Term sometimes applied to the Lower Oxford Clay used for brickmaking in the Peterborough district, England.

Köhn Pipette. Apparatus for particle-size analysis by sedimentation; it has been used more particularly in the study of clay soils. (M. Köhn, *Z. Pflanz. Düng.*, **A11**, 50, 1928.)

Kolene Process. Trade-name: an electrolytic salt-bath treatment of sheet-steel before it is enamelled; (Kolene Corp., Detroit, USA).

Kopecky Elutriator. An elutriator consisting of three cylinders of different diameters; it has been used for the particle-size analysis of clays. (*Tonind. Ztg.*, **42**, 629, 1918.)

Kozeny Equation. An equation relating the rate of flow (q) of a fluid of viscosity μ through a packed bed of particles of depth L and area A, under a pressure difference ΔP, the specific surface of the particles being S and the voids per unit weight being V:

$$q = K \cdot \frac{AV^3}{\mu S^2} \cdot \frac{\Delta P}{L}$$

This equation, due to J. Kozeny (*Sitz. Akad. Wiss. Wien., Math.-Naturw. Kl.*, **136**, 271, 1927) forms a basis for the determination of the specific surface of a powdered ceramic material by the gas-permeability method.

159

Kreüger's Ratio

Kreüger's Ratio. A ratio claimed by H. Kreüger (*Trans. Roy. Swedish Inst., Sci. Res.* No. 24, 1923) to be a criterion of the frost resistance of clay building bricks; it is the ratio of the percentage water absorption after 4 days' immersion in cold water to the total water absorption calculated from the specific gravity (cf. SATURATION COEFFICIENT).

Kreutzer Roof. A design for a furnace roof, particularly for open-hearth steel furnaces. Its feature is the system of transverse and longitudinal ribs, which divide the exterior of the roof into box-like compartments. The design was patented by C. Kreutzer in Germany in 1948.

Kryptol Furnace. The term used on the Continent and in Russia for a granular-carbon resistance furnace; the name derives from the German company, Kryptol Gesselschaft, that originally supplied the carbon granules.

Kühl Cement. A hydraulic cement introduced by H. Kühl (Brit. Pat., 231 535; 31/3/25). It contains less SiO_2 but more Al_2O_3 and Fe_2O_3 (about 7% of each) than does portland cement; its strength properties are similar to those of rapid-hardening portland cement. Also known as BAUXITLAND CEMENT.

Kurlbaum Method. A method for the determination of flame temperature by means of an optical pyrometer (F. Kurlbaum, *Phys. Z.*, **3**, 187, 1902).

Kyanite. A mineral having the same composition (Al_2SiO_5) as sillimanite and andalusite, but with different physical properties. The chief sources are Virginia and S. Carolina (USA), and India. When fired, kyanite breaks down at 1300°C into mullite and cristobalite with a volume expansion of about 10%; it is therefore calcined before use. Calcined kyanite is used in making aluminous refractories.

L

Laced Valley. A form of roof tiling designed to cover a re-entrant corner of a roof. The courses of tiling meet and interlace to provide efficient drainage.

Lacustrine Clay. A clay that was formed by deposition on the bed of a lake; the lacustrine clays of the Vale of York, England, for example, are used in brickmaking.

Ladd Circular Tunnel Kiln. A ROTARY-HEARTH KILN (q.v.) of a type introduced in 1941 by the Ladd Engineering Co., USA (see *Brick Clay Record*, **99**, No. 2, 42, 1941).

Ladder. A horizontal support for glass tubing that is to be cut into lengths (see also WASHBOARD).

Ladle Brick. A refractory brick of suitable shape and properties for use in the lining of a ladle; unless otherwise stated it is assumed that a ladle for the casting of steel is implied. Ladle bricks are almost invariably made of grogged fireclay; the features required are good shape, uniform size, low porosity and, if possible, a permanent expansion when exposed to high temperature. In the USA three qualities of ladle brick are specified in ASTM – C435.

Laitance. The layer of water sometimes formed on the upper surface of freshly placed concrete as a result of the aggregate settling by sedimentation (see also BLEEDING).

Lambert's Law. See under TRANSLUCENCY.

Laminate. A material in sheet form consisting of several different layers united by a ceramic bond. Ceramic laminates have been made to provide lightweight heat insulation. Refractory laminates have been used as bats on which to fire electroceramics and ferrites; these laminates consist of a silicon carbide core with outer layers of alumina or mullite.

Laminated Glass. Safety glass made from alternating and adherent layers of glass and organic plastics, the glass layers being outermost; if broken, the fragments of glass are held in position by the plastics interlayer. For some purposes the interlayer may be made of glass fibre.

Lamination. Textural inhomogeneity in clayware resulting from the shaping process and particularly common in products that have been extruded. In structural clay products, lamination is a source of mechanical weakness and a cause of poor frost-resistance; in refractory products lamination may reduce resistance to slag attack and to thermal shock.

Lamotte Comparator. A pH meter of a type recommended by the US Porcelain Enamel Institute for use in the determination of the acidity or alkalinity of PICKLING (q.v.) solutions. (Lamotte Chemical Products Co., Baltimore, USA.)

Lamp-blown. Glass-ware shaped by means of an oxy-gas or air-gas burner; glass tubing or rod is the usual starting material.

Lancaster Mixer. Trade-name: A counter-current pan-type mixer. (Posey Iron Works Inc., Lancaster, Pa., USA.)

Land-drain Pipe

Land-drain Pipe. See FIELD-DRAIN PIPE.

Lander. See LAUNDER.

Lanthanum Carbide. LaC$_2$ exists in various forms: the tetragonal room-temperature form changes to hexagonal at intermediate temperatures; at 1750°C, through reaction with excess carbon a cubic phase is produced.

Lanthanum Oxide. La$_2$O$_3$; m.p. 2320°C, sp. gr. 6·51. A rare earth sometimes used in special optical glasses.

Lanthanum Titanates. Two compounds have been reported: LaTiO$_3$ and La$_2$Ti$_3$O$_9$; the former has a PEROVSKITE (q.v.) structure and can be synthesized by heating a mixture of La$_2$O$_3$ and Ti$_2$O$_3$ at 1200°C in a vacuum.

Lap. (1) A faulty surface on glass-ware caused by a fold in the glass.

(2) A rotating disk, normally horizontal, carrying abrasive grain or powder and used for the finishing of work-pieces.

(3) The amount by which a roofing tile overlaps the course next but one below it.

Larnite. One of the crystalline forms of CALCIUM ORTHOSILI-CATE (q.v.).

Laterite. Material rich in alumina and/or iron oxide; it is formed by the tropical weathering, *in situ*, of appropriate silicate rocks. The most aluminous laterites are known as BAUXITE (q.v.).

Lath. See CLAY LATH.

Lathe. A machine for turning unfired hollow-ware, e.g. cups or vases.

Lattice Brick. A hollow building brick, or block, in which the cells form a pattern of open lozenges; such bricks are claimed to have a high heat-insulating value because of the extended path of any heat flow through the solid material.

Launder or Lander. An inclined channel, lined with refractory material, for the conveyance of molten steel from the furnace tap-hole to a ladle.

Lawn. Term used in the British pottery industry for a fine-mesh screen or sieve. In the past, makers of these 'lawns' gave them somewhat arbitrary numbers; they are now largely replaced by meshes conforming to British Standards.

L-D Process. A process for the production of steel by the oxygen-blowing of molten iron held in a static vertical vessel. The latter is lined with tarred or fired dolomite refractories; dead-burned magnesite is also sometimes added to the batch. The process was first used, in 1952, at Linz, Austria, and 'L-D' stands for Linzer-Düsenverfahren, i.e. Linz Nozzle Process.

Lea and Nurse Permeability Apparatus. A device for the determination of the specific surface of a powder by measurement of the permeability to air of a prepared bed of the sample; the calculation is based on the CARMAN EQUATION (q.v.). (F. M. Lea and R. W. Nurse, *J. Soc. Chem. Ind.*, **58**, 277, 1939.)

Lead Antimonate (Naples Yellow). This compound, generally used in conjunction with tin oxide or zinc oxide, gives a good on-glaze yellow for pottery decoration; it is also used, with a lead borosilicate flux, as a vitreous enamel colour.

Lead Bisilicate. See LEAD SILICATE.

Lead Chromate. The basic chromate, $PbO.PbCrO_4$, is normally used in ceramics. If the firing temperature is low, it can be used to produce coral reds in pottery glazes; in glassmaking it produces a deep green.

Lead Niobate. $Pb(NbO_3)_2$; a ferroelectric compound having properties that make it useful in high-temperature transducers and in sensing devices. The Curie temperature is 570°C.

Lead Oxide. The more common lead oxides are litharge (PbO) and red lead or minium (Pb_3O_4). These oxides are important constituents of heavy glassware, enamels, and some pottery glazes; because the oxides are poisonous, it is normal (in England compulsory) to form an insoluble lead silicate frit and to employ this non-toxic material as a source of lead for ceramic glazes.

Lead Poisoning. In the UK, lead poisoning has been virtually eliminated from the pottery, glass, and vitreous-enamel industries by successively more stringent factory regulations. The current regulations for the pottery industry are the Pottery (Health and Welfare) Special Regulations of 1947 and 1950.

Lead Silicate. A material obtained by fritting lead oxide with silica in various ratios. The usual ratio is $PbO.2SiO_2$ and a frit of this composition is known as lead bisilicate; so-called lead monosilicate is approx. $PbO.0.7SiO_2$. Tribasic lead silicate, $3PbO.SiO_2$, finds some use in making lead glasses. The bisilicate is used in pottery glazes, enamels, etc.

Lead Solubility Test. The (British) 'Pottery (Health and Welfare) Special Regulations, 1947 and 1950' define this test, as applied to the specification of a Low Solubility Glaze as follows: A weighed quantity of the material which has been dried at 100°C and thoroughly mixed is continuously shaken for 1 h, at the

Lead Stannate

common temperature, with 1000 times its weight of an aqueous solution of HCl containing 0·25% by wt of HCl. This solution is thereafter allowed to stand for 1 h and then filtered. The lead salt contained in the clear filtrate is then precipitated as PbS and weighed as $PbSO_4$.

Lead Stannate. $PbSnO_3$; up to 5% of this material is sometimes added to barium titanate ceramics to reduce their tendency to depolarize when used as piezoelectrics.

Lead Tantalate. $PbTa_2O_6$; a compound believed to have ferroelectric properties and of possible interest as a special electroceramic. The Curie temperature is 260°C.

Lead Titanate. $PbTiO_3$; added in small amounts to barium titanate ceramics to improve their piezoelectric behaviour; a complex lead titanate–zirconate body (P.Z.T.) finds use as a ceramic component in piezoelectric transducers. The Curie temperature is 490°C.

Lead Zirconate. $PbZrO_3$; a ferroelectric material. It is also used in lead titanate–zirconate (P.Z.T.) piezoelectric ceramics.

Leaded Glass. Windows made from small pieces of glass which are held in position by strips of lead, H- or U-shaped in section.

Leadless Glaze. In the UK this is defined in the Pottery (Health and Welfare) Special Regulations of 1950 as: A glaze which does not contain more than 1% of its dry weight of a lead compound calculated as PbO.

Lean Clay. A clay of low plasticity; the adjective 'lean' is also applied to a body of low plasticity.

Leather-hard. A term for clayware that has been partially dried; at this stage of drying, the shrinkage has been largely completed.

LECA. Abbreviation for LIGHTWEIGHT EXPANDED CLAY AGGREGATE (q.v.).

Le Chatelier Soundness Test. A procedure for assessing any expansion of hydraulic cement caused by the presence of excess lime, magnesia or sulphates. The gauged cement is put into a split brass cylinder 30 mm i.d., to which are attached two needles 156 mm long from the centre-line of the mould, one needle on each side of the split in the mould. The cement is allowed to set for 24 h immersed in water at 58–64°F; the distance between the ends of the needles is then measured. The mould is immersed in water and boiled for 1 h. When it has cooled, the distance between the ends of the needles is again measured. The difference

164

between the first and second readings should not exceed 10 mm. The test is included in B.S. 12.

Lee Board. See under HACK.

Leer. The original form of the word LEHR (q.v.).

Lehr. A heated chamber for annealing glass-ware; usually tunnel-shaped, the ware passing through on a conveyor. The spelling *lehr* originated in USA at the end of the 19th century; the older form—*leer*—first appeared in an English text of 1662. The origin of the word is obscure.

Leighton Buzzard Sand. An important source of sand from the Lower Greensand deposits of Bedfordshire, England. The sand is high in silica and well graded; it is used as a refractory foundry sand and as a standard sand for mixing with portland cement for testing according to B.S. 12.

Lens-fronted Tubing. Tubing for liquid-in-glass thermometers made in such a way as to magnify the width of the column of liquid.

Lepidolite. A lithium mineral, approximating to $LiKAlF_2Si_3O_4$; sp. gr. 2·9; m.p. 1170°C. The largest deposits are in S. Rhodesia but it also occurs in USA. Lepidolite is a source of Li for special glasses, pottery bodies and glazes, and vitreous enamels.

Lepol Kiln. See ACL KILN.

Leptize. To disintegrate a material, in the dry state, into very fine particles by impact; at least 50% of the product is $< 50\mu$ and at least 1% is $< 2\mu$. The term was introduced by J. G. Bennett, M. Pirani, and M. W. Thring (Brit. Pat. 549 142; 9/11/42); it has not been generally adopted.

Lessing Rings. A particular shape for chemical stoneware tower fillings.

Leucite. $K_2O . Al_2O_3 . 4SiO_2$; there are two forms stable above and below 620°C respectively; sometimes formed when alkali attacks fireclay refractories, as in a blast furnace stack.

Lew Board. See under HACK.

Libbey–Owens Process. See COLBURN PROCESS.

Libo. Trade-name; a lightweight building material made from very fine sand and lime (Nederlands-Belgische Maatschappij voor Lichte Bouwproducten, Gorinchem, Holland).

Lift. A fault in vitreous enamelware, a relatively large area of enamel coming from the metal; the fault is also known as PEELING. The cause may be inadequate cleaning of the base metal or a defective ground-coat.

Ligand Field Theory

Ligand Field Theory. A theory concerned with the changes in electronic energy levels of ions, especially of transition element ions, which occur when other ions or polar groups (ligands) are brought into their immediate neighbourhood. Variations in the environment of the ion, e.g. in the nature of the ligands or the symmetry of their arrangement, cause variations in the spacings of the energy levels. This forms the basis of the modern explanation of the colours of crystals and of glasses.

Light. As applied to optical glass the term means that the glass has a relatively low refractive index.

Light-extinction Method. See TURBIDIMETER.

Lightweight Expanded Clay Aggregate. A bloated clay aggregate made by the sudden heating of suitable clays either in a rotary kiln (the original method used in Denmark in 1939) or on a sinter-hearth. It is used as an aggregate for making lightweight concrete.

Lignosulphonate. See SULPHITE LYE.

Lime. CaO; m.p. 2570°C; sp. gr. 3·23; thermal expansion (0–1000°C), $13·5 \times 10^{-6}$. Added as WHITING (q.v.) to some pottery bodies and glazes, and as crushed limestone ($CaCO_3$) or magnesian limestone to most glass batches. Limestone is a major constituent of the batch used to make portland cement clinker.

Lime Blowing. The falling away of small pieces from the face of a clay building brick as a result of the expansion (following hydration and carbonation by the atmosphere) of nodules of lime present in the fired brick. Some brick-clays contain nodules of calcite ($CaCO_3$) which are converted into CaO during the kiln firing; these nodules can be rendered innocuous by fine grinding and harder firing. A cure that is more of an expedient is DOCKING (q.v.).

Lime Refractory. Because of its abundance and high m.p. (2570°C), lime would be an attractive basic refractory but for its ready hydration and carbonation when exposed to the air. There has been much research on methods of stabilization, but the only accepted use of lime as a refractory has been as a container material for the melting of the platinum metals.

Lime Saturation Value. The ratio of the actual lime content of a hydraulic cement to that calculated from an equation deduced as representing the amount of lime combined as silicate, aluminate and ferrite. Several such equations have been proposed; all are empirical.

166

Lime-slag Cement. Alternative name for SLAG CEMENT (q.v.).

Limestone. A hard rock consisting of calcium carbonate, $CaCO_3$. It is used as a source of lime in cement manufacture and in glass-making; for the latter use see B.S. 3108.

Linde Flame Plating. See FLAME PLATING.

Lindemann Glass. A lithium beryllium borate glass that is highly transparent to X-rays. It is made from a batch consisting of 10 parts $Li_2B_2O_7$, 2 parts BeO and 3 parts B_2O_3; this glass is difficult to shape and is of low chemical durability. (C. L. Lindemann and F. A. Lindemann, *Z. Roentgenkunde*, **13**, 141, 1911.)

Lindemann–Danielson Test. See DANIELSON–LINDEMANN TEST.

Line Reversal Method. See SODIUM-LINE REVERSAL METHOD.

Linhay. Cornish name for a storage shed for china clay.

Linseis Plastometer. A device designed by M. Linseis for the evaluation of the plasticity of clay on the basis of two parameters: cohesion (measured as tensile strength) and the capacity for relative movement of the clay particles without rupture. The apparatus is made by Netzsch Bros., Selb, Germany.

Lintel or Mantel. The metal supporting structure that takes the weight of the refractories and casing of the stack of a blast furnace.

Liparite. A PEGMATITE (q.v.) that occurs in Italy and is used locally in ceramic bodies.

Liquid Gold. See BRIGHT GOLD.

Liquid-phase Sintering. Strictly, the word SINTERING (q.v.) should not be used if a liquid phase is present; the more correct term in this case is VITRIFICATION (q.v.).

Liquidus. The line on a binary equilibrium diagram (e.g. Fig. 2, p .100) showing the temperature at which any mixture of the two components, when under conditions of chemical equilibrium, becomes completely liquid; conversely, the liquidus shows the temperature at which a liquid of a given composition begins to precipitate solid material. In a ternary, i.e. three-component, system, the liquidus is represented by a curved surface.

Litharge. Lead monoxide, PbO. See LEAD OXIDE.

Lithium Minerals. See AMBLYGONITE, EUCRYPTITE, LEPIDOLITE, PETALITE, SPODUMENE.

Lithium Niobate. $LiNbO_3$; a ferroelectric compound having the ilmenite structure and of potential interest as an electroceramic.

Lithium Tantalate. $LiTaO_3$; a ferroelectric compound of

167

potential value as a special electroceramic. The Curie temperature is above 350°C.

Lithium Titanate. A compound sometimes used as a flux in titanate bodies. It is also characterized by very high thermal expansion: $19 \cdot 5 \times 10^{-6}$ (25–700°C).

Lithium Zirconate. Sometimes used as a flux in more refractory zirconate bodies of the type used as electroceramics.

Lithography (UK); Decalcomania (USA). A method used for the decoration of pottery by means of transfers. A special paper is used but the ceramic colours cannot be printed directly and the actual printing is done in varnish and the colour then dusted on. The Litho or Decal (to use the common contractions of these words) is placed coloured-side down on the sized ware, rubbed firmly on, and the paper then sponged off.

Lithophane. A decorative type of translucent pottery which, when viewed by transmitted light, exhibits a scene or portrait that is present in low relief on the ware. The process originated at Meissen, Germany, in 1828; examples of the ware were also produced in England during the 19th century.

Litre-weight Test. A works' test for the routine control of the firing of portland cement clinker; it was introduced by W. Anselm (*Zement*, **25**, 633, 1936).

Littler's Blue. A royal blue first produced in about 1750 by William Littler of Longton Hall, Stoke-on-Trent, by staining white slip with cobalt oxide, applying the slip to pottery-ware, firing, and salt-glazing.

Littleton Softening Point. See SOFTENING POINT.

Live-hole. A flue left in a CLAMP (q.v.) and filled with brushwood to start the firing process.

Liver Spotting. See MOTTLING (of silica refractories).

Livering. A general term covering curious faults in casting slip which may develop if the slip is allowed to stand; the cause is lack of control in the DEFLOCCULATION (q.v.)

Livesite. A disordered kaolinite, particularly as found in some micaceous fireclays. The term is now little used; it was proposed by K. Carr, R. W. Grimshaw and A. L. Roberts (*Trans. Brit. Ceram. Soc.*, **51**, 339, 1952).

Load. The weight of glass produced by a glass-making furnace in a given time, usually 24 h. Known in USA (and also less commonly in UK) as PULL; other alternative terms are DRAW and OUTPUT.

Load-bearing Tile. US term for a hollow fired-clay building block designed to carry superimposed loads.

Loading. Choking of the pores in the face of an abrasive wheel with debris from the workpiece that is being ground.

LOF–Colburn Process. LOF = Libbey–Owens–Ford Glass Co.; see under COLBURN PROCESS.

L.O.I. Abbreviation sometimes used for LOSS ON IGNITION (q.v.).

London Clay. A tertiary clay used in making building bricks in Surrey, Berkshire, Essex, Suffolk, Dorset, Wiltshire, and Hampshire.

London Stock Brick. See STOCK BRICK.

Long Glass. Glass that is slow-setting.

Longitudinal-arch Kiln. An ANNULAR KILN (q.v.) in which the axis of the arched roof, on both sides of the centre-line of the kiln, is parallel to the length of the kiln (cf. TRANSVERSE-ARCH KILN). Kilns of this type are much used in the firing of building bricks.

Loo Board. See under HACK.

Looping or Loops. See CURTAINS.

Loss on Ignition. The loss in weight of a sample of material expressed as a percentage of its dry weight, when the sample is heated under specified conditions. In the testing of clays and similar materials, the temperature employed is $1050 \pm 50°C$; in the testing of frits, the temperature must be below the fusion point and below the temperature at which there is loss by volatilization.

Lost-wax Process. See INVESTMENT CASTING.

Low-duty Fireclay Brick. Defined in ASTM – C27 as a fireclay refractory having a P.C.E. ⊀ 15 and a modulus of rupture ⊀ 600 p.s.i.

Low-frequency Induction Furnace. See under ELECTRIC FURNACES FOR MELTING AND REFINING METALS.

Low-heat Cement. Hydraulic cement compounded so as to evolve less heat during setting than does normal portland cement. This feature is important when large volumes of concrete are placed, as in dams. It is achieved by reducing the proportions of $3CaO.SiO_2$ and $3CaO.Al_2O_3$. For specification see B.S. 1370.

Low-solubility Glaze. The (British) 'Pottery (Health and Welfare) Special Regulations, 1947 and 1950' define this as a glaze which does not yield to dilute HCl more than 5% of its dry weight of a soluble lead compound, calculated as PbO, when determined by the LEAD SOLUBILITY TEST (q.v.).

169

L.P.G.

L.P.G. Abbreviation for LIQUEFIED PETROLEUM GAS.

Lucalox. Trade-name: a translucent, pure, polycrystalline alumina made by General Electric Co., USA. The translucency results from the absence of micropores. Because the crystals are directly bonded to one another, without either matrix or pores between the crystal boundaries, the mechanical strength is very high: transverse strength, 50 000 p.s.i.; modulus of elasticity, 56×10^6 p.s.i.

Ludwig's Chart. A diagram proposed by T. Ludwig (*Tonind. Ztg.*, **28**, 773, 1904) to relate the refractoriness of a fireclay with its composition, which is first recalculated to a molecular formula in which Al_2O_3 is unity, i.e. $xRO.Al_2O_3.ySiO_2$. In the chart, x is plotted as ordinate and y as abscissa; a series of diagonal lines indicates compositions of equal refractoriness. The diagram is inevitably a rough approximation only.

Lumenized. Lenses coated with MgF_2. (Trade-name—Kodak Ltd.)

Luminous Wall Firing. Term sometimes used in USA for kiln- or furnace-firing by SURFACE COMBUSTION (q.v.).

Lumnite Cement. Trade-name: a HIGH-ALUMINA CEMENT (q.v.) made by Universal Atlas Cement Division of U.S. Steel Corp., New York.

Lump. (1) A heap of unmelted batch floating on the molten glass in a tank furnace.

(2) The most defective saleable pottery-ware remaining after the sorting process; cf. FIRSTS, SECONDS.

Lump Man. A workman in a glassworks whose job is to observe and control any lumps of floating batch in a glass-tank furnace.

Lunden Conductive Tile Flooring. A method of using ceramic tiles in an ANTI-STATIC (q.v.) floor: electrically-conducting 'dots' of metal, or of special ceramic material that has been made conducting by loading the body with carbon, are interspersed among the normal tiles. The method was devised by S. E. Lunden, a Californian architect, in 1950.

Lurgi Process (for Cement Manufacture). A method for the production of hydraulic cement by sintering the charge on a grate.

Lustre. A form of metallic decoration on pottery or glass introduced in the Middle East before 900 AD. Lustres for both materials can be made from metallic resinates, e.g. of Cu, Mn, or Co, in a solvent. The firing conditions must be reducing, this

170

being achieved by the presence of reducing agents in the lustre composition and/or by adjustment of the kiln atmosphere.

L.W.F. Abbreviation for Luminous Wall Firing, a term sometimes used in USA for kiln- or furnace-heating by SURFACE COMBUSTION (q.v.).

Lynch Machine. A machine for the manufacture of glass bottles; it is based on the original design introduced by J. Lynch in 1917 and operated on the 'blow-and-blow' principle. 'Press-and-blow' Lynch machines are also widely used.

Lytag. Trade-name: a lightweight aggregate for concrete made by sintering pulverized fuel ash (Lytag Ltd, Boreham Wood, England).

M

MacAdam System. A method of colour notation in which non-spherical regions of equal perceptual differences are inscribed in the CIE (q.v.) space. (W. R. J. Brown and D. L. MacAdam, *J. Opt. Soc. Amer.*, **33**, 18, 1943.)

Mack's Cement. A quick-setting plaster made by adding Na_2SO_4 or K_2SO_4 to plaster of Paris.

Mackenzie-Shuttleworth Equation. Relates to the progress of shrinkage, or densification, during the sintering of a compact, the latter being considered as a continuous matrix containing uniformly distributed spherical pores:

$$d\sigma/dt = K(1 - \sigma)^{2/3}\sigma^{1/3}$$

where

$$K = \frac{3}{2}\left(\frac{4\pi}{3}\right)^{1/3}\frac{\gamma n^{1/3}}{\eta}$$

In these equations σ is the relative density, t the time in seconds, η the viscosity in poises and n the number of pores per c.c of real material. (J. K. Mackenzie and R. Shuttleworth, *Proc. Phys. Soc.*, B, **62**, 833, 1949.)

Mäckler's Glaze. A type of AVENTURINE GLAZE (q.v.); (H. Mäckler, *Tonind. Zeit.*, **20**, 207, 219, 1896).

MacMichael Viscometer. A rotation-type viscometer designed for the testing of clay slips. (R. F. MacMichael, *Trans. Amer. Ceram. Soc.*, **17**, 639, 1915.)

Maerz–Boelens Furnace. A type of open-hearth steel furnace

Magdolo

with back and front walls sloping inwards, thus reducing the span of the roof; the design was introduced by E. Boelens, in Belgium, in 1952. (Also known as a PORK-PIE Furnace.)

Magdolo. Trade-name: a half-way product in the extraction of magnesia from sea-water, used in Japan as a refractory material for the L-D PROCESS (q.v.). It contains 55–62% MgO and 30–35% CaO, together with minor amounts of SiO_2, Al_2O_3 and Fe_2O_3.

Magnesia; Magnesium Oxide. A highly refractory oxide, MgO; m.p. 2800°C. The crystalline form is known as PERICLASE (q.v.).

Magnesio-chromite. See PICROCHROMITE.

Magnesio-ferrite. $MgFe_2O_4$; m.p. 1750°C; sp. gr. 4·20. This spinel is formed in some basic refractories by reaction between the periclase (MgO) and the iron oxide present.

Magnesite. Magnesium carbonate, $MgCO_3$; the natural mineral often contains some $FeCO_3$ (if more than 5% the mineral is known as breunnerite). The main sources are Russia, Austria, Manchuria, USA, Greece and India. After it has been dead-burned at 1600–1700°C, magnesite is used as raw material for the manufacture of basic refractories (see also SEA-WATER MAGNESIA).

Magnesite Refractory. A refractory material, fired or chemically-bonded, consisting essentially of dead-burned magnesite; the MgO content usually exceeds 80%. Such refractories are used in the hearths and walls of basic steel furnaces, mixer furnaces and cement kilns, for example.

Magnesite-chrome Refractory. A refractory material made from dead-burned magnesite and chrome ore, the magnesite being present in the greater proportion. Such refractories may either be fired or chemically-bonded and are used principally in the steel industry. A typical chemical analysis is (per cent): SiO_2, 3–4; Fe_2O_3, 4–6; Al_2O_3, 4–6; Cr_2O_3, 8–10; CaO, 1–2; MgO, 70–75. In Western Europe, P.R.E. define a magnesite-chrome refractory as containing 55–80% MgO.

Magnesium Aluminate. $MgAl_2O_4$; m.p. 2135°C; sp. gr. 3·6; thermal expansion (100–1000°C), $9·0 \times 10^{-6}$. This compound is the type mineral of the SPINEL (q.v.) group.

Magnesium Fluoride. MgF_2; m.p. 1255°C. Hot-pressed (30 000 p.s.i. at 650°C) compacts had the following properties: sp. gr. 3·185; hardness, Knoop 585, Rockwell C52; Young's Modulus, $10·8 \times 10^6$ p.s.i.; thermal expansion, $11·5 \times 10^{-6}$ (20–500°C). The hot-pressed material is suitable for use in infra-red components operating under severe conditions.

172

Magnesium Oxide. See MAGNESIA.

Magnesium Oxychloride. See SOREL CEMENT.

Magnesium Stannate. $MgSnO_3$; when added to titanate electro-ceramic bodies, the dielectric constant is usually somewhat increased up to the Curie point.

Magnesium Sulphate. $MgSO_4.7H_2O$; Epsom salt; used in the vitreous-enamel industry to improve the suspension of the enamel slip.

Magnesium Titanate. $MgTiO_3$; an electroceramic having a low dielectric constant (13–17). It is sometimes added to other titanate bodies to reduce their dielectric constants.

Magnesium Zirconate. $MgZrO_3$; m.p. 2150°C. This compound is sometimes added in small amounts (up to 5%) to other electro-ceramic bodies to lower their dielectric constant at the Curie point.

Magnet Thickness Gauge. A non-destructive device for determining the thickness of a coating of vitreous enamel; it depends on the principle that the force required to pull a permanent magnet off the surface of enamelware is inversely proportional to the thickness of the non-magnetic enamel layer on the magnetic base metal.

Magnetic Ceramics. See FERRITES.

Magnetic Separator. Equipment for removing iron particles from clay or other ceramic material; in the design commonly used in the pottery industry, the prepared body is allowed to flow as a fluid slip over a series of permanent magnets, which are periodically removed and cleaned.

Magnetite. Fe_3O_4; a spinel; m.p. 1594°C; sp. gr. 5·14. This magnetic iron oxide is present in many ferruginous slags and in their reaction products with refractory materials.

Magnit. Trade-name: A tarred magnesitic-dolomite refractory made by Vereinigte Öesterreichische Eisen und Stahlwerke, Linz, Austria, and used by them in the L-D PROCESS (q.v.) of steel-making. The composition of the calcined raw material varies as follows: 65–80% MgO, 10–25% CaO, 2–5% SiO_2, 1% Al_2O_3, 4–6% Fe_2O_3, Loss on ignition, 0·5–2%.

Magnum. A 2-quart wine bottle.

Main Arch. (1) The refractory blocks forming the part of a horizontal gas-retort comprising the DIVISION WALLS (q.v.) and the roof that covers the retorts and the recuperators.

(2) General term for the central part of a furnace roof, particularly used as a synonym of the Crown of a glass-tank furnace.

Maiolica; Majolica

Maiolica; Majolica. Originally a porous type of pottery, with a glaze opacified with tin oxide, from the island of Majorca; this pottery was first made in the 16th century under the combined influence of Hispano-Moresque and Near Eastern wares; it was essentially similar, technically, to delft-ware and the original faience. As applied to present-day pottery, the term signifies a decorated type of earthenware having an opaque glaze, usually fired at a comparatively low temperature (900–1050°C).

Malm. (1) A synonym of Marl in its original sense of a calcareous clay.

(2) The best quality of clamp-fired London Stock Bricks.

Manchester Kiln. A type of longitudinal-arch kiln that was introduced in the Manchester district of England for the firing of building bricks. It is top-fired. A distinctive feature is the flue-system, with horizontal damper-plates, in the outside wall. The Manchester kiln usually has a hot-air system.

Mandrel. (1) A former used in the production of blown glass-ware from tubing or of precision tubing.

(2) A refractory tube used in the production of glass tubing or rod.

Mangalore Tile. A clay roofing tile of the interlocking type as made in the Mangalore district of India.

Manganese-Alumina Pink. A ceramic stain, particularly for the colouring of pottery bodies; it is produced by calcination of a mixture of $MnCO_3$, hydrated alumina and borax. This stain will produce a strong, clean colour over a wide firing range but the glaze must be rich in Al_2O_3; this requirement makes it difficult to produce a smooth glaze.

Manganese Oxide. There are several manganese oxides, the commonest being MnO_2 (pyrolusite). It is used as a colouring oxide (red or purple); mixed with the oxides of Co, Cr, Fe, it produces a black. This oxide is also used to colour facing bricks, and to promote adherence of ground-coat vitreous enamels to the base-metal.

Mangle. A type of dryer used in the tableware section of the pottery industry. The ware, while still in the plaster mould, is placed on trays suspended between two endless chains that pass over wheels above and below the working opening in the dryer; by means of these chains the trays, when filled, are moved vertically through the hot air current in the dryer until they arrive at a second opening where the moulds and dried ware are removed.

Mansard Roof. A tiled roof having a steeper pitch towards the eaves than towards the ridge; the term is also applied to a flat-topped roof with steeply-pitched tiling towards the eaves. (F. Mansard, 17th century French architect.)

Manson Effect. The BOILING (q.v.) of vitreous enamel that occurs when heated cast-iron is DREDGED (q.v.) with only a thin coating of enamel, which is allowed to fuse before a further coating of enamel has been applied. The effect was first observed by M. E. Manson (*Bull. Amer. Ceram. Soc.*, **11**, 204, 1932).

Mantel See LINTEL.

Mantle Block. (1) A refractory block for use in the gable wall of a glass-tank furnace above the DOG-HOUSE (q.v.).

(2) The upper course of refractory blocks, in a sheet-glass furnace, enclosing the Fourcault drawing pit.

Manx Stone. See CHINA STONE.

Marbleized. A vitreous-enamel finish resembling marble and, in fact, obtained by an offset process from natural marble.

Marbling. A procedure sometimes used by the studio potter: the marble effect is obtained by covering the piece of previously dried ware with coloured slips and then shaking the ware (while the slips are still wet) to make the colours run into one another.

Marcasite. A form of iron sulphide, FeS_2, sometimes found as particles or nodules in clays, forming dark spots when the clay is fired.

Margin. See GAUGE.

Mariotte Tube. An orifice-type viscometer designed by E. Mariotte (Paris, 1700). One such instrument with a tube diameter of $2\frac{1}{8}$ in and an $\frac{1}{8}$ in orifice has been used for the assessment of the fluidity of clay slips (*J. Amer. Ceram. Soc.*, **27**, 99, 1944).

Marl. Strictly, an impure, calcareous clay; the word is used in N. Staffordshire to denote a low-grade fireclay, e.g. saggar-marl. See also ETRURIA MARL, KEUPER MARL.

Marlow Kiln. A tunnel kiln fired with producer gas and pre-heated air; it was first used, in Stoke-on-Trent, England, for the biscuit- and glost-firing of wall tiles (J. H. Marlow, *Trans. Brit. Ceram. Soc.*, **21**, 153, 1922).

Marquardt Porcelain. A mullitic porcelain introduced early in the present century by the State Porcelain Factory, Berlin, chiefly for pyrometer sheaths and furnace tubes. Typically, the body consists of 55% clay, 22·5% feldspar and 22·5% quartz; it is fired

Marseilles Tile

at 1400–1450°C for a sufficiently long time for all the quartz to dissolve so that the fired product consists of mullite in a glassy matrix.

Marseilles Tile. A clay roofing tile of the interlocking type, particularly of the pattern made in the Marseilles district of France. These tiles are made in a REVOLVER PRESS (q.v.).

Martin's Cement. A quick-setting plaster made by adding K_2CO_3 to plaster of Paris.

Marver. A small metal or stone table on which a gather of glass is rolled or shaped by hand.

Mason's Ironstone China. A vitrified type of earthenware introduced by C. J. Mason, Stoke-on-Trent, England. According to his patent (Brit. Pat. 3724, 1813) the batch composition was: 4 pts. china clay, 4 pts. china stone, 4 pts. flint, 3 pts. prepared ironstone, and a trace of cobalt oxide. It is now known that the body did not contain ironstone, the name merely being a highly successful method of indicating to the public that the ware was very strong.

Masonry. Bricks (or other structural units of a similar type) bonded together and depending for their structural stability mainly on the strength of the bricks and of the mortar, but also on the bonding and weight. See also PREFABRICATED MASONRY; REINFORCED BRICKWORK.

Master Mould. See under MOULD.

Match Mark. See JOINT LINE.

Matching Piece. A short refractory channel between the spout of a glass-tank furnace and the pot spout for a revolving pot; also known as an INTERMEDIATE PIECE.

Matrix. That part of a ceramic raw material or product in which the larger crystals or aggregates are embedded. A fired silica refractory, for example, consists of crystalline silica (quartz, cristobalite and tridymite) set in a glassy matrix; a fireclay refractory may consist of quartz and mullitic GROG (q.v.) set in a largely amorphous matrix. (Also sometimes known as the GROUND-MASS.)

Matt Blue. A colour for pottery decoration depending on the formation of cobalt aluminate; zinc oxide is usually added, a quoted recipe being 60% Al_2O_3, 20% CoO, 20% ZnO.

Matt Glaze. A ceramic glaze that has partially devitrified; the effect is deliberately achieved, for example, on some types of glazed wall tile. The glazes used for the purpose are usually

176

leadless and devitrification is encouraged by the introduction, into the glaze batch, of such oxides as TiO_2, CaO or ZnO.

Maturing Temperature or Maturing Range. (1) The firing temperature at which a ceramic body develops a required degree of vitrification.

(2) The firing temperature at which the constituents of a glaze have reacted to form a glass that, when the ware has been cooled, appears to the eye to be homogeneous and free from bubbles.

Mazarine Blue or Royal Blue. A ceramic colour, for on-glaze or under-glaze, based on the use of cobalt oxide (40–60%) together with a flux.

Mechanical Analysis. A term sometimes used as a synonym of PARTICLE-SIZE ANALYSIS (q.v.).

Mechanical Boy. A device used for controlling the mould during the hand-blowing of glass-ware.

Mechanical Press. A press in which bricks (building or refractory) are shaped in a mould by pressure applied mechanically, e.g. by a toggle mechanism, as distinct from the procedure in a hydraulic press.

Mechanical Shovel. The original form of excavator, still much used in the getting of brick-clays. It consists of a bucket (or 'dipper') on the end of an arm pivoted to a boom; the bucket scoops upwards, the excavator operating below the clay face.

Mechanical Spalling. SPALLING (q.v.) of a refractory brick or block caused by stresses resulting from pressure or impact.

Mechanical Water or Mechanically-held Water. The water that is present in moist clay and is not chemically combined. When the clay is heated, the mechanically-held water is virtually all removed by the time that the temperature has reached 150°C.

Medina Quartzite. A US quartzite used as a raw material for silica refractories. A quoted chemical analysis is (per cent): SiO_2, 97·8; Al_2O_3, 0·9; Fe_2O_3, 0·85; alkalis, 0·4.

Medium-duty Fireclay Brick. Defined in ASTM – C27 as a fireclay refractory having a P.C.E. $\nleqslant$ 29 and a modulus of rupture $\nleqslant$ 500 p.s.i.

Meldon Stone. A CHINA STONE (q.v.) of low quality from Cornwall, England. A quoted analysis is (per cent): SiO_2, 70; Al_2O_3, 18; Fe_2O_3, 0·4; CaO, 0·6; Na_2O, 4; K_2O, 4·5; loss on ignition, 2·5.

Melting. See CONGRUENT MELTING; INCONGRUENT MELTING. For melting points of various compounds and minerals, see under name of compound or mineral.

177

Melting End

Melting End. The part of a glass-tank furnace where the batch is melted and the glass is refined.

Membrane Theory of Plasticity. Attributes the plasticity of clay to the compressive action of a postulated surface envelope of water around the clay particles. (F. H. Norton, *J. Amer. Ceram. Soc.*, **31**, 236, 1948).

Mendheim Kiln. A gas-fired chamber kiln designed by G. Mendheim, of Munich, in about 1910 for the firing of refractories at high temperature. The gas enters at the four corners of each chamber, and burns within BAG WALLS (q.v.) which direct the hot products of combustion towards the roof; they then pass downward through the setting and are exhausted through the chamber floor.

Merch Bricks. Term sometimes used in USA for building bricks that come from the kiln discoloured, warped or off-sized.

Mercury Penetration Method. A procedure for the determination of the range of pore sizes in a ceramic material. It depends on the fact that the volume of mercury that will enter a porous body at a pressure of P dynes/cm² is a measure of the volume of pores larger than a radius r cm where $r = -2\sigma \cos\theta/P$, σ being the surface tension of mercury in dynes/cm and θ being the contact angle between mercury and the ceramic. A development of the method has been described by R. D. Hill (*Trans. Brit. Ceram. Soc.*, **59**, 198, 1960).

Mereses. The ribs found on the stems of some wine-glasses.

Merriman Test or Sugar Test. A quality test (now discarded) for hydraulic cement. The sample is shaken with a solution of cane sugar and the amount of cement dissolved is determined by titration with HCl (T. Merriman, *J. Boston Soc. Civil Eng.*, **26**, No. 1, 1, 1939).

Merwinite. $3CaO \cdot MgO \cdot 2SiO_2$; melts incongruently at 1575°C. It may occur in, or be formed during service in, dolomite refractories.

Metahalloysite. A clay mineral of the kaolinite group and of the same approximate composition as kaolinite—$Al_2O_3 \cdot 2SiO_2 \cdot 2H_2O$. It is formed by the partial dehydration of HALLOYSITE (q.v.).

Metakaolin. An intermediate product formed when kaolinite is heated at temperatures between about 500° and 850°C; the layer structure of the parent kaolinite persists in modified form but collapse of the layers destroys any periodicity normal to the layers. At higher temperatures (925°C) metakaolin transforms to a cubic

178

phase with a spinel-type structure; at 1050–1100°C mullite is formed.

Metal. In the glass industry, the molten glass is called 'metal'.

Metal Blister. A defect that may occur in vitreous enamelware if there is any gas trapped in a fault in the base metal.

Metal-cased Refractory. A basic refractory with a thin sheet-metal casing on three or four sides, leaving the ends of the brick exposed; the refractory material itself (generally magnesite or chrome-magnesite) is usually chemically bonded. Such bricks are used chiefly in steel furnaces; during use, the metal case at the hot end of the brick oxidizes and 'knits' each brick to its neighbours. There are many variants of this type of basic refractory but all derive from the steel tubes packed with magnesite as patented in 1916 by N. E. MacCallum of the Phoenix Iron Co., USA.

Metal Line. See FLUX LINE.

Metal-marking. See SILVER MARKING.

Metal Protection (at high temperature). Metals can be protected from oxidation at high temperature by various types of ceramic coating, the commonest being flame-sprayed alumina and refractory enamels. Such coatings have found particular use on the exhaust systems of aircraft.

Metallizing. Electroceramics are metallized when it is required to join the ceramic to a metal or form a seal. An alumina ceramic, for example, can be metallized by painting it with a powdered mixture of Mo and Fe and firing it in a protective atmosphere to bond the metals to the surface; the metallized area is then plated with Cu or Ni. This two-stage procedure is known as the TELE-FUNKEN PROCESS (Telefunken G.m.b.H., Berlin). Single stage metallizing is possible if the hydride of Ti or Zr is used, together with a hard solder; this is sometimes known as the BONDLEY PROCESS (R. J. Bondley, *Electronics*, **20**, (7), 97, 1947).

Metalloceramic. One of the old, now obsolete, names of CERMET (q.v.).

Metameric Colours. Colours that appear to be the same under one type of illuminant but, because they have different spectral reflectivity curves, will not match under a different illuminant. Two white vitreous enamels may be metameric, for example, if one is opacified with titania in the anatase form whereas the other contains titania in the form of rutile.

Methuselah. A 9-quart wine bottle.

Mettlach Tile

Mettlach Tile. A vitreous floor tile (especially of the multi-coloured type) as first made by Villeroy and Boch at Mettlach, in the Saar. The term (often mis-spelt 'Metlach') is now commonly used on the Continent and in Russia for any vitreous floor tile.

Mg Point. See TRANSFORMATION POINTS.

Mica Convention. See under RATIONAL ANALYSIS.

Mica (Glass-bonded). An electroceramic made by the bonding of mica (natural or synthetic) with a glass of high softening point. Some products of this type can be used up to 500°C. They have good dielectric properties. Transverse strength, 13 000–15 000 p.s.i. Thermal expansion 11×10^{-6}.

Mica (Natural). Group name for a series of silicates, the most important of which, to the ceramist, is MUSCOVITE (q.v.), a common impurity in clays.

Michigan Slip Clay. A clay occurring in Ontonagon County, Michigan; it is similar to ALBANY CLAY (q.v.).

Micoquille. Thin glass similar to COQUILLE (q.v.) but with a radius of curvature of 7 in.

Microcharacter Hardness. A scratch hardness test in which a loaded diamond is used; it has been applied to the testing of glazes. The 'Microcharacter' instrument was developed by C. H. Bierbaum (*Trans. Amer. Soc. Mech. Engrs.*, p. 1099, 1920; p. 1273, 1921).

Microcline. See FELDSPAR.

Micro-glass. Thin glass for making cover-slips for use in microscopy.

Microlite. Trade-name: a dense corundum, consisting of uniform micro-crystals $(1–3\mu)$, introduced in 1951 by the Moscow Hard-Alloy Combine. It contains 99% Al_2O_3 and a small amount of MgO to inhibit crystal growth; the primary use is in tool tips.

Micromeritics. The science of small particles. This term, which has not achieved popularity, was proposed by M. J. DallaValle in his book *Micromeritics* (Pitman, 1943).

Micromerograph. Trade-name: an instrument for particle-size analysis depending on the rate of fall of particles in air; it is suitable for use in the range $1–250\mu$. (Sharples Corp. Research Labs., Bridgeport, Pa., USA.)

Micron. One-thousandth of a millimetre; the symbol for this unit is μ. (In German literature the symbol μm is sometimes used.)

Micronizer. Trade-name: a type of fluid-energy mill used for very fine grinding, e.g. talc to 5μ.

180

Microscopy. Books on the use of the microscope in ceramics are: *Thin-Section Mineralogy of Ceramic Materials* by G. R. Rigby (British Ceramic Research Assoc., 2nd Ed., 1953) and *Microscopy of Ceramics and Cements* by H. Insley and D. Frechette (Academic Press, New York, 1955).

Midfeather. A dividing wall between two flues, in a gas-retort or glass-tank furnace for example; in the latter, the wall may also be called a TONGUE.

Mil. One-thousandth of an inch.

Milford-Astor Machine. A simple printing machine that has been used for the small-scale production of transfers for the decoration of pottery. (Brittains Ltd., Stoke-on-Trent, England.)

Milk Glass. Opaque white glass (cf. ALABASTER GLASS and OPAL GLASS).

Mill Addition. Any of the materials, other than the frit, charged to the ball mill during the preparation of a vitreous enamel slip. Such materials include clay, opacifier, colouring oxide and setting-up agent.

Millefiori. Decorative glassware, particularly paperweights, made by setting multi-coloured glass cane to form a design within a clear glass matrix. This type of glass-ware originated in Italy and derives its name from Italian words meaning 'A Thousand Flowers'.

Mineral Wool or Rock Wool. There are three main types of mineral wool: *Rock Wool*, made from natural rocks or a mixture of these; *Slag Wool* made from metallurgical slag; *Glass Wool*, made from scrap glass or from a glass batch. All three varieties are used as heat-insulating materials. US Dept. Commerce Standard CS 117–49 defines 5 classes of mineral wool having the following max. temperature limits of use: Class A—600°F; Class B—1000°F; Class C—1200°F; Class D—1600°F; Class E—1800°F.

Mineralizer. A substance that, even though present in only a small amount, assists the formation and/or crystallization of other compounds during firing. A small amount of alkali, for example, mineralizes the conversion of quartz to cristobalite; boric oxide acts as a mineralizer in the synthesis of spinel ($MgAl_2O_4$); the presence of iron compounds facilitates the growth of mullite crystals.

Minium. Red lead, Pb_3O_4. See LEAD OXIDE.

Minton Oven. A down-draught type of pottery BOTTLE OVEN (q.v.) patented by T. W. Minton (Brit. Pat. 1709, 10th May 1873).

Mirac Process

Mirac Process. Trade-name: a process for the treatment of steel claimed to permit one-coat enamelling; (Pemco Corp., Baltimore, Md., USA).

Mismatch. If the properties, particularly the thermal expansion, of the components of a heterogeneous solid are mutually incompatible, there is said to be 'mismatch'. A typical example is the case of a ceramic material consisting of crystals set in a glassy matrix; it may be important that there is no mismatch between these two components.

Mitre Seam. See JOINT LINE.

Mix. A BATCH (q.v.) after it has been mixed.

Mixer. In the clay industries the usual types of mixer are:

(1) *Batch-type Mixer;* operates by rotating arms.

(2) *Shaft Mixer;* a continuous mixer for wet or plastic material which is fed into an open trough along which it is propelled and mixed by one or two rotating shafts carrying blades.

(3) *Pug Mill;* a shaft mixer with a closed barrel instead of an open trough; the term PUG MILL should not be confused with PUG (q.v.).

Mixer Furnace. See HOT-METAL MIXER.

Mobilometer. See GARDNER MOBILOMETER.

Mocca or Mocha Ware. A type of pottery having a brown moss-like or dendritic decoration similar in appearance to the Mocha, or Mecca, stones (brown agates) used in Arabia for brooches. The effect is produced by allowing differently coloured slips to flow over the surface of the ware.

Mock Acid Gold. See ACID GOLD.

Modular Co-ordination; Module. A system for the standardization of the dimensions of building components on the basis of multiples of one or more modules, i.e. basic units of length. The British Standards Institution has issued (1963) two proposals: *A.* A single module of 4 in., *B.* Four modules—$1\frac{1}{2}$, 3, 4 and 12 in. Proposal *A* would approximate closely to the metric module of 10 cm; proposal *B* is more flexible and would permit retention of the present British size of building brick.

Modulus of Elasticity. The ratio of stress to strain within the elastic range. See YOUNG'S MODULUS.

Modulus of Rigidity. The MODULUS OF ELASTICITY (q.v.) in shear.

Modulus of Rupture. The maximum transverse breaking stress, applied under specified conditions, that a material will withstand before fracture. The modulus of rupture, M, of a test-piece of rectangular cross-section is given by: $M = 3Wl/2bd^2$, where W

is the breaking load, l the distance between the knife-edges in the transverse-strength test, b the breadth of the test-piece and d its depth. For a cylindrical test-piece of diameter d, the equation is: $M = 8Wl/\pi d^3$.

Mohammedan Blue. See CHINESE BLUE.

Mohs' Hardness. A scale of scratch hardness based on the following series of minerals in ascending order of hardness: 1 Talc, 2 Rock Salt, 3 Calcite, 4 Fluorspar, 5 Apatite, 6 Feldspar, 7 Quartz, 8 Topaz, 9 Corundum, 10 Diamond. (Named from Mohs who proposed this scale in the early 19th century.)

Moil. The excess glass remaining on a shaping tool or in a glass-making machine; it is re-used as CULLET (q.v.).

Moisture Content. In chemical analysis the moisture content is determined by drying the sample at 110°C until the weight is constant. For works' control of the moisture content of a powder or of clay, instruments are available based on measurement of the pressure developed by the evolution of acetylene from a mixture of the sample with calcium carbide. Methods depending on the variations of the electrical properties of a material with changes in its moisture content are not applicable to clays owing to the great influence of exchangeable bases and soluble salts. The neutron absorption method is more promising.

Moisture Expansion. See ADSORPTION.

Moler. A deposit of DIATOMITE (q.v.) of marine origin occurring on the island of Mors, Denmark; it has been worked since 1912 for use as a heat-insulating material, as a constituent of special cements and for other purposes.

Möller and Pfeiffer Dryer. A multi-track tunnel dryer with con-current air flow and heat recuperation from the hot air at the exit end. This design of dryer is well suited to heavy clay products requiring gentle initial drying.

Molochite. Trade-name: a pre-fired china clay. Composition (per cent): SiO_2, 54–55; Al_2O_3, 42–43; Fe_2O_3, 0·7; Alkalis, 1·5–2·0. P.C.E. > 1770°C. Thermal expanson $4·7 \times 10^{-6}$ (0°–1000°C). It is used as a CHAMOTTE (q.v.) in blast-furnace stack refractories, kiln furniture, refractory cements, etc. (English Clays Lovering Pochin & Co. Ltd, St. Austell, England).

Molten-cast Refractory. A refractory made by FUSION CASTING (q.v.); also, and in the UK more commonly, known as a FUSION-CAST REFRACTORY.

Molybdenum Borides. Five compounds have been reported.

Molybdenum Carbides

Mo_2B; m.p. 2120°C; sp. gr. 9·3. Mo_3B_2; dissociates at 2250°C. MoB; exists in two crystalline forms—α-MoB, m.p. 2350°C, sp. gr. 8·8; β-MoB, m.p. 2180°C, sp. gr. 8·4. Mo_2B_5; dissociates at approx. 1600°C; sp. gr., 7·5. MoB_2; m.p. 2100°C; sp. gr., 7·8; thermal expansion, $7·7 \times 10^{-6}$.

Molybdenum Carbides. MoC; m.p. 2692°C; sp. gr. 8·5. Mo_2C; m.p. 2687°C; sp. gr. 8·9.

Molybdenum Disilicide. $MoSi_2$; m.p. approx. 2000°C; hardness 80–87 Rockwell A; thermal expansion 8×10^{-6}. This special refractory material resists oxidation up to 1700°C. Because of its relatively high electrical conductivity, it has found some use as a material for furnace heating elements.

Molybdenum Enamel ('Moly' Enamel). A vitreous enamel containing about 7·5% molybdic oxide, which improves the adherence and acid resistance.

Molybdenum Silicide. See MOLYBDENUM DISILICIDE.

Monazite. Strictly, the mineral $(La,Ce)PO_4$; the term is commonly used for the naturally occurring sands rich in rare earths.

Monel. Trade name (Henry Wiggin & Co. Ltd., Birmingham, England). A nickel-copper alloy having good resistance to acids and therefore used for equipment in the PICKLING (q.v.) process in the vitreous-enamel industry.

Monitor. Equipment for directing a high-pressure (150 p.s.i.) jet of water against a clay face in the hydraulic process of winning china clay.

Monk-and-Nun Tile. Popular name for SPANISH TILE (q.v.).

Monnier Kiln. A tunnel kiln designed for the firing of building bricks; it is mechanically fired from the top with coal, which burns among the bricks as in a HOFFMANN KILN (q.v.). Although the design is of French origin (Patented by J. B. Monnier) the first such kiln was built in Kent, England, in 1929.

Monolithic. The word is applied to furnace linings that are rammed or cast in place, as opposed to being built of jointed brickwork.

Monometer Furnace. Trade-name: Monometer Mfg. Co. Ltd, London. An oil-fired rotary furnace, particularly for the melting of cast iron; a rammed refractory lining is used, generally similar to that in a CUPOLA (q.v.).

Monticellite. The complex orthosilicate of calcium and magnesium, $CaO.MgO.SiO_2$; this compound dissociates at 1485°C with the formation of a liquid phase. Monticellite is formed in basic refractories when appreciable amounts of lime and silica are present; it is for this reason that specifications for chrome ore and dead-burned magnesite usually stipulate low maxima for the SiO_2 and CaO contents.

Montmorillonite. A magnesian clay mineral approximating in composition to $5Al_2O_3 . 2MgO . 24SiO_2 . 6H_2O$; some of the H^+ groups are usually replaced by Na^+ or by Ca^{2+}. Montmorillonite is the principal constituent of bentonite; Na-bentonite is characterized by considerable swelling when it takes up water, whereas little, if any, swelling occurs with Ca-bentonite. The mineral was originally found (1847) at Montmorillon, France.

Moore Bin Discharger. Trade-name: a system of vibrating fins arranged to assist the discharge of clay, or other 'sticky' material, from a storage bin. (Conveyors (Ready Built) Ltd., Stroud, Glos., England.)

Moore–Campbell Kiln. An early type of electric tunnel kiln; its principal feature was the free suspension of the heating elements on refractory knife-edges. (B. J. Moore and A. J. Campbell, Brit. Pat. 270 035, 5/5/27.)

Morehouse Mill. A mill of the attrition type consisting essentially of a vertically-driven shaft, with horizontal milling elements, made of sintered alumina. It has been used for the preparation of vitreous-enamel slips and ceramic glazes.

Morgan–Marshall Test. A sand-blast test developed by Morgan Refractories Ltd., and Thos. Marshall & Co. Ltd., England, for the evaluation of the abrasion resistance of refractory bricks. In the UK the original apparatus (*Trans. Brit. Ceram. Soc.*, **54**, 239, 1955), modified in a few details, has been generally adopted by the refractories industry.

Mortar. Material for the joining of structural brickwork. (In USA the term is also applied to refractory jointing material for furnace brickwork.) The usual types of building mortars, and the proportions (by vol.) recommended in B.S. Code of Practice 121.101 are:

Type of Mortar	Cement	Lime	Fine Aggregate
Cement Mortar	1	$0-\frac{1}{4}$	3
Cement-Lime Mortar	1	1	5–6
Cement-Lime Mortar	1	2	8–9
Cement-Lime Mortar	1	3	10–12
Lime Mortar	—	1	2–3

Morted Ware. A fault in pottery-ware resulting from a small local concentration of soluble salts forming on the surface of the green ware or the biscuit ware. In the green state the fault may result from a drop of water falling on the ware, during fettling for example. In the biscuit state, the fault arises from moisture condensing on the ware during the early stages of firing.

Mosaic. Small pieces of coloured glass or clay tile arranged to form a pattern or picture and cemented to a wall or floor. (The word is cognate with Muse, a Greek goddess of the arts.) In the USA the largest facial area for a ceramic mosaic tile is 6 in².

Mottling. (1) A form of decoration applied to vitreous enamelware, to glazed ceramic tiles, and to some studio pottery. For tiles the effect is produced by hand, with a sponge, or by a rubber roller.

(2) The staining that sometimes develops during the firing of silica refractories; the stains are irregular in shape and vary from red to brown. The cause is precipitation of Fe_2O_3 from solution in the matrix of the brick. These stains are also sometimes referred to as LIVER SPOTTING or FLOWERS.

Mould (for the Shaping of Pottery). The plaster moulds used for the casting, jiggering and jolleying of pottery-ware are made as follows. *Stage* 1, *Model*: the prototype of the piece of ware that is to be made is usually shaped of clay, occasionally of wood or plaster. *Stage* 2, *Master Mould*: plaster is cast round the model; depending on the shape of the latter, the mould, when set, may need to be cut into several pieces. *Stage* 3, *Case Mould*: a replica of the original model, usually made by casting plaster in the master mould, which is first coated with a separating agent, e.g. soft soap. More recently, organic plastics have been used in the making of case moulds; they are costly but, owing to their hardness, have a long life. *Stage* 4, *Working Mould*: as many moulds as may be required are made by casting plaster around the case mould.

Mould Brick. A fireclay brick, or a heat-insulating brick, shaped to fit in the top of an ingot mould to help to maintain the top of the ingot molten until the main part of the ingot has solidified.

Mould Mark. A fault on glass-ware caused by an uneven mould surface, or, in the case of curved glass sheets, by the dies or frame used in its shaping (see also JOINT LINE).

Mould Plug. A piece of refractory material, or metal, for sealing the bottom of an ingot mould in the casting of steel. (See also PLUG.)

Mould Runner. A man or youth who, in old pottery factories, carried filled moulds from the making department to the drying room and empty moulds back again to the maker; this work has now been eliminated by the introduction of dryers of various types placed adjacent to the making machines.

Mouldable Refractory. A graded refractory material, moistened ready for use and intended for ramming into position in a furnace lining; such a material may be air-hardening. Also known as a PLASTIC REFRACTORY.

Mounted Point. A small abrasive unit, variously shaped, permanently mounted on a spindle.

Mouth (of a Converter). See NOSE.

Mouthpiece. There is a difference in usage of this term as applied to a PUG (q.v.) or AUGER (q.v.): more commonly it refers to the die only, but sometimes to the die-plus-spacer.

Move. In the hand-made glass industry, a stipulated output per CHAIR (q.v.) for an agreed rate of pay.

Moving-fire Kiln. See ANNULAR KILN.

Mud Hog. A US machine for the disintegration of dry or moist plastic clay. It consists of a rotating swing hammer operating close to a series of anvils linked together to form a steeply inclined slat conveyor.

Muffle. A refractory internal shell, forming a permanent part of some fuel-fired kilns. Ware placed in a muffle is protected against contact with the combustion gases. In small laboratory kilns the muffle is usually a single refractory shape; in industrial kilns the muffle is built up with refractory brickwork. In a SEMI-MUFFLE the refractory protection acts only as a shield against direct flame impingement, the combustion gases subsequently being free to pass through the setting.

Muller Mixer. A WET PAN (q.v.) in which the mullers are suspended out of contact with the pan bottom; by this means the material charged to the pan is mixed without further grinding taking place.

Mullet. A knife-shaped piece of iron used in severing a hand-made glass bottle from the blow-pipe.

Mullite. The only compound of alumina and silica that is stable at high temperatures. The composition is $3Al_2O_3 \cdot 2SiO_2$ but it can take a small amount of Al_2O_3 into solid solution. It melts congruently at 1850°C. Mullite is formed when any refractory clay is fired and is an important constituent of aluminosilicate

Multi-bucket Excavator

refractories, aluminous porcelains, etc. In USA, mullite refractories are defined (ASTM – C467) as containing 56–79% Al_2O_3.

Multi-bucket Excavator. A type of mechanical excavator sometimes used in the getting of brick-clays. Buckets are carried by endless chains round a steel boom (known as a 'ladder') and as they travel they scoop up clay. This type of excavator normally gets clay from below the level on which the truck of the machine is standing.

Multi-passage Kiln. A kiln consisting of a number of adjacent tunnels, each of small cross-section, through which the ware is pushed on refractory bats. These kilns may be electric or gas-fired. The typical electric multi-passage kiln has 24 passages arranged 4 across and 6 high; the ware is pushed in opposite directions in adjacent passages. Gas-fired multi-passage kilns usually have only 4 passages and the ware is pushed in the same direction in each passage.

Multiple-hearth Furnace. A furnace designed primarily for the roasting of ores in the non-ferrous metal industry, but which has more recently been adapted to the calcination of fireclay to produce CHAMOTTE (q.v.). The furnace is in the form of a large, squat, vertical cylinder and consists of several hearths around a central shaft on which rabble arms are mounted. The material to be calcined is charged to the top hearth and slowly descends, hearth by hearth, to the bottom, where it is discharged.

Munsell System. A method of colour notation based on evaluation of three parameters: lightness, hue and saturation. (A. H. Munsell, *A Colour Notation*, Munsell Color Co., Inc., Baltimore, USA.)

Murgatroyd Belt. That part of the side wall of a glass container in which stresses are concentrated following thermal shock; the belt normally extends from the base to about three-quarters of the height of the wall. (J. B. Murgatroyd, *J. Soc. Glass Tech.*, **27**, 77, 1943.)

Murray–Curvex Machine. A device for off-set printing from an engraved copper plate on to pottery flatware by means of a solid convex pad of gelatine. Cold-printing colours are used, the medium being formulated from synthetic resins and oils. Output is about 1200 pieces/day. (G. L. Murray, Brit. Pat., 735 637, 24/8/55; 736 312, 7/9/55.)

Muscovite. A mica approximating in composition to $(Na, K)_2O \cdot 3Al_2O_3 \cdot 6SiO_2 \cdot 2H_2O$; it is in this form that most of the alkali is

188

present in many clays. Thermal expansion (20–800°C) 9.6×10^{-6}. When heated, mica begins to lose water at about 250°C but, at normal rates of heating, dehydration is not complete until about 1100°C; the product is a glass containing mullite and some free Al_2O_3.

Musgrave–Harner Turbidimeter. An instrument developed in 1947 by J. R. Musgrave and H. R. Harner of the Eagle-Picher Co., Missouri, USA; it has been used for the particle-size analysis of clays and other ceramic raw materials in the range 0.25–60μ.

Mussel Gold. An old form of prepared gold for use in the decoration of pottery. It was made by rubbing together gold leaf, sugar (or honey) and salt; the paste was then washed free from soluble material and, traditionally, stored in mussel shells.

Mycalex. Trade-name: a particular type of MICA (GLASS-BONDED) (q.v.).

Mylius Test. See IODOEOSIN TEST.

N

n, n_D, etc. Symbol for refractive index; the subscript indicates the spectral line to which the quoted value refers, e.g. n_D = the refractive index for a wavelength equivalent to the sodium D line.

v-**value.** Alternative name for ABBE VALUE (q.v.).

Nacrite. $Al_2O_3 . 2SiO_2 . 2H_2O$. A rare, exceptionally well-crystallized, clay mineral of the kaolinite group.

Naples Yellow. See LEAD ANTIMONATE.

Natch. The key (a hemispherical hollow and complementary protrusion) in a two- or multi-part plaster mould for pottery making; the purpose is to ensure that the parts fit correctly. Occasionally known as a JOGGLE.

Natural Clay Tile. US term defined (ASTM – C242) as a tile made by either the dust-pressed method or the plastic method, from clays that produce a dense body having a distinctive, slightly textured appearance.

Navvy. Old name for a MECHANICAL SHOVEL (q.v.)—from its original use in canal excavation.

Neat Work. In a brick structure, the brickwork set out at the base of a wall.

Nebuchadnezzar. A 20-quart wine bottle (the largest size).

Neck

Neck. A stack of unfired bricks as set in a CLAMP (q.v.)

Neck Mould or Neck Ring. The mould that shapes the neck of a glass bottle. (Also known as a FINISH MOULD or RING MOULD.)

Needle or Plunger. A refractory part of a feeder in a glass-making machine; it forces glass through the orifice and then pulls it upwards after the shearing operation.

Neodymium Oxide. Nd_2O_3, a rare earth that imparts a violet colour to glass.

Nepheline Syenite. An igneous rock composed principally of nepheline ($K_2O.3Na_2O.4Al_2O_3.8SiO_2$) and feldspar. There are large deposits in Ontario (Canada), and Stjernoy (Norway). It is used as a flux in whiteware bodies, and as a constituent of some glasses and enamels.

Nernst Body. A sintered mixture of thoria, zirconia and yttria together with small amounts of other rare-earth oxides. After it has been preheated to about 2000°C this body becomes sufficiently electrically conducting for use as a resistor in high-temperature laboratory furnaces. (W. Nernst, Brit. Pat. 6135, 12/3/1898.)

Network-forming Ion. One of the ions in a glass that form the network in the glass structure as postulated by Zachariasen (see under GLASS). The ratio of the ionic radius of the network-forming ion to that of the oxygen ion must lie between 0·155 and 0·225 for triangular co-ordination, or between 0·225 and 0·414 for tetrahedral co-ordination; such ions include B^{3+}, Al^{3+}, Si^{4+} and P^{5+}. See also NETWORK-MODIFYING ION.

Network-modifying Ion. One of the ions in a glass which, according to Zachariasen's theory (see under GLASS), do not participate in the network. They must have a rather large radius and a low valency, e.g. the alkali metals and the alkaline earths.

Neutral Glass. A name sometimes applied to glass that is resistant to chemical attack (cf. NEUTRAL-TINTED GLASS).

Neutral Refractory. A refractory material such as chrome ore that is chemically neutral at high temperatures and so does not react with either silica or basic refractories.

Neutral-tinted Glass. A grey glass, usually of the borosilicate type; these glasses are used in light filters to reduce the trans-mission without (as far as possible) selective absorption of particular wavelengths (cf. NEUTRAL GLASS).

Neutralizer. A dilute solution of alkali or of sodium cyanide used in the treatment of the base-metal for vitreous enamelling after the PICKLING (q.v.) process. *Alkali Neutralizer* consists of a warm (65–70°C) solution of soda ash, borax or trisodium phosphate, the strength being equivalent to 0.3–0.4% Na_2O. *Cyanide Neutralizer* is a 0.10–0.15% solution of sodium cyanide; other ingredients may be present to neutralize the hardness of the water.

Neutron-absorbing Glass. For thermal neutron-absorption a glass must contain cadmium, which is the only common glass constituent having a high neutron-capture cross-section; it is used to form a cadmium borate glass, which is made more chemically durable by the addition of TiO_2 and ZrO_2.

New Mine Fireclay. A fireclay (there are five seams) occurring below the OLD MINE FIRECLAY (q.v.) in the Stourbridge district, England.

Newcastle Kiln. A type of intermittent kiln formerly popular in the Newcastle-on-Tyne area. In its original form it is a rectangular kiln with two or three fireboxes at one end and openings for the exhaustion of waste gases at the base of the other end-wall, which incorporates a chimney. In a later design the kiln is of double length, there are fireboxes at each end, and the waste gases are removed from the centre of the kiln. Such kilns found particular use in the firing of refractories and salt-glazed ware.

Nib. (1) A protrusion at one end of a roofing tile serving to hook the tile on the laths in the roof.

(2) A fault in flat glass, in the form of a protrusion at one corner, caused during cutting.

Nibbed Saggar. A SAGGAR (q.v.) with internal protrusions to support a bat (thus permitting the placing of two separate layers of ware) or to allow a cover to be placed inside the top of the saggar.

Nickel Dip. Also sometimes known as NICKEL FLASHING or NICKEL PICKLING. A process for the treatment (after PICKLING (q.v.)) of base-metal for vitreous enamelling to improve adherence and reduce the likelihood of COPPERHEADS (q.v.) and FISH-SCALING (q.v.). A solution of a nickel salt, e.g. the sulphate or double ammonium sulphate, is used; boric acid and ammonium carbonate are also added in small amounts to give the correct pH.

Nickel Oxide. There are two common oxides: grey, NiO; black, Ni_2O_3. They are used (about 0.5% of either type) in ground-coat enamels for sheet steel to promote adherence.

Niobium Borides. Several compounds have been reported,

Niobium Carbide

including the following. NbB_2; m.p. 3050°C; sp. gr. 7·0; thermal expansion, $5·9 \times 10^{-6}$ parallel to a and $8·4 \times 10^{-6}$ parallel to c. NbB; m.p., 2300°C; sp. gr., 7·6. Nb_3B_4; melts incongruently at 2700°C; sp. gr., 7·3.

Niobium Carbide. NbC; m.p. 3500 ± 125°C; sp. gr. (theoretical), 7·5 g/ml; modulus of rupture (25°C), 35 000 p.s.i.

Niobium Nitrides. Three nitrides have been reported: NbN, Nb_2N and Nb_4N_3. During reaction between Nb and N_2 at 800–1500°C the product generally consists of more than one compound. Most of the phases are stable at least to 1500°C.

Nip. A small glass bottle for mineral water.

Nitre. See POTASSIUM NITRATE.

Nitrides. A group of special ceramic materials: see under the nitrides of the following elements: Al, Be, Ga, Hf, Nb, Si, Ta, Th, Ti, U, Zr.

Nobel Elutriator. An early type of multiple-vessel elutriator for particle-size analysis; it is described in *Z. Anal. Chem.*, **3**, 85, 1864.

No-fines Concrete. Concrete made with aggregate from which all the fine material, i.e. less than $\frac{3}{8}$ in., has been removed; the proportion of cement to aggregate varies from 1:8 to 1:12 (vol.). The strength is low (1000–1200 p.s.i.) but so too are heat and moisture transfer.

Nodular Fireclay. An argillaceous rock containing aluminous and/or ferruginous nodules.

Nominal Dimension. As applied to hollow clay building blocks, this term is defined in the USA (ASTM – C43) as: A dimension that may be greater than the specified masonry dimension by the thickness of a mortar joint but not by more than $\frac{1}{2}$ in.

Nomogram or Nomograph. Two or more scales drawn as graphs and arranged so that a calculation involving quantities represented by the scales can be reduced to a reading from the diagram. Examples of the use of nomograms in ceramics include the determination of porosity, the solids concentration in slips, the drying and firing shrinkage of clay pipes.

Non-load Bearing Tile. US term for a hollow fired-clay building block for use in masonry carrying no superimposed load.

Non-metallic Inclusion. A non-metallic particle that has become embedded in steel during its processing. Tests, particularly those employing radioactive tracers, have shown that (contrary to earlier belief) these inclusions rarely originate from the refractory materials of the furnace, ladle or casting-pit.

Non-stoichiometric. A chemical compound is said to be non-stoichiometric if the ratio of its constituents differs from that demanded by the chemical formula. This may happen with oxides that are readily reducible, or with compounds containing an element of variable valency, or when interstitial atoms are present in the lattice. Non-stoichiometric ceramics are of interest as being semi-conducting.

Non-vitreous or Non-vitrified. These synonymous terms are defined in USA (ASTM – C242) in terms of water-absorption: the term non-vitreous generally implies that the ceramic material has a water absorption above 10%, except for floor tiles and wall tiles which are considered to be non-vitreous if the water absorption exceeds 7%.

Norman Slabs. A type of glass for stained windows; it is made by blowing bottles of square section and cutting slabs of glass from the four sides.

Nose or Mouth. The constricted circular opening at the top of the refractory lining of a Converter (q.v.); through this opening molten pig-iron is charged and steel is subsequently poured.

Nose-ring Block. A block of refractory material specially shaped for building into the discharge end of a rotary cement kiln; also known as a Discharge-end Block.

Nostril. The term used for the refractory gas- or air-port in a gas retort.

Nostril Blocks. See Rider Bricks.

Notch Test. A test proposed by R. Rose and R. S. Bradley (*J. Amer. Ceram. Soc.*, **32**, 360, 1949) for the assessment of the low-temperature spalling tendency of fireclay refractories. A notch is made in a transverse-strength test-piece and the effect of this on the strength at 800°C is determined; if the transverse strength is but little reduced by the notch, the low-temperature spalling resistance of the fireclay refractory will be good.

Nozzle. A cylindrical fireclay shape traversed by a central hole of uniform diameter; the top of the nozzle is contoured to form a seating for a Stopper (q.v.). Nozzles are fitted in the bottom of ladles used in the teeming of steel. In the USA, three qualities are specified in ASTM – C435.

Nuclear Reactor Ceramics. These special ceramics include: fuel elements—UO_2, UC and UC_2; control materials—rare-earth oxides and B_4C; moderators—BeO and C.

Nucleation; Nucleating Agent. Nucleation is the process of

Ochre

initiation of a phase change by the provision of points of thermo-dynamic non-equilibrium; the points are provided by 'seeding', i.e. the introduction of finely-dispersed nucleating agents. The process has become of great interest to the ceramist since DEVITRIFIED GLASS (q.v.) became a commercial product.

O

Ochre. Impure hydrated iron oxide used to a limited extent as a colouring material for coarse pottery and structural clayware.

Ocrate Process. The treatment of concrete with gaseous SiF_4 to transform any free CaO into CaF_2. The treated concrete has improved resistance to chemicals and to abrasion. (Ocrietfabriek NV, Baarn, Holland.)

Octahedrite. See ANATASE.

Oddments or Fittings. Special glazed clayware shapes used in conjunction with glazed pipes. These shapes include Bends, Junctions, Tapers, Channels, Street Gullies, Syphons, Interceptors and Yard Gullies; the first five of these are made partly by machine and partly by hand, the last three items have to be entirely hand-moulded.

Oden Balance. Apparatus for particle-size analysis; one of the pans of a delicate balance is immersed in the settling suspension and the change in weight as particles settle on the pan is measured. (S. Oden, *Internat. Mitt. Bodenk.*, **5**, 257, 1915.)

Off-hand. Hand-made glass-ware produced without the aid of a mould.

Off-hand Grinding. See FREEHAND GRINDING.

Offset Finish. See FINISH.

Offset Punt. Term applied to the bottom of a bottle if it is asymmetric to the axis.

OH Furnace. See OPEN-HEARTH FURNACE.

Oil Spot. A surface fault, seen as a mottled circle, on electric lamp bulbs or valves; it is caused by carbonization of a contaminating drop of oil.

Old Mine Fireclay. A fireclay occurring in the Brierley Hill district, near Stourbridge (England); it usually contains 56–64% SiO_2, 25–30% Al_2O_3 and $\not> 2.5\%$ Fe_2O_3. A smooth plastic fireclay formerly much used for making glass-pots, but now largely worked out.

Olivine. A mineral consisting principally of a solid solution of

FORSTERITE (q.v.) and FAYALITE (q.v.). Rocks rich in olivine find some use as a raw material for the manufacture of forsterite refractories and, when crushed, as a foundry sand.

Once-fired Ware (UK) or Single-fired Ware (USA). Ceramic whiteware to which a glaze is applied before the ware is fired, the biscuit firing and glost firing then being combined in a single operation. Because the glaze must mature at a relatively high temperature, it is usually of the leadless type. Sanitary-ware is the principal type of ware made in this way.

One-fire Finish. US term for vitreous enamelware produced by a single firing.

O'Neill Machine. A machine for making glass bottles originally designed by F. O'Neill in 1915; it was based on the use of blank-moulds and blow-moulds. The principle in modern O'Neill machines is the same; they are electrically operated and are capable of high outputs.

On-glaze Decoration. Decoration applied to pottery after it has been glazed; the ware is again fired and the colours fuse into the glaze, the decoration thus becoming durable. Because the decorating fire can be at a lower temperature with on-glaze decoration, a more varied palette of colours is available than with under-glaze decoration.

Ooms-Ittner Kiln. An annular, longitudinal-arch kiln divided into chambers by permanent walls and with a flue system designed for the salt-glazing of clay pipes; the fuel can be coal, gas or oil. (Ooms-Ittner Co., Braunsfeld, Cologne, Germany.)

Opacifier. A material added to a batch to produce opal glass, or to a glaze or enamel to render it opaque. Glass is usually opacified by the addition of a fluoride, e.g. fluorspar or cryolite; for glazes and enamels the oxides of tin, zirconium or titanium are commonly used.

Opacity. Defined in ASTM – C286 as: The property of reflecting light diffusely and non-selectively. For a definition relating specifically to vitreous enamel see CONTRAST RATIO.

Opal. Non-crystalline hydrated silica.

Opal Glass. A translucent to opaque glass that is generally fiery, as distinct from ALABASTER GLASS (q.v.).

Opaque Ceramic Glazed Tile. US term (ASTM – C43) for a hollow clay facing block the surface faces of which are covered by an opaque ceramic glaze which is coloured and which has a bright satin or gloss finish.

Open Handle. A cup handle of the type that is attached to the cup at the top and bottom only, the side of the cup itself forming part of the finger-opening (cf. BLOCK HANDLE).

Open-hearth Furnace. A large rectangular furnace in which steel, covered with a layer of slag, is refined on a refractory hearth; it is heated by gas or oil, and operates on the regenerative principle which was first applied to this type of furnace by Sir Wm. Siemens in 1867 (hence the earlier name SIEMENS FURNACE). The type of refractory lining depends on the particular steelmaking process used: see ACID OPEN-HEARTH FURNACE, BASIC OPEN-HEARTH FURNACE and ALL-BASIC FURNACE.

Opening Material. Term occasionally used for a non-plastic that is added to a clay to decrease the shrinkage and increase the porosity; such materials include grog, chamotte, sand and pitchers.

Optical Blank. A piece of optical glass that has been pressed to approximate to the shape finally required; also called a PRESSING.

Optical Crown Glass. Any glass of low dispersion used for optical equipment (cf. FLINT GLASS). There are many varieties, their names indicating their characteristic composition, e.g. barium crown, borosilicate crown, fluor-crown, phosphate crown, zinc crown.

Optical Flint Glass. See under FLINT GLASS.

Optical Glass Classification. A system by which an optical glass is classified according to its refractive index, n_D, and its Abbe Value, v. Standard borosilicate crown glass, for example, has $n_D = 1 \cdot 510$ and $v = 64 \cdot 4$; its classification by this system is 510644. Further identification is often provided by letters, preceding the number; e.g. BSC = Boro-Silicate Crown; LF = Light Flint, etc.

Optical Pyrometer. A device for the measurement of high temperatures depending ultimately on visual matching. The type most used in the ceramic industry is the DISAPPEARING FILAMENT PYROMETER (q.v.); a less accurate type is the WEDGE PYROMETER (q.v.).

Orange Peel. A surface blemish, adequately described by its name, sometimes occurring on vitreous enamelware, glass-ware and glazed ceramics. In vitreous enamelling, the fault can usually be prevented by an alteration in the spraying process and/or in the specific gravity of the enamel slip.

196

Organic Bond. (1) A material such as gum or starch paste that can be incorporated in a ceramic batch to give it dry-strength; the organic bond burns away during the firing process, which develops a ceramic bond or causes sintering.

(2) Some abrasive wheels have an organic bond such as synthetic resin, rubber or shellac, and are not fired.

Orifice Ring. See under BUSHING.

Orthoclase. See FELDSPAR.

Orton Cones. PYROMETRIC CONES (q.v.) made by the Edward Orton Jr. Ceramic Foundation, Columbus, Ohio. They are made in two sizes: $2\frac{1}{2}$ in. high for industrial kiln control, and $1\frac{1}{8}$ in. high for P.C.E. testing. For nominal equivalent softening temperatures see Appendix 2.

Osborn-Shaw Process. See SHAW PROCESS.

Osmosis. See ELECTRO-OSMOSIS.

Oven. Obsolete term for a kiln, more particularly the old type of BOTTLE-OVEN (q.v.).

Overburden. The soil and other unwanted material that has to be removed before the underlying clay or rock can be worked by the open-cast method.

Overflush. A fault in glass-ware caused by the flow of too much glass along the line of a joint (cf. FIN).

Over-glaze Decoration. See ON-GLAZE DECORATION.

Over-glazed. Pottery-ware having too thick a glaze layer, particularly on the bottom; this thick glaze is likely to be crazed. Causes of this fault are incorrect dipping, the use of slop glaze of too high a density, or biscuit ware that is too porous, i.e. under-fired.

Over-pickling. PICKLING (q.v.) for too long a period or in too strong a solution; this causes blisters in vitreous enamelware.

Overpress. A fault, in glass-ware, in the form of an inside FIN (q.v.).

Overspray. (1) In the application, by spraying, of enamel slip to base-metal or of slop glaze to ceramic whiteware, that proportion of slip or glaze that is not deposited on the ware; it is normally collected for re-use.

(2) In vitreous enamelling, the application of a second (usually thinner) coat of enamel slip over a previous layer of enamel that has not yet been fired.

Owens Machine. A suction-type machine for making glass bottles; the original machine was designed by M. J. Owens between 1897 and 1904. Since then, there have been many improvements but the basic principle remains the same.

Ox Gall. This material (an indefinite mixture of fats, glycocholates and taurocholates) has been added to glaze suspensions to prevent crawling.

Oxford Clay. A clay of the Upper Jurassic system providing raw material for 30% of the building bricks made in the UK; the Fletton brick industry of the Peterborough area is based on this clay, which contains so much carbonaceous material that the dry-pressed bricks can be fired with very little additional fuel.

Oxidation Period. The stage in the firing of clayware during which any carbonaceous matter is burned out, i.e. the temperature range 400–850°C. It is important that all the carbon is removed before the next stage of the firing process (the vitrification period) begins, otherwise a black core may result.

Oxide Ceramics. Special ceramics made from substantially pure oxides, usually by dry-pressing or slip-casting followed by sintering at high temperature. The most common oxide ceramics are Al_2O_3, BeO, MgO, ThO_2, and ZrO_2.

P

Packing Density. The BULK DENSITY (q.v.) of a granular material, e.g. grog or crushed quartzite, when packed under specified conditions. A common method of test, particularly for foundry sands, involves the use of an AFA RAMMER (q.v.).

Pad. The refractory brickwork below the molten iron at the base of a blast furnace.

Paddling. The process of preliminary shaping of a piece of glass, while it is in a furnace and soft, prior to the pressing of blanks of optical glass.

Padmos Method. A comparative method for the determination of the coefficient of thermal expansion of a glass; the glass being tested is fused to a glass of known expansion and similar transformation temperature. From the birefringence resulting from the consequent stress at the junction of the two glasses, the difference in expansion can be calculated. (A. A. Padmos, *Philips Res. Repts.*, **1**, 321, 1946.)

Pale Glass. Glass of a pale green colour.

Pall Ring. A type of ceramic filling for towers in the chemical industry. The 'ring' is a small hollow cylinder with slots in the sides and projections in the core. The design was introduced by Badische Anilin und Soda Fabrik—see *Industr. Chemist*, **35**, 36, 1959.

Pallet. A board, small platform or packaging unit sometimes used, for example, in the transport of refractories or building bricks; cf. STILLAGE.

Pallette. See BATTLEDORE.

Palygorskite. A fibrous, hydrated magnesium aluminium silicate; structurally, this mineral is the same as ATTAPULGITE (q.v.). There are no known deposits of economic size.

Pan Mill. Term sometimes applied to an EDGE-RUNNER MILL (q.v.) but more properly reserved for the old type of mill (also known as a BLOCK MILL) for grinding flint for the pottery industry. It consists of a circular metal pan paved with chert stones over which are moved heavy blocks of chert (RUNNERS) chained to paddle arms.

Pandermite. A hydrated calcium borate that has been used as a component of glazes.

Panel Brick. A special, long, silica refractory shape laid stretcher-fashion in the wall of a coke-oven.

Panel Spalling Test. A test for the spalling resistance of refractories that was first standardized in 1936, in USA. A panel of the bricks to be tested is subjected to a sequence of heating and cooling cycles and the loss in weight due to the spalling away of fragments is reported. General details of the apparatus and test are given in ASTM – C38. The procedures for individual types of refractory are: High-duty Fireclay Refractory, ASTM – C107; Super-duty Fireclay Refractory, ASTM – C122; Plastic Refractory, ASTM – C180.

Pannetier's Reds. Brilliant on-glaze ceramic colours, based on iron oxide, developed in the early 19th century by Pannetier in Paris.

Pantile. A single-lap roofing tile having a flat S-shape in horizontal section. Details of clay pantiles are specified in B.S.1424; the minimum pitch recommended is 35°.

Parallel-plate Plastometer. See WILLIAMS' PLASTOMETER.

Parian. A white, vitreous, type of pottery made from china clay and feldspar (in the proportion of about 1:2) generally in the form of figures—hence the name, from Paros, the Aegian island

199

from which white marble was quarried for sculpture in Classical Greece. Parian was created by the firm of Spode, England; it was subsequently made by several other firms and was popular between about 1840 and 1860. (PARIAN CEMENT or PLASTER is an alternative name for KEENE'S CEMENT (q.v.).)

Paris White. See WHITING.

Parison. A piece of glass that has been given an approximate shape in a preliminary forming process ready for its final shaping. (From the French *paraison*, derived from *parer*—to prepare.)

Parker Pre-namel. A process for the treatment of steel prior to enamelling (US Pat., 2 809 907, 15/10/57).

Particle Size or (preferably) Particle Mean Size. A concept used in the study of powders and defined as: 'The dimensions of a hypothetical particle such that, if a material were wholly composed of such particles, it would have the same value as the actual material in respect of some stated property' (cf. EQUIVALENT PARTICLE DIAMETER).

Particle-size Analysis. The process of determining the proportions of particles of defined size fractions in a granular or powdered sample; the term also refers to the result of the analysis. The methods of determination available include: ADSORPTION, in which the particle size is assessed on the basis of surface area; the DIVER METHOD; the use of a CENTRIFUGE; ELUTRIATION; TURBIDIMETER; the ANDREASEN PIPETTE; and, for coarser particles, a SIEVE. (For further details see under each heading.)

Parting Line. See JOINT LINE.

Parting Wheel. See CUTTING-OFF WHEEL.

Partition Tile. US term for a hollow fired-clay building block for use in the construction of interior partitions but not carrying any superimposed load.

Paste. A literal translation of the French word *Pâte* in the sense of ceramic body; see HARD PASTE and SOFT PASTE.

Paste Mould. A metal mould lined with carbon and used wet in the blowing of glass-ware.

Pat Test. A qualitative method for assessing the soundness of hydraulic cement. Pats of cement are made about 3 in. dia., $\frac{1}{2}$ in. thick at the centre but with a thin circumference. They are immersed in cold water for 28 days or in boiling water or steam for 3–5 hours. Unsoundness is revealed by distortion or cracking.

Pâte-sur-Pâte. A method of pottery decoration in which a bas-relief is built up by hand-painting with clay slip as the paint.

The method is more particularly suited to the decoration of vases in the classical style; fine examples were made by M. L. Solon at Mintons Ltd., England, in the late 19th century.

Paul Floc Test or Paul Water Test. See FLOC TEST.

Paver. US term for a dust-pressed, unglazed, relatively thick, floor tile having a superficial area of at least 6 in².

Paving Brick. A dense, well-vitrified clay brick for use as a paving material. Properties are specified in ASTM – C7.

Paviour. A term applied to clamp-fired STOCK BRICKS (q.v.) that are not of first quality but are nevertheless hard, well-shaped and of good colour.

PCE. Abbreviation for PYROMETRIC CONE EQUIVALENT (q.v.).

Peach Bloom. A glaze effect on pottery produced by the Chinese; it is achieved by the addition of copper oxide to a high-alkali glaze but requires very careful control of the kiln atmosphere. The 'bloom' results from incipient devitrification of the glaze surface.

Peacock Blue. A ceramic colour made from a batch such as: 33% Cobalt oxide, 7% STANDARD BLACK (q.v.), 45% China Stone, 15% Flint.

Pearl-ash. Potassium carbonate, K_2CO_3; sometimes used in glass and glaze batches.

Pearson Air Elutriator. A down-blast type of ELUTRIATOR (q.v.) designed by J. C. Pearson (*US Bur. Stand. Tech. Paper No.* 48, 1915) and used for determining the fineness of portland cement.

Pebble Heater. A heat exchanger in which refractory 'pebbles' (which may be made of mullite, alumina, zircon or zirconia) are used as heat carriers. One type of pebble heater consists of two refractory-lined chambers joined vertically by a throat; both chambers are filled with 'pebbles', which descend at a steady rate, being discharged from the bottom of the lower chamber and returned to the top of the upper chamber. In the latter they are heated by a countercurrent of hot gases; in the lower chamber they give up this heat to a second stream of gas or air.

Pebble Mill. A BALL-MILL (q.v.) in which flint pebbles are used as grinding media.

Pedersen Process. A process devised in 1944 by H. Pedersen, a Norwegian, for the extraction of alumina from siliceous bauxite; the bauxite is first melted in an electric furnace with limestone and coke, the reaction product then being leached with NaOH.

Peeling or Shivering. (1) The breaking away of glaze from

Pegmatite

ceramic ware in consequence of too high a compression in the glaze layer; this is caused by the glaze being of such a composition that its expansion coefficient is too low to match that of the body (a certain degree of compression in the glaze is desirable, however).

(2) A similar effect sometimes occurs on the slagged face of a refractory.

(3) A fault in vitreous enamelling that is also known as LIFT (q.v.).

Pegmatite. A coarsely crystalline rock occurring as veins, which may be several hundred feet thick. Feldspars are a common constituent of pegmatites, which are of ceramic interest for this reason.

PEI. Porcelain Enamel Institute, Washington D.C., USA.

Pencil Edging. The process of rounding the edges of flat glass.

Pencil Ganister. A GANISTER (q.v.) with black markings, as though pencilled, formed by decomposition of the rootlets that were present in the original seat-earth from which the ganister was formed.

Pencil Stone. Term sometimes applied to PYROPHYLLITE (q.v.).

Pennvernon Process. See PITTSBURGH PROCESS.

Pentacalcium Trialuminate. A compound formerly believed to have the composition $5CaO . 3Al_2O_3$ and to occur in high-alumina hydraulic cement; it is now known that the compound in question is DODECACALCIUM HEPTALUMINATE ($12CaO . 7Al_2O_3$) (q.v.).

Peptize. Term used in colloid chemistry for the process that, in ceramic technology, is known as DEFLOCCULATION (q.v.).

Perforated Brick. A building brick made lighter in weight by its being pierced with numerous, relatively small ($\frac{1}{4}$–$\frac{1}{2}$ in. dia.) holes, usually in the direction of one of the two short axes. In the UK, the total volume of the perforations does not usually exceed 15% of the volume of the brick; on the Continent of Europe the proportion is higher and in USA it is normally over 25% (cf. HOLLOW CLAY BLOCKS).

Periclase. Crystalline magnesium oxide, MgO; m.p. 2800°C; sp. gr. 3·58; thermal expansion (20–1000°C), $13·8 \times 10^{-6}$. Periclase, generally with a little FeO in solid solution, is the main constituent of magnesite refractories. SEA-WATER MAGNESIA (q.v.) can be produced as a nearly-pure periclase. It has been shown that large single crystals of periclase have measurable ductility; this research aims at an understanding of the factors

202

determining ductility and how this property might be achieved in ceramic products.

Periodic Kiln. See INTERMITTENT KILN.

Peripheral Speed. The rate of movement of a point on the edge of a rotating disk or cylinder; it is the product of the circumference and the rate of revolution. The peripheral speed is of importance, for example, in the operation of abrasive wheels and ball mills.

Perish. To disintegrate as a result of slow hydration on exposure to moist air; calcined dolomite disintegrates in this manner if stored for more than a short period.

Perkiewicz Method for Preventing Kiln Scum. A process in which clay bricks, prior to their being set in the kiln, are coated with a combustible, e.g. tar or a mixture of gelatine and flour; should sulphur compounds condense on the bricks during the early stages of firing, the deposit will fall away when the combustible coating subsequently burns off. (M. Perkiewicz, Brit. Pat. 3760, 15/2/04).

Perlite. A rock of the rhyolite type that is of interest on account of its intumescence when suddenly heated to a temperature of about 1000°C. Perlite occurs in economic quantity in USA, Hungary, Turkey and New Zealand. The expanded product has a bulk density of 7–12 lb/ft^3 and is used as a heat-insulating material.

Permanent Linear Change. A preferred term including the two terms AFTER-CONTRACTION (q.v.) and AFTER-EXPANSION (q.v.).

Permeability. The rate of flow of a fluid (usually air) through a porous ceramic material per unit area and unit pressure gradient. This property gives some idea of the size of the pores in a body, whereas the measurement of POROSITY (q.v.) evaluates only the total pore volume. From the permeability of a compacted powder the SPECIFIC SURFACE (q.v.) of the powder can be deduced (see CARMAN EQUATION; LEA AND NURSE PERMEABILITY APPARATUS; RIGDEN'S APPARATUS).

Perovskite. CaTiO$_3$; m.p. 1915°C; sp. gr. 4·10. This mineral gives its name to a group of compounds of similar structure, these forming the basis of the titanate, stannate and zirconate dielectrics. Extensive solid solution is possible so that the potential range of compositions and ferroelectric properties is very great. R. S. Roth (*J. Res. Nat. Bur. Stand.*, **58**, 75, 1957) has classified the perovskites on the basis of ionic radii of the constituent ions; a graph of this type can be divided into orthorhombic, pseudocubic and cubic fields with an area of ferroelectric and antiferroelectric compounds superimposed on the cubic field. In a development of this classification a three dimensional graph includes the polarizability of the ions as the third dimension.

Perpend. An alignment of cross-joints in brickwork.

Perrit. A support, made of heat-resisting alloy, designed to carry vitreous enamelware through an enamelling furnace.

Perthite. A mineral consisting of a lamellar intergrowth of feldspars, especially of albite and orthoclase.

Petalite. A lithium mineral, $Li_2O.Al_2O_3.8SiO_2$; sp. gr. 2·45; m.p. 1350°C. The chief sources are S. Rhodesia and S.W. Africa. It finds some use in glasses, ceramic bodies and glazes.

Petersen Air Elutriator. An up-blast ELUTRIATOR (q.v.) that has found some use in determining the fineness of portland cement.

Petuntse, Petunze. The Chinese name of CHINA STONE (q.v.).

PFA. Abbreviation for PULVERIZED FUEL ASH (q.v.).

pH. Symbol for the acidity or alkalinity of a solution; it is a number equivalent to the logarithm, to the base 10, of the reciprocal of the concentration of hydrogen ions in an aqueous solution. The point of neutrality is pH7; a solution with a pH below 7 is acid, above 7 is alkaline. The pH of a casting slip is of practical importance in determining its rheological properties.

Phase Diagram. See EQUILIBRIUM DIAGRAM.

PHI Scale. A particle-size series introduced by W. C. Krumbein (*J. Sed. Petrology*, **8**, 84, 1938). The particle-sizes are expressed logarithmically in phi (φ) units, where $\varphi = -log_2 d$, d being the diameter of the particles in mm.

Phosphides. As yet, research on the phosphide ceramics has been very limited; they include AlP, BP, Be_3P_2, InP and TiP. Some of these compounds have potential use as semiconductors.

Phosphor. A fluorescent material of the type used, for example, in the coating of the screen of a cathode-ray tube. Ceramic phosphors have been made from Ca silicate, stannate or phosphate activated with Sn, Mn or Bi; some of the Ca can be replaced by Sr, Zn or Cd.

Phosphoric Acid. A syrupy liquid of variable composition sometimes used as a binder for refractories (see ALUMINIUM PHOSPHATE).

Photoelectric Pyrometer. A device for measuring high temperatures on the basis of the electric current generated by a photoelectric cell when it receives focused radiation from a furnace or other hot body. Instruments have been designed for the measurement of temperatures exceeding 2000°C, but others are available for use at lower temperatures, e.g. for checking the temperature of the molten glass used in lamp-bulb manufacture.

Photosensitive Glass. A glass containing small amounts ($< 0\cdot1\%$) of gold, silver or copper, which render the glass sensitive to light in a manner similar to photographic films; the image is developed by heat treatment.

Piano-wire Screen. A screen formed by piano wires stretched tightly, lengthwise, on a frame 2–3 ft wide and 4–8 ft high. The screen is set up at an angle of about 45° and crushed material is fed to it from above. The mesh size varies from about 4 to 16. Because there are no cross-wires, and because the taut wires can vibrate, there is less tendency for 'blinding', but some elongated particles inevitably pass the screen.

Pick-up. The amount of vitreous enamel slip (expressed in terms of dry weight per unit area) after the dipping and draining process. For ground-coats, the pick-up is usually $1\cdot2$–$1\cdot5$ oz/ft^2.

Pickings. Term applied to clamp-fired STOCK BRICKS (q.v.) that are soft though of good shape.

Pickle Basket. A container for vitreous enamelware during PICKLING (q.v.); the basket is made from a corrosion-resistant metal, e.g. Monel.

Pickle Pills. Capsules, containing Na_2CO_3 and an indicator such as methyl orange, for use in works' tests of the strength of the PICKLING (q.v.) bath employed in vitreous enamelling.

Pickling. The chemical treatment of the base metal used in vitreous enamelling to ensure a satisfactory bond between the enamel and metal. The various methods may be classified as follows: (1) the use of reducing acids, e.g. H_2SO_4, HCl and acid salts; (2) the sulphur compounds, H_2S and $Na_2S_2O_3$; (3) metal coatings, e.g. Ni, As and Sb; (4) oxidizing acids and salts, e.g. H_3PO_4, HNO_3, solutions of ferric salts in acids, or $NaNO_3 + H_2SO_4$.

Picotite. A complex spinel of general formula (Fe,Mg)O. $(Al,Cr)_2O_3$; the proportion of Al exceeds that of Cr and the Fe:Mg ratio is from 3 to 1. Picotite may occur in slagged basic refractories.

Picrochromite. Magnesium chromite, $MgO.Cr_2O_3$; m.p. 2250°C; sp. gr. 4·41. This spinel can be synthesized by heating a mixture of the two oxides at 1600°C; it is formed (usually with other

Piezoelectric

spinels in solid solution) in fired chrome-magnesite refractories. Picrochromite is highly refractory but when heated at 2000°C the Cr_2O_3 slowly volatilizes.

Piezoelectric. A material is stated to be 'piezoelectric' when an applied stress results in the setting up of an electric charge on its surface; conversely, if the material is subjected to an electric field, it will expand in one direction and contract in another direction. Typical piezoelectric materials are the titanate and zirconate ceramics.

Pigskin. A defect in the surface texture of vitreous enamelware and of glazed sewer-pipes; it is caused by variations in the thickness of the enamel or glaze.

Pilkington Twin Process. A continuous process for the production of polished plate glass; a continuous ribbon of glass is rolled, annealed and then simultaneously ground on both faces. The process was introduced by Pilkington Bros. Ltd., England, in 1952. (For a more recent process introduced by this firm see FLOAT GLASS PROCESS.)

Pillar. (1) A column of brickwork; for example the refractory brickwork between the doors of an open-hearth steel furnace.

(2) An item of KILN FURNITURE (q.v.) forming one of the upright parts of a CRANK (q.v.); cf. POST.

Pin. An item of KILN FURNITURE (q.v.); it is a small refractory SADDLE (q.v.) for use in conjunction with a CRANK (q.v.). See Fig. 4, p. 158.

Pin Disk Mill. A type of rotary disintegrator sometimes used (more particularly on the Continent and in USA) for the size-reduction of clay.

Pin-hole. (1) A fault in vitreous enamelware. It is the result of a blister that has burst and partially healed; the usual sources of the gas that gave rise to the blister are a hole in the base-metal or a speck of combustible foreign matter in the cover-coat.

(2) Pin-holes in glazes also result from burst bubbles; here, most of the gas has its origin in the air between the particles of powdered glaze becoming trapped as the glaze begins to mature.

(3) A frequent source of pin-holes in pottery biscuit-ware, and in subsequent stages of processing, is air occluded in the clay during its preparation.

(4) Pin-holes in plaster moulds originate in air attached to the particles of plaster during blending; this can be eliminated by blending the plaster in a vacuum.

Pin Mark or Point Mark. A fault in vitreous enamelware caused by the imprint of the supports used during the firing process.

Pinch Effect. Term applied by J. W. Mellor (*Trans. Brit. Ceram. Soc.*, **31**, 129, 1932) to the crazing of wall tiles as a result of contraction of the cement, used to fix the tiles on a wall, while it is setting.

Pinite. A rock consisting largely of sericite. The composition is (per cent): SiO_2, 46–50; Al_2O_3, 35–37; K_2O, 8–10; loss on ignition, 4–6. A deposit in Nevada, USA, was worked from 1933–1948 as a raw material for making dense refractories for rotary frit kilns and for the cooler parts of cement kilns where high abrasion-resistance is required.

Pinning. The process of fixing PINS (q.v.) in kiln furniture.

Pip. An item of KILN FURNITURE (q.v.). A pip is a small refractory button with a point on its top surface; ware is supported on the point.

Pipe. (1) See FIELD-DRAIN PIPE; SEWER PIPE.
(2) A cavity at the top of a fusion-cast refractory resulting from the contraction of the molten material as it cools.

Pipe Blister. A large bubble sometimes produced on the inside of hand-made glass-ware by impurities or scale on the blow-pipe.

Pipeclay. A white-firing, siliceous, clay of a type originally used for making tobacco pipes.

Pipette Method. A method for the determination of particle size: see ANDREASEN PIPETTE.

Piston Extruder. See STUPID.

Pit. A fault in vitreous enamelware similar to, but smaller than, a DIMPLE (q.v.).

Pitch. The slope of a tiled roof. For clay roofing tiles the common pitch, measured internally at the ridge between the two sides of the roof, is 105°. Less common are the SQUARE PITCH (90°), GOTHIC PITCH (75°) and SHARP PITCH (60°).

Pitch Polishing. The polishing of glass with a polishing agent supported on pitch instead of the more usual felt.

Pitchers. Pottery that has been broken in the course of manufacture. Biscuit pitchers are crushed, ground and re-used, either in the same factory or elsewhere; the crushed material is also used in other industries as an inert filler. Because of the adhering glaze, glost pitchers find less use.

Pittsburgh Process. A process for the vertical drawing of sheet glass invented by the Pittsburgh Plate Glass Co. in 1921. The sheet is drawn from the free surface of the molten glass, the drawing slot being completely submerged; the edges of the sheet are formed by rollers. (This process has also been referred to as the PENNVERNON PROCESS.)

Place Brick. A clamp-fired STOCK BRICK (q.v.) of very poor quality—of use only for temporary erections.

Placer. A man who sets pottery-ware in SAGGARS (q.v.) or with KILN FURNITURE (q.v.) ready for firing in a kiln. The man in charge of a team of placers is known in the N. Staffordshire potteries as a COD PLACER.

Placing Sand. Fine, clean silica sand used in the placing of earthenware in SAGGARS (q.v.); calcined alumina is used for this purpose in the bedding of bone china. With the increased use of open setting in tunnel kilns and top-hat kilns, placing sand is less used than formerly.

Plagioclase. See FELDSPAR.

Plain Tile. See under ROOFING TILE; WALL TILE (Fig. 6, p. 307).

Plaining (of Glass). See REFINING.

Plasma Spraying. The process of coating a surface (of metal or of a refractory) by spraying it with particles of oxides, carbides, silicides or nitrides that have been made molten by passage through the constricted electric arc of a PLASMA GUN; the temperature of the plasma arc 'flame' can be as high as 30 000°C. The usual purpose of a refractory coating applied in this way is to protect a material, e.g. Mo or C, from oxidation when used at high temperature.

Plaster of Paris. Calcium sulphate hemihydrate, $CaSO_4 \cdot \frac{1}{2}H_2O$; prepared by heating GYPSUM (q.v.) at 150–160°C. There are two forms: α, produced by dehydrating gypsum in water or saturated steam; β, produced in an unsaturated atmosphere. Plaster usually contains both forms. It is used for making MOULDS (q.v.) in the pottery industry.

Plaster-base Finish Tile. US term (ASTM – C43) for a hollow clay building block the surfaces of which are intended for the direct application of plaster; the surface may be smooth, scored, combed or roughened.

Plastic Making. This term includes all processes of shaping clay in the plastic condition, i.e. pressing, extrusion, jiggering, jolleying and throwing.

Plastic Pressing. The shaping of ceramic ware in a die from a plastic body by direct pressure.

208

Plastic Refractory. See FIRECLAY PLASTIC REFRACTORY; MOULDABLE REFRACTORY.

Plastic Shaping. See PLASTIC MAKING.

Plasticity. The characteristic property of moist clay that permits it to be deformed without cracking and to retain its new shape when the deforming stress is removed. Plasticity is associated with the sheet structure of the clay minerals and with the manner in which water films are held by the clay particles.

Plastometer. An instrument for the evaluation of plasticity; see, for example, BINGHAM PLASTOMETER; LINSEIS PLASTOMETER; WILLIAMS' PLASTOMETER.

Plat. Cornish term for the overburden of a china-clay mine.

Plate Glass. Flat, transparent, glass both surfaces of which have been ground and polished so that they give undistorted vision. Plate glass is used, for example, in shop windows.

Platting. See under SCOVE.

Plauson Mill. The original 'colloid mill' designed by H. Plauson, in Germany, and covered by a series of patents beginning with Brit. Pat. 155 836, 24/12/20. The novel feature was the high speed of operation (up to 100 ft/s) and the concentration of the grinding pressure at a small number of points.

Plique-à-Jour. One type of vitreous-enamel artware: a pattern is first made of metal strips and coloured enamels are fused into the partitions thus formed; after the object has been polished on both sides, it can be held to the light to give a stained-glass window effect. (French words: 'open-work plait'.)

Pluck, Plucking. (1) A surface blemish sometimes found on pottery where the glaze has been removed by the pointed supports for the ware used in the firing process.

(2) A surface fault on rolled glass arising if the glass sticks to the rollers.

Plug. Part of a glass-blowing machine for hollow-ware; it moves in a blank-mould with a reciprocating action, forming a cavity for blowing; this part is sometimes referred to as a PLUNGER. (See also MOULD PLUG.)

Plugging Compound. See FILLER.

Plumbago. A refractory material composed of a mixture of fireclay and graphite; some silicon carbide may also be included. It finds considerable use as a crucible material for foundries. In USA refractories of this type are termed CARBON-CERAMIC REFRACTORIES.

209

Plunger

Plunger. See NEEDLE; PLUG.

Ply Glass. Glass-ware, particularly for lamp shades and globes, made by covering opal glass (usually on both sides) with transparent glass of matched thermal expansion (cf. CASED GLASS).

Pneumatolysis. The breakdown of the minerals in a rock by the action of the volatile constituents of a magma, the main body of the magma having already become solid. The china clays of Cornwall, England, were formed by the pneumatolysis of the feldspar present in the parent granite.

Pneumoconiosis. Disability caused by the inhalation, over a long period, of various dusts; the form of the disease encountered most commonly in the ceramic industry is SILICOSIS (q.v.).

Pocket Clay. A highly siliceous clay sometimes found in large pockets in Carboniferous Limestone. In England it is found, typically, in Derbyshire and is worked as a refractory raw material at Friden, about 10 miles S.W. of Bakewell; a quoted composition is (per cent): SiO_2, 78; Al_2O_3, 13·5; Fe_2O_3, 2·0; TiO_2, 0·7; CaO, 0·2; MgO, 0·3; K_2O, 1·6; Na_2O, 0·1; loss on ignition, 3·9.

Pocket Setting. See BOXING-IN.

Podmore Factor. A factor proposed by H. L. Podmore (*Pottery Gazette*, **73**, 130, 1948) to indicate the intrinsic solubility of a frit:

Podmore factor = (solubility $\times 100$) $\div$ (specific surface). The solubility of any given frit is approximately proportional to the surface area of frit exposed to the solvent, i.e. to the fineness of grinding; the Podmore factor is independent of this fineness.

Poge. A tool formerly used for lifting the 'balls' of ball-clay.

Point. See MOUNTED POINT.

Point Bar. One type of support for vitreous enamelware used during the firing process.

Point Mark. See PIN MARK.

Poling. A method of removing bubbles from molten glass in a pot; the glass is stirred with a wooden pole, gases sweep through the molten glass and remove the smaller bubbles, which would otherwise rise to the surface only sluggishly. If a block of wood is used, the process is called BLOCKING.

Polished Section. A section of material that has been ground and plane-polished on one face for examination, under a microscope, by reflected light (cf. THIN SECTION).

Polishing. (1) A finishing process for plate glass and optical glass. Plate glass is polished by a series of rotating felt disks with rouge (very fine hydrated iron oxide) as the polishing medium.

210

For the polishing of optical glass cerium oxide, rouge or (less commonly) zircon are used.

(2) The grinding away of small surface blemishes from the face of glazed pottery-ware.

Polyphant Stone. An impure SOAPSTONE (q.v.) from the village of Polyphant, near Launceston, Cornwall, England. Composition (per cent): SiO_2, 28–36; Al_2O_3, 6–9; $FeO + Fe_2O_3$, 10–12; CaO, 4–5, MgO, 23–27; loss on ignition, 10–12. The m.p. is 1300–1400°C. Blocks of this stone are used as a refractory material in alkali furnaces.

Pontil. See PUNTY.

Pooles Tile. An interlocking clay roofing tile of a design specified in B.S. 1424.

Popper. A fault sometimes occurring during vitreous enamelling on sheet steel, small disks of ground-coat becoming detached and rising into the first cover-coat.

Popping. Term sometimes used for (1) LIME BLOWING (q.v.); (2) the heat treatment of PERLITE (q.v.) to cause expansion.

Porcelain. One type of vitreous ceramic whiteware. In the UK the term is defined on the basis of composition: a vitreous whiteware made from a feldspathic body (typified by the porcelain table-ware made in Western Europe and containing 50% kaolin, 25% quartz and 25% feldspar). In the USA the term is defined on the basis of use: a glazed or unglazed vitreous ceramic whiteware used for technical purposes, e.g. electrical porcelain, chemical porcelain, etc. Note, however, that the term electrical porcelain is also used in the UK. The firing of porcelain differs from that of earthenware in that the first firing is at a low temperature (900–1000°C), the body and the feldspathic glaze being subsequently matured together in a second firing at about 1350–1400°C.

Porcelain Enamel. The term used in USA for VITREOUS ENAMEL (q.v.) and defined in ASTM – C286 as: A substantially vitreous or glassy inorganic coating bonded to metal by fusion at a temperature above 800°F (427°C).

Porcelain Tile. US term defined (ASTM – C242) as a ceramic mosaic tile or PAVER (q.v.) that is generally made by dust-pressing and of a composition yielding a tile that is dense, fine-grained, and smooth, with sharply-formed face, usually impervious. The colours of such tiles are generally clear and bright.

Pore-size Distribution. The range of sizes of pores in a ceramic product and the relative abundance of these sizes. This property

Pork-pie Furnace

is usually expressed as a graph relating the percentage volume porosity to the pore diameter in microns. Pore-size distribution is difficult to determine, the most generally satisfactory methods are the MERCURY PENETRATION METHOD (q.v.) and the BUBBLE-PRESSURE METHOD (q.v.).

Pork-pie Furnace. See MAERZ-BOELENS FURNACE.

Porosimeter. A term variously applied to apparatus for the measurement of POROSITY (q.v.) or of PORE-SIZE DISTRIBUTION (q.v.).

Porosity. A measure of the proportion of pores in a ceramic material, defined as:

Apparent Porosity. The ratio of the open pores to the BULK VOLUME (q.v.), expressed as a percentage.

Sealed Porosity or Closed Porosity. The ratio of the volume of the sealed pores to the bulk volume, expressed as a percentage.

True Porosity. The ratio of the total volume of the open and sealed pores to the bulk volume, expressed as a percentage.

Port. An opening through which fuel gas or oil, or air for combustion, passes into a furnace; ports are usually formed of refractory blocks or monolithic material.

Port Walker. A workman who observes the sheet of glass issuing from a FOURCAULT (q.v.) tank-furnace and gives warning of faults.

Portland Blast-furnace Cement. A hydraulic cement made by mixing portland cement clinker with up to 65% blast-furnace slag and grinding them together (see B.S. 146 and cf. SLAG CEMENT).

Portland Cement. A hydraulic cement produced by firing (usually in a rotary kiln) a mixture of limestone, or chalk, and clay at a temperature sufficiently high to cause reaction and the formation of calcium silicates and aluminates; the proportion of $3CaO.SiO_2$ is about 45% and that of $2CaO.SiO_2$ about 25%, the other major constituents being $3CaO.Al_2O_3$ and $4CaO.Al_2O_3$. Fe_2O_3. The chemical composition is (per cent): SiO_2, 17–24; Al_2O_3, 3–7; Fe_2O_3, 1–5; CaO, 60–65; MgO, 1–5; alkalis, 1; SO_3, 1–3. Properties are specified in B.S. 12 and ASTM – C150.

Post. (1) The gather of glass after it has received preliminary shaping and is ready to be hand-drawn into tube or rod.

(2) An item of KILN FURNITURE (q.v.). Posts, also known as PROPS or UPRIGHTS, support the horizontal bats on which ware is set on a tunnel-kiln car. (See Fig. 4, p. 158.)

(3) A discrete portion of bond between abrasive grains in a grinding wheel or other abrasive article. When the abrasive grain held by a post has become worn, the post should break to release the worn grain so that a fresh abrasive grain will become exposed.

Post Clay. A term sometimes used in N.E. England for an impure siliceous fireclay.

Pot. A large fireclay crucible used for melting and refining special types of glass. A pot may either be open (the surface of the glass being exposed to the furnace gases) or closed by an integrally moulded roof, a mouth being left for charging and gathering; a CLOSED POT is also known as a COVERED or HOODED POT.

Pot Arch. A furnace for preheating the POTS (q.v.) used in one method of glass-making.

Pot Bank. The old term in N. Staffordshire for a pottery factory.

Pot Clay. A somewhat siliceous fireclay that is used in the manufacture of glass-pots or crucibles for use in melting steel; the term is also applied to a mixture of such a clay with grog, so that it is ready for use.

Pot Furnace. A furnace in which glass is melted and refined in POTS (q.v.), or in which a frit is melted for use in a glaze or in a vitreous enamel.

Pot Ring. See RING.

Pot Spout. A refractory block used in the glass industry to connect the working end of a glass-tank furnace to a revolving pot.

Potassium Cyanide. KCN; used as a neutralizer in vitreous enamelling.

Potassium Niobate. $KNbO_3$; a ferroelectric compound having a perovskite structure at room temperature. The Curie temperature is 420°C.

Potassium Nitrate. KNO_3; used to some extent as an oxidizing agent in glass-making. The old name was NITRE.

Potassium Tantalate. $KTaO_3$; a ferroelectric material having a dielectric constant exceeding 4000 at the Curie temperature (-260°C).

Potassium Titanate. This compound, which approximates in composition to $K_2Ti_6O_{13}$ and melts at 1370°C, can be made into fibres for use as a heat-insulating material.

Potette. A refractory shape used in the glass industry; it is partly immersed in the molten glass and protects the gathering point from the furnace gases and from any scum that is floating on the glass.

Potsherd. A piece of broken pottery. The word is not now used in the industry; the N. Staffs. dialect word is 'shord'.

Potter's Horn or Kidney. A thin kidney-shaped piece of horn or metal used, until the early 20th century, by pottery pressers. To make dishes, a bat of prepared body was placed on a plaster mould and hand-pressed to shape with a piece of fired ware; the horn was used for final smoothing of the surface.

Potter's Red Cement. A POZZOLANA (q.v.) type of cement consisting of crushed fired clay mixed with portland cement; (C. J. Potter, *J. Soc. Chem. Ind.*, **28**, 6, 1909).

Pottery. This term is generally understood to mean domestic ceramic ware, i.e. tableware, kitchenware and sanitaryware, but the pottery industry also embraces the manufacture of wall- and floor-tiles, electroceramics and chemical stoneware. There are subdivisions within these main groups of pottery-ware, e.g. tableware may be earthenware, bone china or porcelain; similarly, sanitaryware may be sanitary fireclay, sanitary earthenware or vitreous china sanitaryware. In USA the term WHITEWARE (q.v.) is used.

Potting Material. A material for the protection of electrical components, e.g. transformers. The use for this purpose of sintered alumina powder, within hermetically sealed cans, has proved successful.

Pouring-pit Refractories. Alternative term (particularly in USA) for CASTING-PIT REFRACTORIES (q.v.).

Powder Blue. See SMALT.

Powder Density. See TRUE DENSITY.

Powdering. A process sometimes used for the on-glaze decoration of the more expensive types of pottery-ware. Colour mixed with an oil medium is first brushed on the ware, dry powdered colour then being applied to achieve a stippled effect.

Pozzolana. A material that, when ground and mixed with lime and water, will react with the former to produce compounds having hydraulic properties. There are both natural and artificial pozzolanas. The original Natural Pozzolana was a volcanic tuff worked at Pozzoli (or Pozzuoli), Italy, in Roman times. One of the principal types of Artificial Pozzolanas is produced by firing clay or shale at 600–1000°C.

PPG Ring-roll Process. PPG = Pittsburgh Plate Glass Co.; see RING-ROLL PROCESS.

Prall Mill. An impact mill consisting of an impeller rotating clockwise at 1000 rev/min, a baffle plate moving anti-clockwise at 1500 rev/min, and a second baffle plate that is stationary.

Praseodymium Oxide. A rare earth that, together with zirconia and silica, produces a distinctive and stable yellow colour for pottery decoration.

Praseodymium Yellow. A ceramic colour made by calcining a stoichiometric mixture of ZrO_2 and SiO_2 with 5% PrO_2. This clean, bright yellow can be used in glazes firing at any temperature from about 1100–1300°C; the presence of zircon acts as a stabilizer.

PRE. The West European refractories federation: Produits Réfractaires Européen, Via Corridoni 3, Milan, Italy.

Precision-bore Tubing. Special glass tubing made by heating ordinary glass tubing until it is soft and then shrinking it on to a steel mandrel.

Prefabricated Masonry. Panels or beams of specially designed hollow clay blocks formed by the insertion of metal reinforcing wires through a number of the blocks, laid side by side, followed by infilling with cement. Many designs of such prefabricated panels and sections have been developed to accelerate building construction.

Preform. Term used in the glass industry, and particularly in the making of glass-to-metal seals, for a compact of powdered glass that has been fired to a non-porous state.

President Press. Trade-name: a dry-press brickmaking machine of a type that gives two pressings on both the top and bottom of the mould, a short pause between the pressings allowing any trapped air to escape. (W. Johnson & Son (Leeds) Ltd., England.)

Press-and-Blow Process. A method of shaping glass-ware, the PARISON (q.v.) is pressed and then blown to the final shape of the ware.

Press Cloth. See FILTER CLOTH.

Pressed Glass. Glass-ware shaped by pressing molten glass between a mould and a plunger.

Pressing. In the pottery industry, pressing was an old method of shaping ware by hand-pressing a prepared piece of body between two halves of a mould. See also DRY PRESSING; OPTICAL BLANK.

Pressure Casting. Slip-casting, in plaster moulds, under pressure. Experiments have shown that with slip under a pressure of 30 p.s.i. the rate of casting is increased very significantly; such a

215

Pressure Cracking

pressure exposes the mould to a stress higher than the usual plaster mould can withstand. As yet, the process has not been applied industrially to any extent.

Pressure Cracking. Cracking of the compacted semi-dry powder immediately after it has been shaped in a dry-press; the cause is sudden expansion of air that has been trapped and compressed in the pores of the compact. The fault has been largely eliminated by designing presses so that the plunger descends twice to its lowest level, the trapped air having time to escape between the first and second pressings.

Pressure Sintering. This term has the same meaning as HOT PRESSING (q.v.).

Preston Density Comparator. An instrument designed in 1950 at the Preston Laboratories, USA, for use in the routine quality-control of glass on the basis of an observed relationship, for any specific glass, between any change in composition and the associated change in density; the sink-float method is employed.

Primary Air. The air that mingles with the fuel and effects the initial stages of its combustion. In a ceramic kiln of a type fired by solid fuel on a grate, the primary air passes through the grate and the fuel bed (cf. SECONDARY AIR and TERTIARY AIR).

Primary Boiling. Gas evolution during the initial firing of vitreous enamel; this may result in faults in the ware.

Primary Clay or Residual Clay. A clay still remaining in the geographical location where it was formed; in the UK such a clay is typified by the CHINA CLAY (q.v.) of Cornwall (cf. SEDIMENTARY CLAY).

Primary Crusher. A heavy crusher suitable for the first stage in a process of size reduction, the product being about 1 to 3 in. in size. Gyratory and jaw crushers, and Kibbler rolls, fall into this class.

Primary Phase. The first crystalline phase to appear when a liquid is cooled. For example, in the Al_2O_3-SiO_2 system (see Fig. 2, p. 100), if a liquid is formed containing 40% Al_2O_3, and if this liquid is then cooled, the primary phase, when solidification begins, will be mullite; if the liquid contains 80% Al_2O_3, however, it can be seen that the primary phase will be corundum.

Prince Rupert's Drops. Drops of glass that have been highly stressed by quenching; when the 'tail' of one of these drops is broken the glass explodes to dust, but the drop itself is immensely strong. These drops were first made by Prince Rupert, nephew of King Charles I.

Printer's Bit. An item of KILN FURNITURE (q.v.). It is a small piece of refractory material for use as a distance-piece in the stacking of decorated pottery-ware before and during firing.

Prismatic Glass. Glass that has been pressed or rolled to produce a pattern of prisms; these refract light passing through the glass. (The term is also sometimes, erroneously, applied to LENS FRONTED TUBING (q.v.).)

Process Fish-scaling. A fault in vitreous enamelling, FISH-SCALING (q.v.) occurring during the drying or firing of the cover-coat.

Proctor Dryer. The original 'Proctor Dryer' of the early 1920s was a tunnel dryer for heavy-clay and refractory bricks; drying was achieved by air recirculating over heated steam coils, or over pipes carrying hot waste gases. Today, Proctor dryers still operate on the same principle but are made in a variety of types suitable for all kinds of ceramic product. The name derives from the manufacturers: Proctor and Schwartz, Philadelphia, USA.

Prokaolin. A term that has been applied to an amorphous intermediate product in the process of KAOLINIZATION (q.v.).

Prop. See POST.

Prouty Kiln. A tunnel kiln of small cross-section suitable for rapid firing of pottery-ware, which is carried through the kiln on bats. (T. C. and W. O. Prouty, US Pat., 1 676 799, 10/7/28.)

Proximate Analysis. See RATIONAL ANALYSIS.

Prunt. A crest, or other device, that has been fused on glass-ware subsequent to the general shaping process.

PSD. Abbreviation for PORE SIZE DISTRIBUTION (q.v.).

p.s.i. Abbreviation for pounds per square inch.

Psychrometric Chart. A graphical representation of the relation-ship between the relative humidity, specific volume, weight ratio of moisture to air, dry-bulb temperature, vapour pressure, total heat, and dew-point of moist air. The chart finds use in the ceramic industry, particularly in the control of dryers.

Pucella. A tool for widening the top of a wine glass in the hand-making process; from Italian word meaning a virgin.

Pug. A machine for consolidating plastic clay or body into a firm column. It consists of a steel cylinder ('barrel') which tapers at one end to a die, through which the clay or body is forced by knives mounted on a shaft which rotates centrally to the barrel (cf. AUGER and MIXER).

Pull. See LOAD.

217

Pulled Stem. See STEMWARE.

Pulverized Fuel Ash. Finely divided ash carried over from coal-fired power-station boilers. It has found some use in the manufacture of building materials, e.g. THERMALITE YTONG (q.v.), LYTAG (q.v.), and to a less extent in clay building bricks. The composition of the ash is (per cent): SiO_2, 43–50; Al_2O_3, 24–28; Fe_2O_3, 6–12; CaO, 2–4; MgO, 2–3; alkalis, 3–5; loss on ignition, 2–10. In the UK, the Central Electricity Generating Board Specification is: $\lessdot$ 35% SiO_2, $\gg$ 4% MgO, $\gg$ 3% SO_3, $\gg$ 12% loss on ignition.

Pulverizer. See FINE GRINDER.

Punch Test. A simple test to determine whether the glaze on a piece of fired pottery is in tension or compression. A steel centre-punch with a blunt end $\frac{1}{32}-\frac{1}{16}$ in. dia. is placed on the glazed surface and hit sufficiently hard with a hammer to break the glaze. If the latter was in tension, one or more cracks will be found to have radiated from the point of impact; if the glaze was in compression a circular crack will have formed round the punch mark or a conical piece of glaze will have become detached.

Punch Ware. US term for thin, hand-blown, glass tumblers, etc.

Punt. The bottom of a glass container.

Punt Code. A mark of identification on the base of a glass bottle.

Punty. An iron rod used in the hand-making of glass-ware to hold the base of the ware during its manipulation; from Latin *puntellum*, the word used in 16th century manuscripts on glass-making.

Pup. A brick (particularly a refractory brick) of a shape and size based on that of a standard square but with the end faces square (or nearly square). This shape of brick is also known as a SOAP or CLOSER. See Fig. 1, p. 37.

Purple of Cassius. Colouring material produced by adding a mixture of tin chlorides to a dilute solution of gold chloride; the latter is reduced to the metal and hydrated stannic oxide is precipitated, the colour resulting from the finely-divided gold on the SnO_2 particles.

Purpling. A fault liable to occur with CHROME-TIN PINK (q.v.) ceramic colour if the amount of alkali and borax is too high and the amount of lime too low.

Push-down Cullet. A fault occasionally found in sheet glass as a result of the presence of CULLET (q.v.) in the zone of the furnace from which the glass was drawn.

Push-up. Alternative name for PUSHED PUNT (q.v.).

Pushed-bat Kiln or Sliding-bat Kiln. A tunnel kiln, of small cross-section, through which the ware is conveyed on sliding bats instead of on the usual cars; when there are a number of such tunnels in a single kiln, it is known as a MULTI-PASSAGE KILN (q.v.).

Pushed Punt. The concave bottom of a glass wine-bottle or other container.

Putnam Clay. A fine-grained, plastic, Florida kaolin; it fires to a good white colour.

Puzzuolana. Obsolete spelling of POZZOLANA (q.v.).

Pycnometer. A small bottle having an accurately-ground glass stopper pierced with a capillary. The bottle is of known volume and is used for the accurate determination of the specific gravity of a material.

Pyrex. Trade name. A borosilicate glass introduced by the Corning Glass Co., USA in 1915 and made in England by J. A. Jobling & Co. Ltd., since 1922. This glass has high thermal endurance and is much used for making chemical ware.

Pyrite, Pyrites. Iron sulphide, FeS_2; sometimes present in clays, causing slaggy spots on the fired product.

Pyroceram. Trade-name (Corning Glass Works, USA) for the original commercially available DEVITRIFIED GLASS (q.v.); the equivalent material made in the UK under licence is known as PYROSIL (q.v.).

Pyrolusite. MnO_2; see MANGANESE OXIDE.

Pyrolytic Coatings. Pyrolysis is the decomposition of a material by heat; a pyrolytic coating is a thin coating produced by the breakdown of a volatile compound on a hot surface. Some types of resistor are made by the pyrolytic coating of an electro-ceramic rod with carbon. Pyrolytic coatings of BN, SiC, and Si_3N_4 have been applied to components for their protection during exposure to high temperatures.

Pyrometer. A device for the measurement of high temperature, e.g. a thermocouple or a radiation pyrometer.

Pyrometric Cone. A Pyramid with a triangular base and of a defined shape and size; the 'cone' is shaped from a carefully proportioned and uniformly mixed batch of ceramic materials so that, when it is heated under stated conditions, it will bend due to softening, the tip of the cone becoming level with the base at a definite temperature. Pyrometric cones are made in series, the temperature interval between successive cones usually being 20°C. The best known series are Seger Cones (Germany). Orton Cones (USA), Staffordshire Cones (UK); for nominal temperature equivalents see Appendix 2.

Pyrometric Cone Equivalent. A measure of refractoriness: the identification number of the standard pyrometric cone that bends at the temperature nearest to that at which the test-cone bends under the standardized conditions of test. These conditions are specified in the following national standards: Britain, B.S. 1902; USA, ASTM – C24; France, B49–102; Germany, DIN 51 063.

Pyrophyllite. A natural hydrated aluminium silicate, Al_2O_3. $4SiO_2 . H_2O$. It occurs in N. Carolina, Newfoundland, Japan, and S. Africa, and finds some use in whiteware bodies. The massive material can be turned on a lathe and then fired with no appreciable change in dimensions.

Pyroplastic Deformation. The irreversible deformation suffered by many ceramic materials when heavily stressed at high temperatures. The term has been applied more particularly to the slow deformation of fireclay refractories when loaded at high temperatures.

Pyroscope. A device that, by a change in shape or size, indicates the temperature or, more correctly, the combined effect of time and temperature (which has been called HEAT-WORK). The best-known pyroscopes are PYROMETRIC CONES (q.v.), BULLERS' RINGS (q.v.); HOLDCROFT BARS (q.v.); and WATKIN HEAT RECORDERS (q.v.).

Pyrosil. Trade-name (J. A. Jobling & Co. Ltd., Sunderland, England) for ceramic ware consisting of DEVITRIFIED GLASS (q.v.); it is the British equivalent of PYROCERAM (q.v.).

Pythagoras Ware. Trade-name: W. Haldenwanger, Berlin-Spandau. A gas-tight mullitic porcelain that can be used at temperatures above 1400°C. P.C.E. 1825°C. Thermal expansion (20–1000°C), $5·7 \times 10^{-6}$. Tensile strength, 7000 p.s.i.; crushing strength, 98 000 p.s.i.

PZT. Lead zirconate-titanate, $Pb(Zr,Ti)O_3$; used as a ceramic component in piezoelectric transducers.

Q

Quarl Block. A refractory shape forming the whole, or a segment, of a gas- or oil-fired burner, particularly in a boiler furnace or glass-tank furnace.

Quarry. (1) An open-cast working of rock.

(2) See Floor Quarry.

Quartz. The common form of silica, SiO_2. Quartz is the principal constituent of silica sand, quartzite and ganister, and it occurs as an impurity in most clays. The room-temperature form, α-quartz (sp. gr. 2·65) undergoes a reversible crystalline change to β-quartz at 573°C; this inversion is accompanied by a linear expansion of 0·45%. At 870°C quartz ceases to be stable but, in the absence of fluxes, does not alter until a much higher temperature is reached, when it is converted into cristobalite and/or tridymite, depending on the temperature and nature of the fluxes present.

Quartz Glass. In UK, a non-recommended alternative term for Vitreous Silica (q.v.); in USA the term is reserved for vitreous silica that has been made from vein quartz.

Quartzite. A silica rock of a type in which the quartz grains, as originally deposited, have grown at the expense of the siliceous cement. Its principal use in the ceramic industry is as raw material for the manufacture of silica refractories. For this purpose the quartzite should contain as little alkali and alumina compounds as possible; the porosity should be low ($< 3\%$) and the rock should consist of small quartz grains set in a chalcedonic or opaline matrix. Quartzite for silica refractories is worked in Britain in N. and S. Wales, the Sheffield area and the N. Pennines (cf. Ganister; Silcrete).

Quebracho Extract. An extract of tannin from the S. American tree *Quebracho Colorado*. It finds some use, particularly as sodium tannate, as a deflocculant for pottery slips.

Queen's Ware. The name given by Queen Charlotte, wife of George III, to the fine white earthenware introduced by Josiah Wedgwood in 1763.

Quenching of Frit. Molten glaze-frit or enamel-frit is quenched to break it up, thus making it easier to grind. The simplest method is to allow the stream of molten frit to fall into water, but this does not give uniform quenching and fracture. Better methods are to expose the stream of molten frit to a blast of air and water, or to pass the stream between water-cooled rolls; the latter process gives a flaky product.

Quoin. An external corner of brickwork.

R

R-value. The partial dispersion ratio of a glass expressed as $(n_D-n_C)(n_F-n_C)^{-1}$ where n_C, n_D and n_F are the refractive indices at wavelengths equivalent to the spectral lines C, D and F. (cf. ABBE NUMBER.)

Rack Mark. A surface blemish on rolled glass resulting from a mechanical defect in the drive actuating the forming roller.

Radial Brick. A brick with the two end faces curved to form parts of concentric cylinders (cf. CIRCLE BRICK and see Fig. 1, p. 37).

Radio-frequency Heating. See under INDUCTION HEATING.

Radome. A protective cone for the radar equipment in the nose of an aircraft, rocket or space vehicle. At high speeds the skin temperature may exceed 500°C and ceramic radomes become necessary; alumina and devitrified glass ceramics have been used.

Raku Ware. A type of pottery used in Japan for the Tea Ceremony. The ware is thick-walled, is covered with a very soft lead borosilicate glaze, and is once-fired at about 750°C.

Ram Pressing. A process for the plastic shaping of tableware and sanitaryware by pressing a bat of the prepared body between two porous plates or mould units; after the pressing operation, air is blown through the porous mould parts to release the shaped ware. The process was patented in 1952 by Ram Incorporated (US Pat., 2 584 109/10).

Ramming. The forming of refractory shapes or monolithic linings by means of a portable device (usually pneumatic) by which the material is compressed by repeated impact from a blunt-ended tool (cf. TAMPING).

Ramming Material. A coarsely-graded refractory batch which, when moistened, can be rammed into position, in a furnace hearth for example. The rammed mass becomes strong and monolithic by vitrification or sintering *in situ.*

Ramp. That part of the roof of an open-hearth steel furnace that rises from the end of the main roof to the top of the end wall; as the furnace is symmetrical, there is a ramp at each end.

Rankine Temperature Scale. The Fahrenheit scale displaced downwards so that Absolute Zero ($-492°F$) is $0°$ Rankine; $32°F = 492°$ Rankine; etc.

Rankinite. $3CaO.2SiO_2$; melts incongruently at $1464°C$ to form dicalcium silicate and a liquid. It is *not* found in portland cement.

Rapid-hardening Portland Cement. The UK equivalent of the US term HIGH EARLY-STRENGTH CEMENT. It is more finely ground than ordinary portland cement; B.S. 12 stipulates a specific surface $\nless$ 3250 cm²/g compared with $\nless$ 2250 cm²/g for ordinary cement. This extra fineness results in the 1-day strength being as high as the 3-day strength of ordinary cement.

Rare Earths. Oxides of the rare earth elements of Group III in the Periodic System. Those used in ceramics are the oxides of LANTHANUM, CERIUM, PRASEODYMIUM, NEODYMIUM, and YTTRIUM (q.v.).

Raschig Ring. A thin-walled hollow cylinder made of chemical stoneware, glass, carbon or metal, for the packing of absorption towers. These rings are made in various sizes from about $\frac{1}{2} \times \frac{1}{2}$ in. (10 000/ft³) to 2×2 in. (170/ft³) and even up to 8×8 in.

Rational Analysis. The mineralogical composition of a material as deduced from the chemical analysis. With materials whose general mineralogical composition is not already known, the calculation is made only after micro- and/or X-ray examination has shown what minerals are present and their approximate proportions. With pottery clays the calculation is made without such guidance, either on the FELDSPAR CONVENTION or the MICA CONVENTION. The former is the older procedure and assumes that the minerals present are kaolinite, feldspar and quartz. It is now known, however, that the alkali impurity in English pottery clays is present as mica and according to the mica convention the rational analysis is calculated on the basis of kaolinite, mica and quartz. (The term PROXIMATE ANALYSIS is sometimes applied to the rational analysis of clays.)

Rattler Test. A test introduced in 1913 by ASTM, as a method for evaluating the resistance of paving bricks to impact and abrasion; it is now standard (ASTM C7 – 42). A sample of 10 bricks is subjected to the action of 10 cast-iron balls 3·75 in. dia. and 245–260 balls 1·875 in. dia. in a drum (28 in. dia; 20 in. long) rotating at 30 rev/min for 1 h. The severity of abrasion and impact is reported as a percentage loss in weight.

Raw Glaze. A glaze in which none of the constituents has been previously fritted.

RE, REL, REO, REOL, RER, RES, REX, REXH, REXR.

Reaction-sintering

Symbols for various shapes of ceramic wall tiles and fittings (see Fig. 6, p. 307).

Reaction-sintering. A process in which SINTERING (q.v.) and chemical reaction between two or more components take place simultaneously. For example, Si and C can be reaction-sintered to produce SiC; compacted Si powder can be reaction-sintered in N_2 to form Si_3N_4.

Ream. A fault in flat glass resulting from the presence of a heterogeneous layer within the glass.

Rearing. The setting of pottery flatware on edge for the firing process (cf. DOTTLING).

Réaumur Porcelain. A fritted porcelain ('soft paste') made from 75% frit, 17% chalk and 8% clay; the frit contained 60% SiO_2, the remainder being K_2O, Na_2O, Al_2O_3 and CaO.

Réaumur Temperature Scale. A little-used scale in which the temperature interval between the freezing point and boiling point of water is divided into 80 degrees.

Reboil. The appearance of bubbles in glass or vitreous enamel during the process of manufacture. Molten glass may reboil after it has apparently freed itself from gases; in vitreous enamelling, bubbles may appear on the surface of sheet-steel ground-coats during the second or third firing process.

Recessed Wheel. An abrasive wheel having a contoured central recess in one or both sides.

Reckna Clay Beams. Prefabricated beams made of specially designed hollow clay building blocks reinforced with steel rods; the load-bearing capacity is about 40 lb/ft². The design originated in Germany (*Ziegelindustrie*, **3**, 60, 1950).

Reclaim. Overspray (enamel or glaze) that, after it has been reconditioned, is suitable for re-use. In vitreous enamelling, reclaim is generally used only in first-coats.

Rectangular Kiln. An intermittent kiln, rectangular in plan, with fireboxes at intervals along the two long sides. Such kilns find limited use in the heavy-clay and refractories industries.

Recuperator. A continuous heat exchanger in which heat from furnace gases is transferred to incoming air or gas through walls (usually tubes) of metal or refractory material.

Red Clay. A term applied in the pottery industry to any of the ferruginous clays used for making tea-pots, floor tiles and plant pots; in England these clays are typified by the ETRURIA MARLS (q.v.).

224

Red Edge. A fault in plate glass characterized by numerous small cavities, containing rouge from the polishing, around the edges.

Red Lead. Pb_3O_4; see LEAD OXIDE.

Red Ware. See ROCKINGHAM WARE.

Redston-Stanworth Annealing Schedule. A procedure for determining the optimum conditions for annealing small glass articles; it is based on the Maxwell model of stress release, modified to take account of the variation of viscosity with time. (G. D. Redston and J. E. Stanworth, *J. Soc. Glass Tech.*, **32**, 32, 1948.)

Reel Cutter. A type of CUTTING-OFF TABLE (q.v.) in which the wires are stretched within a large circular frame, the axis of which is slightly above the line of the extruding clay column. This type of cutter can operate on a stiff clay and, if powered and automatic, has a high output.

Rees–Hugill Flask. A 250 ml flask, with a neck graduated directly in specific gravity units, designed expressly for the determination of the specific gravity of silica and other refractories. It was introduced by W. Hugill and W. J. Rees (*Trans. Brit. Ceram. Soc.*, **24**, 70, 1925) and has since become the subject of a British Standard (B.S. 2701).

Refining. The process in glass-making during which the molten glass becomes virtually free from bubbles; this is effected in the REFINING ZONE of a glass-tank furnace, or in a pot.

Reflex Glass. Glass strip, for use in vessels containing liquids, made with prismatic grooves to facilitate the reading of the liquid level.

Refractoriness. The ability of a material to withstand high temperatures; it is evaluated in terms of the PYROMETRIC CONE EQUIVALENT (q.v.).

Refractoriness under Load. The ability of a material to withstand specified conditions of load, temperature and time. Details of variations in this test will be found in the following national standards: Britain—B.S. 1902; USA—ASTM – C16; Germany—DIN 1064; France—AFNOR B49–105. It is usual to show the result of this test as a curve (Fig. 5) relating the expansion/contraction to the temperature. In the British version of the test, the temperature is reported at which the test-piece shears or at which 10% subsidence has occurred. In the ASTM method the test-piece is a whole brick; unless the brick fails by shear, the subsidence caused by the test is determined by direct

Refractory Cement

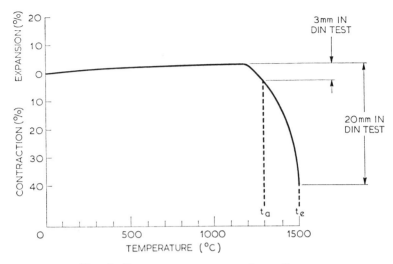

FIG. 5. REFRACTORINESS-UNDER-LOAD CURVE.

measurement of the brick after it has cooled to room temperature. The German procedure reports two temperatures: t_a, at which the curve has fallen 3 mm from its highest point, and $t_e - 20$ mm subsidence from its highest point.

Refractory Cement. A finely-ground refractory material, or mixture, used for the jointing of furnace brickwork. The term is often used to denote all types of such cement but should more properly be reserved for materials that harden only as a result of ceramic bonding at high temperature.

Refractory Coating. A refractory material for the protection of the surface of refractory brickwork or of metals (e.g. pyrometer sheaths or aircraft exhaust systems). Examples of such coatings include Al_2O_3, $ZrSiO_4$, $MoSi_2$ and (for the protection of metals) refractory enamels. Refractory coatings for furnace brickwork are sometimes known as WASHES.

Refractory Concrete. A concrete made from a graded aggregate of crushed refractory material together with high-alumina hydraulic cement, e.g. Ciment Fondu or Lumnite. The concrete may be cast in position in a furnace lining or it can be made into blocks and special shapes. With an aggregate of crushed 40%–Al_2O_3 firebrick, such a concrete can be used up to about 1300°C.

Insulating concrete can be made for use up to 900°C; its thermal conductivity is about 2–2·5 Btu/ft²/h/°F/in.

Refractory Material. Also known as a REFRACTORY. Broadly, a material that will withstand high temperatures sufficiently to permit its use in a furnace lining or other location where it will be exposed to severe heating. The official definition (B.S. 3446) requires that a refractory material shall have a PCE (q.v.) not less than 1500°C. Typical refractory materials are fireclay, silica, magnesite, chrome-magnesite, sillimanite; special refractory materials include sintered alumina, beryllia, boron nitride, etc.

Regenerator. A periodic heat exchanger in which refractory checker bricks alternately receive heat from furnace gases and give up heat to incoming air or gas; this is achieved by reversing the direction of gas flow.

Rehoboam. A 6-quart wine bottle.

Reinforced Brickwork or Reinforced Masonry. Brickwork that is reinforced by steel rods or wire inserted into continuous cavities through the brickwork, the space between steel and brick then being filled with poured concrete. Such brickwork can withstand tensile, as well as the usual compressive, stresses. Rules for the design of such brickwork are included in B.S. Code of Practice CP111.

Relieving Arch. An arch built into structural or refractory brickwork to take from that part of the wall below the arch the weight of brickwork above the arch.

Repressing. A second pressing sometimes given to bricks (both building and refractory) to improve their final shape; with building bricks the purpose can also be to provide some special surface effect.

Re-setting (of Portland Cement). The property of portland cement, after it has been mixed with water and allowed to set, to set for a second time (although giving a greatly reduced strength) if it is again finely ground. The effect is due to the fact that the hydration causing setting is confined to the surface (10μ) of the cement particles, some of which are as large as 100μ.

Residual Clay. See PRIMARY CLAY.

Resinous Cement. A term used in chemical engineering for an acid-proof cement, based on synthetic resin, for jointing chemical stoneware or acid-resisting bricks; the cement, when set, is impervious and very hard.

Resist. (1) A fluoride-resisting layer used during acid etching

Retarder

to protect those parts of glass or ceramic ware not required to be etched. Wax, asphalt, resin, etc., dissolved in turpentine are suitable for this purpose.

(2) A method for the on-glaze decoration of pottery, particularly for use with lustres. The decoration is painted on the ware with glycerine, or other resist; the ware is then coated with the lustre solution, dried, the painted area washed clear with water, re-dried and fired. Alternatively, an infusible resist can be used, e.g. china clay, and the ware can be fired before the resist is gently rubbed away with a soft abrasive powder.

Retarder. (1) A material, such as keratin, that can be added to plaster to retard setting.

(2) Retarders for the setting of portland cement generally consist of cellulose derivatives (e.g. carboxymethyl-cellulose), starches or sugars; these are additional to the 1–3% of gypsum added to all ordinary portland cements to retard the setting time to within convenient limits (see INITIAL SET and FINAL SET).

Retort. A structure, generally built of refractory material, into which raw materials are charged to be decomposed by heat, e.g. a zinc retort or gas retort.

Revergen Kiln. Trade-mark: See DAVIS REVERGEN KILN.

Revolver Press. A type of press for the shaping of clay roofing tiles. It consists of a pentagonal or hexagonal drum, mounted on a horizontal shaft and carrying bottom press-moulds on each flat surface. As the drum rotates, discontinuously, each mould in turn is brought beneath a vertical plunger which consolidates a clot of clay in the mould. While the drum is stationary (during pressing) a clot of clay is fed to the mould next due to arrive beneath the plunger.

Revolving Pot. A shallow refractory pot, which is slowly rotated and from which molten glass is gathered by a suction machine.

Revolving Screen or Trommel. A cylindrical screen, usually of the perforated-plate type, set at an inclined angle and made to rotate about its axis.

Rheology. The science of the deformation and flow of materials, e.g. the study of the viscosity of a glass, glaze or enamel, the study of the plasticity of clay.

Rheopexy. The complex behaviour of some materials, including a few ceramic bodies, which show DILATANCY (q.v.) under a small shearing stress followed by THIXOTROPY (q.v.) under higher stress.

228

Rib. A wooden or metal tool for smoothing the outside of a vase or bowl while it is being thrown.

Rib Marks. The marks found on the surface of broken glass; they are in the form of raised arcs perpendicular to the direction in which fracture occurred.

Ribbed Roof. A furnace roof (particularly of an open-hearth steel furnace) in which some of the refractory bricks, while conforming with the smooth internal surface of the roof, project outwards to form continuous ribs across the furnace; these ribs, because they remain cool, confer strength on the roof even when it has worn thin.

Ribbon. An ornamental course of tiles on a roof (cf. RIBBON COURSES).

Ribbon Courses. Successive courses of roofing tiles laid to alternately greater and lesser exposures.

Ribbon Process. A glass-making process in which the glass, after it has been melted and refined, is delivered as a continuous ribbon.

Rice Hulls or Rice Husks. The waste husks from rice consist, when calcined, of about $96 \cdot 5\%$ SiO_2, $2 \cdot 2\%$ CaO and minor amounts of other oxides. Because this composition is highly refractory, and because the porosity of the calcined material is nearly 80%, this waste product has been used, particularly in Italy where rice is an important crop, as a raw material for the manufacture of insulating refractories.

Rider Bricks. Refractory bricks, which may be solid, perforated or arched, used in the base of a regenerator chamber to form a support for the checker bricks; in a coke-oven regenerator, also known as SOLE-FLUE PORT BRICKS or NOSTRIL BLOCKS; in a glass-tank regenerator, also known as the RIDER ARCH, BEARER ARCH or SADDLE ARCH.

Ridge Tile. A special fired-clay or concrete shape for use along the ridge of a pitched roof. There are a number of varieties, e.g. SEGMENTAL, HOG'S BACK, WIND RIDGE, etc.

Rigden's Apparatus. An air-permeability apparatus for the determination of the specific surface of a powder; air is forced through a bed of the powder by the pressure of oil displaced from equilibrium in a U-tube. (P. J. Rigden, *J. Soc. Chem. Ind.*, **62**, 1, 1943.)

Ring. (1) A refractory ring that floats on the molten glass in a pot to keep any scum from the gathering area within the ring (cf. FLOATER).

Ring Mould

(2) The part of the mould that forms the rim of pressed glass-ware.

(3) An arch of refractory bricks forming part of a furnace roof and unbonded with the adjacent arches.

(4) A SAGGAR (q.v.) without a bottom, also sometimes known as a RINGER.

Ring Mould. See NECK MOULD.

Ring-roll Crusher. A grinding unit consisting, typically, of a steel casing in which is housed a free-running grinding ring, within which revolve three crushing rolls; the top roll is mounted on the driving shaft while the two lower rolls are carried on floating shafts supported by springs. The three rolls are pressed against the inner surface of the ring by these springs and material is ground as it passes between the ring and the rolls. This type of crusher has been used successfully with ceramic raw materials such as bauxite.

Ring-roll Process. A method for the production of blanks for plate-glass manufacture. Molten glass passes between a heated 'ring-roll' casting table, of large diameter, and a smaller forming roll.

Ring Test. (1) GLAZE FIT: a test first used by H. G. Schurecht and G. R. Pole (*J. Amer. Ceram. Soc.*, **13**, 369, 1930). The test-pieces are hollow cylindrical rings 2 in. dia. glazed on the out-side only with the glaze to be tested. The glazed ring is fired and two grooves or holes are then cut in one edge of the ring, approx. $\frac{1}{4}$ in. apart and large enough to hold glass capillary tubes $\frac{1}{6}$ in. long and $\frac{1}{32}$ in. dia.; these capillaries provide sharp reference marks, the distance between which is measured with a micrometer microscope. The ring is then cut open with a diamond saw and the distance between the reference marks is again measured. Similar measurements are made on unglazed rings so that the true expansion or contraction caused by partially releasing the stress between glaze and body can be determined. If the glaze is in tension the ring will expand when cut open; if in compression it will contract.

(2) THERMAL-SHOCK: a test proposed by W. R. Buessem and E. A. Bush (*J. Amer. Ceram. Soc.*, **38**, 27, 1955). A stack of ceramic rings, each 2 in. O.D. and 1 in. I.D., and $\frac{1}{2}$ in. long, are heated from the inside by a heating element and cooled from the outside by a calorimetric chamber. Both thermal conductivity and thermal-shock resistance can be evaluated (cf. BRITTLE-RING TEST).

Ring Wall. See SHELL WALL.

Ringed Roof. A furnace roof consisting of arches of bricks unbonded with adjacent arches (cf. BONDED ROOF).

Ringelmann Chart. A chart divided into five (Nos. 0–4) shades of darkness introduced in the late 19th century by Professor Ringelmann, of Paris, as a means of designating the blackness of smoke emitted from industrial chimneys. The charts have been standardized (B.S. 2742) and are used in the ceramic industry in compliance with the Clean Air Act of 1955 in which dark smoke is defined as equal to, or denser than, Shade 2 on the Ringelmann Chart.

Rinman's Green. A colouring material consisting of a solid solution of CoO.ZnO in ZnO; it finds limited use as a ceramic colour. (S. Rinman, *Kong. Vet. Akad. Handl.*, p. 163, 1780.)

Rippled. See DRAGGED.

Rise. (1) In a sprung arch, the vertical distance between the level of the springer and the highest point of the arch at its inner surface.

(2) The vertical distance through which the centre of the sprung roof of a furnace rises as a result of expansion when the furnace is heated to its operating temperature.

Riser Brick or End Runner. A RUNNER BRICK (q.v.) with a hole near one end of its upper face and (generally) a short tubular projection from this hole to lead molten steel into the bottom of an ingot mould.

Rittinger's Law. A law relating to the theoretical amount of energy required in crushing or grinding: the energy necessary for the size-reduction of a material is directly proportional to the increase in total surface area. (P. R. von Rittinger, Berlin, 1867.)

Rittinger Ratio. The $\sqrt{2}$ ratio used in some series of sieves, e.g. Tyler sieves (see Appendix 1). The ratio was proposed by P. R. von Rittinger in 1867.

RO Fusion-cast Refractory. A fusion-cast refractory made in an inclined mould designed to concentrate the shrinkage cavity in one corner of the block; such blocks are made in France (RO = Retassure Orientée, i.e. oriented cavity).

Robertson Kiln. Several types of tunnel kiln were designed by H. M. Robertson but that most commonly associated with his name is a tunnel kiln for salt-glazing; the salt is introduced via fireboxes in the side walls and the fumes are extracted in the cooling zone. (Brit. Pats., 331 224 and 331 225, 9/5/28.)

Robey Oven

Robey Oven. A down-draught type of pottery BOTTLE OVEN (q.v.) patented by C. Robey (Brit. Pat., 970, 17th March, 1873).

Rocaille Flux. An alternative (now less common) name for STRASS (q.v.).

Rock Cement. US term for ROMAN CEMENT (q.v.).

Rock Crystal. Large, naturally occurring, transparent crystals of quartz. In USA the term is also used for CRYSTAL GLASS (q.v.).

Rock Wool. See MINERAL WOOL.

Rocker. A glass bottle that has a faulty, convex, bottom.

Rocket Nozzle. The flame temperature in a rocket nozzle exceeds 2500°C; in addition, thermal shock is very severe. Special refractory materials that have been used in rocket nozzles include silicon carbide (usually with graphite additions), silicon nitride, boron nitride, beryllia and various refractory carbides.

Rockingham Ware. The term originally referred to the ornate porcelain made at a pottery at Swinton, Yorkshire, England, on the estate of the Marquis of Rockingham during the years 1826–1842. Ware with a brown manganese glaze was also produced and it is this type of glaze, which in the UK is usually applied to tea-pots made from red clay, to which the term Rockingham Ware now refers. In the USA, Rockingham Ware was made at Bennington.

Rockwell Hardness. A criterion of hardness based on indentation, either by a steel ball or by a diamond cone. Details are given in B.S. 891. It has been but little used in the ceramic industry.

Rod Cover or Sleeve. A cylindrical fireclay shape having an axial hole and terminating in a spigot at one end and a socket at the other. These refractory sleeves are used to protect the metal stopper-rod in a steel-casting ladle. In the UK six sizes of rod cover are standardized in B.S. 2496. In the USA three qualities are specified in ASTM – C435.

Rod Mill. A steel cylinder into which is charged material that is to be finely ground together with metal rods as grinding media. This type of mill will reduce material from 1 in. to 10 mesh in one stage and the product tends to be granular. Rod mills are employed for the preparation of the materials used in the manufacture of sand-lime bricks.

Rokide Process. Trade-name: (Norton Company, USA): a process for the production of a refractory coating, on metal or on ceramic, by atomizing directly from the fused end of a rod of the coating material, e.g. Al_2O_3 or ZrO_2; the molten particles are blown against the cool surface that is to be coated. The original patent was granted to W. M. Wheildon (US Pat., 2 707 691, 3/5/55).

Rolands' Cement. Trade-name: a HIGH-ALUMINA CEMENT (q.v.) made by Rolandshütte A.G., Lubeck, Germany.

Roll Quenching. See under QUENCHING OF FRIT.

Rolled Edge. The edge of a plate or saucer is said to be 'rolled' if its diameter is greater than the general thickness of the rim of the ware.

Rolled Glass. (1) Flat glass that has been rolled so that one surface is patterned or textured; cf. ROUGH CAST.

(2) In USA the term is also applied to optical glass that has been rolled into plates at the time of manufacture, as distinct from TRANSFER GLASS (q.v.).

Roller-bat Machine. A machine for making, from stiff-plastic clay, bats for a final pressing process in one method of roofing-tile manufacture.

Roller Elutriator. An air-type elutriator designed for the determination of fineness of portland cement. (P. S. Roller, *Proc. ASTM*, **32**, Pt. 2, 607, 1932.)

Roller's Equation. A relationship between the percentage weight and the size of powders: $w = a\sqrt{d.} \exp(-b/d)$, where w is the weight percent of all material having diameters less than d, and a and b are constants. Other equations were deduced relating to specific surface and to the number of particles per gram of powder. (P. S. Roller, *J. Franklin Inst.*, **223**, 609, 1937.)

Roller's Plasticity Test. A method for the assessment of plasticity on the basis of the stress/deformation relationship when clay cylinders are loaded. (P. S. Roller, *Chem. Industries*, **43**, No. 4, 398, 1938.)

Roller-head Machine. A machine for the shaping of pottery flatware on a rotating mould, as in a JIGGER (q.v.), but with a rotary shaping tool instead of a fixed profile. The rotary tool is in the form of a shallow cone of the same diameter as the ware and shaped to produce the back of the article being made. The ware is completely shaped in one operation at the rate of about 12 pieces per minute. The machine, developed from earlier attempts to improve on the use of a fixed tool, was patented by T. G. Green & Co. and H. J. Smith (Brit. Pat., 621 712, 14/4/49) with subsequent improvements by J. A. Johnson (Brit. Pat., 765 097, 2/1/57; 895 988, 9/5/62). It is made by Service (Engineers) Ltd., Stoke-on-Trent, England.

Roller-hearth Kiln. A tunnel kiln through which the ware, placed on bats, is carried on rollers. Such kilns are rare in the ceramic industry.

Roller Mark or Roller Scratch. A surface blemish on vertically drawn sheet glass caused by contact with the rollers.

Rolling (of Glaze). A term sometimes used instead of CRAWLING (q.v.).

Rolls. See CRUSHING ROLLS.

Roman Brick. US term for a building brick of nominal size $12 \times 4 \times 2$ in.

Roman Cement. The misleading name given in the early 19th century to a naturally-occurring mixture of clay and limestone after its calcination; the product had hydraulic properties and was the forerunner of PORTLAND CEMENT (q.v.). A small quantity of Roman Cement is still made. In USA it is known as ROCK CEMENT.

Roman Tile. A roofing tile that is channel-shaped and tapered; the roof is built of alternate descending lengths of tiles placed channel-wise and tiles placed ridge-wise over the junction of adjacent channels.

Roofing Tile. A tile for the covering of a pitched roof. Most of the roofing tiles now being made in England are concrete; these must meet the requirements of B.S. 473 (plain roofing tiles) and B.S. 550 (interlocking roofing tiles); clay roofing tiles are dealt with in B.S. 402 (plain) and B.S. 1424 (single-lap).

Rosin–Rammler Equation. An equation relating to fine grinding: for most powders that have been prepared by grinding, the relationship between R, the residue remaining on any particular sieve, and the grain-size in microns (x) is exponential:

$$R = 100e^{-bx^n}$$

where e is the base of the natural logarithm and b and n are constants. (P. Rosin, E. Rammler and K. Sperling, *Ber.* *C*52 *Reichs-Kohlenstaubs*, Berlin, 1933.)

Rosso Antico. A red unglazed stoneware made by Josiah Wedgwood; it was a refinement of the red ware previously made in North Staffordshire by the Elers.

Rotary Disk Feeder. See DISK FEEDER.

Rotary-hearth Kiln. A circular tunnel with a slowly rotating platform for conveyance of the ware through the kiln. An early example was the WOODHALL-DUCKHAM KILN (Brit. Pat., 212 585,

234

23/2/23). The principle has since been adapted in the CLARK CIRCLE SYSTEM of brickmaking introduced in Australia in 1953, and in some modern pottery-decorating kilns. (This type of kiln is sometimes also known as a ROTATING-PLATFORM KILN.)

Rotary Kiln. A kiln in the form of a long cylinder, usually inclined, and slowly rotated about its axis; the kiln is fired by a burner set axially at its lower end. Such kilns are used in the manufacture of portland cement and in the dead-burning of magnesite, calcination of fireclay, etc.

Rotary Smelter. A batch-type cylindrical furnace which can be rotated about its horizontal axis while frit is being melted in it; when the process of melting is complete the furnace is tilted so that the molten frit runs out. Such furnaces are fired by gas or oil and their capacity varies from about 1–10 cwt.

Rotary Vane Feeder. See VANE FEEDER.

Rotating-platform Kiln. See ROTARY-HEARTH KILN.

Rotolec. Trade-name: a circular electric decorating kiln for pottery-ware. (Gibbons Bros. Ltd., Dudley, England.)

Rotomixair. Trade-name: a reciprocating air-jet system of drying bricks and tiles; the system originated in Italy in 1958.

Rotor Process. A process for the production of steel by the oxygen-blowing of molten iron held in a horizontal, rotating vessel; this is usually lined with tarred dolomite refractories, but tarred magnesite refractories have also been used.

Rouge. Very fine ($< 1\mu$) ferric oxide used for polishing glass.

Rouge Flambé. A red glaze first made by the Chinese; the colour is due to colloidal copper produced in the glaze by firing under reducing conditions.

Rough-cast. A descriptive term for glass that has been rolled so that one surface is textured (cf. ROLLED GLASS).

Round Edge Tile. For the various types of wall tiles with round edges see Fig. 6, p.307.

Round Kiln or Beehive Kiln. An intermittent kiln, circular in plan, with fireboxes arranged around the circumference. Such kilns find use in the firing of blue engineering bricks, pipes, some refractory bricks, etc.

Rouse-Shearer Plastometer. See R & S PLASTOMETER.

Rowlock. A form of brickwork which includes courses of HEADERS (q.v.) laid on edge.

Royal Blue. See MAZARINE BLUE.

R & S Plastometer. A device for the assessment of the

Rubber

rheological properties of a clay slip in terms of the time taken for a given volume of the slip to flow through a tube of known diameter; it was designed by Rouse and Shearer Inc., a firm of ceramic consultants in Trenton, N.J., USA. (*J. Amer. Ceram. Soc.*, **15**, 622, 1932.)

Rubber. A building brick made from a sandy clay and lightly fired so that it can be readily rubbed to shape for use in gauged work. The crushing strength of such a brick is about 1000 p.s.i.

Rubbing Block. A shaped block of abrasive material used in the grinding of blocks of marble or other natural stones.

Rubbing Stone. A block of fine-grained abrasive, e.g. corundum, for the STONING (q.v.) of vitreous enamel.

Rubbing Up. When flatware is placed in a bung for biscuit firing, the spaces between the rims of the ware are filled with sand (if the ware is earthenware) or calcined alumina (if the ware is bone china); this is done by taking handfuls of sand, or alumina, from a heap around the bung and allowing it to fall between the rims of the ware. The process is known in N. Staffordshire as 'rubbing up'.

Ruby Glass. Glass having a characteristic red colour resulting from the presence of colloidal gold, copper or selenium. To produce GOLD RUBY a batch containing a small quantity of gold is first melted and cooled; at this stage it is colourless, but when gently heated it develops a red colour as colloidal gold is formed. For COPPER RUBY, produced in the same manner but with copper substituted for the gold, the batch must contain zinc and must be melted in a non-oxidizing atmosphere. The most common ruby glass today is SELENIUM RUBY; a recommended batch contains 2% Se, 1% CdS, 1% As_2O_3 and 0·5% C; the furnace atmosphere must be reducing.

RuL. Abbreviation for REFRACTORINESS UNDER LOAD (q.v.).

Rumbling. See SCOURING.

Run-down. A fault in vitreous enamelling resulting from an excessive amount of cover-coat becoming concentrated in one area of the ware.

Runner. (1) The channel in which molten iron or slag flows from a blast furnace when it is tapped. The runner is usually lined with fireclay refractories which are then covered with a layer of refractory ramming material, e.g. a mixture of fireclay, grog and carbon. Refractory concrete has also been used to line runners.

(2) A large block of chert as used in a PAN MILL (q.v.).

Runner Brick. A fireclay shape, square in section and about 1 ft long, with a hole about 1 in. dia. along its length and terminating in a spigot at one end and a socket at the other end. A number of such refractory pieces, when placed together, form a passage through which, during the bottom-pouring of steel, the molten metal can pass from the CENTRE BRICK (q.v.) to the base of the ingot mould. Two standard sizes are specified in B.S. 2496.

Running-out Machine. Name sometimes applied to a batch-type extrusion machine of the type more commonly called a STUPID (q.v.).

Rust-spotting. A fault in vitreous enamelling: the appearance of rust-coloured spots in ground-coat enamels during drying. The fault is most frequent with ground-coats of high $Na_2O:B_2O_3$ ratio; it can often be cured by the addition of $0.05-0.10\%$ of sodium nitrite to the enamel slip.

Rustic Brick. A fired clay facing brick with a rough textured surface; this effect can be obtained by stretching a wire across the top of the die so that it removes a thin slice of clay from the moving column.

Rusticating. A mechanical process for the roughening of the facing surface of a clay building brick, while it is still moist, to give it a 'rustic' appearance when fired; wire combs or brushes, or roughened rollers are used.

Rutile. A form of TITANIUM OXIDE (q.v.) occurring in the beach sands of Australia, Florida and elsewhere; used as an opacifying agent in enamels and glazes—the rutile break in pottery (particularly tile) glazes is a crystalline effect obtained by the addition of rutile to a suitable basic glaze composition. Rutile is also used in the production of titania and titanate dielectrics; pure rutile has a dielectric constant of 89 perpendicular to the principal axis and 173 parallel to this axis; the value for a polycrystalline body is 85–95 at 20°C.

S

S-crack. Lamination, in the form of a letter 'S', in a clay column from a pug having a poorly designed mouthpiece. The crack develops from the central hole in the clay column formed at the end of the shaft in the pug, the hole being deformed to an 'S' in the rectangular mouthpiece.

Sacrificial Red

Sacrificial Red. Old Chinese name for the ceramic colours known as ROUGE FLAMBÉ (q.v.) and SANG DE BOEUF (q.v.).

Saddle. (1) An item of KILN FURNITURE (q.v.). It is a piece of refractory material in the form of a bar of triangular cross-section. See Fig. 4, p. 158.

(2) The upper part of a two-piece low-tension porcelain insulator of the Callender–Brown type.

Saddle Arch. See RIDER BRICKS.

Sadler Clay. An ALBANY CLAY (q.v.).

Saggar. A fireclay box, usually oval (23×17 in.), in which pottery-ware can be set in a kiln. The object is to protect the ware from contamination by the kiln gases and the name is generally thought to be a corruption of the word 'safe-guard'. Since the bottle-oven has become obsolete as a kiln for the firing of pottery and since clean fuels and electricity have become increasingly used in the industry, the use of saggars has been largely displaced by the setting of ware on lighter pieces of KILN FURNITURE (q.v.).

Saggar-maker's Bottom-knocker. The man whose job it was to beat out, by means of a heavy wooden tool, a wad of grogged fireclay to form a bottom for a SAGGAR (q.v.). To the regret of folklorists, the species is now virtually extinct.

Sagging. (1) The permanent distortion, by downward bending, of vitreous enamelware that is inadequately supported during firing.

(2) The flow of enamel on vertical surfaces of vitreous enamel ware while it is being fired; the fault is visible as roughly horizontal lines or waves.

(3) A method of shaping glass-ware by heating the glass above a mould and allowing it to sag into the contours of the mould.

Salamander. See BEAR.

Salmanazar. A 12-quart wine bottle.

Salmon. US term for an underfired building brick (in allusion to its colour).

Salt Cake. Commercial grade sodium sulphate, Na_2SO_4; it is added to glass batches, together with carbon, to prevent scumming; it is also added to facilitate melting and refining but to a less extent than formerly.

Salt Glaze. In the 17th century and 18th century salt-glazing was used in the manufacture of domestic pottery. Now, except for its use by some studio potters, the process is applied only to the production of salt-glazed sewer-pipes. The pipes (which

238

should be made from a siliceous clay to ensure a good glaze) are fired in the usual way up to 1100–1200°C and salt is then thrown on the fires, where it volatilizes; the salt vapour passes through the setting of pipes and reacts with the clay to form a sodium aluminosilicate glaze. Borax is sometimes added with the salt in the ratio of about 1:4. (See also SEWER-PIPE.)

Saltpetre. See SODIUM NITRATE.

Samarium Oxide. Sm_2O_3; m.p. 2350°C; thermal expansion (100–1000°C) $9·9 \times 10^{-6}$.

Samian Ware. See TERRA SIGILLATA.

Sampling. The choice of procedure for the obtaining of a representative sample of a consignment of ceramic raw material or product, prior to testing, is of great importance. In USA a procedure for the sampling of clays is laid down in ASTM – C322; in the UK, B.S. 812 deals with the sampling of mineral aggregates and sands, and B.S. 3406 deals with the sampling of powders for particle-size analysis. Details of the number of bricks that should be taken as a sample for testing are given in B.S. 1902 (Refractories) and B.S. 1257 (Building Bricks).

Sand. Typically, sand consists of discrete particles of quartz; some beach sands contain a concentration of heavy minerals, e.g. rutile, ilmenite and zircon. Silica sand occurs widely and is used as a major constituent of glass, as a refractory material for furnace hearths and foundry moulds, and for mixing with portland cement to make cement mortar and concrete. The purest sand is used for glassmaking; a specification was issued by the Society of Glass Technology, England, in 1954. For specifications for building sands see B.S. 812, 1198, 1199 and 1200. Data sheets on refractory sands for steel foundries were issued in 1957 by the British Steel Castings Research Assn., Sheffield.

Sand-creased. A texture produced on the surface of clay facing bricks. In the hand-made process the clot of clay is first sprinkled with, or rolled in, sand; where the sand keeps the clay away from the mould a crease is produced. An imitation of this texture can be achieved in machine-made bricks by impressing the surface of the green bricks with a tooled steel plate.

Sand-faced. A clay building brick that has been blasted with sand while in the GREEN (q.v.) state; alternatively, with extruded bricks, the sand-facing can be applied by gravity to the column of clay. A wide range of surface textures can be obtained by sand-facing.

Sand-gritting

Sand-gritting. The process of roughening those parts of the glaze of an electrical porcelain insulator where cement will be applied in the final assembly. The 'sand' (a specially-prepared mixture of coarse-graded porcelain together with powdered glaze) is applied to areas of the glazed insulator before it is fired.

Sand-lime Brick. A building brick shaped from a mixture of sand (or crushed siliceous rock) and lime and hardened by exposure to steam under pressure. In the UK, four qualities are specified in B.S. 187; for the relevant US specification see ASTM – C73.

Sand Seal. A device for preventing hot gases from reaching the undercarriage of the cars in a tunnel kiln. It consists of two troughs running parallel to the internal walls of the kiln and filled with sand; into this sand dip vertical metal plates (APRONS) which are fixed to the sides of each kiln car.

Sanding. (1) The sprinkling of sand or finely crushed brick between courses of bricks as they are being set in a kiln; the object is to make even any irregularity in level and to prevent sticking.

(2) The treatment, during manufacture, of the surface of a facing brick to give it a pleasing texture and colour.

(3) The BEDDING (q.v.) of earthenware in sand.

Sandstone. A sedimentary rock which may be of several types, e.g. siliceous, calcareous, argillaceous, depending on whether the grains of silica in the rock are bonded with secondary silica, with lime or with clay. Siliceous and argillaceous sandstones both find some use in the refractories industry; crushed sandstone is used as a source of silica in glass manufacture.

Sandwich Kiln. A tunnel kiln designed for rapid firing; the height of the setting is small compared with the width and the firing is from above and below the setting.

Sang de Boeuf. A distinctive red glaze originally produced in China during the Sung dynasty; the colour is due to metallic copper formed in a glaze containing copper and fired under reducing conditions.

Sanitary Earthenware. A type of sanitaryware made from white-firing clays but often covered with a coloured glaze; the body itself has a water absorption of 6–8% (cf. VITREOUS-CHINA SANITARY-WARE). The body is made from a batch containing 22–24% ball clay, 24–26% china clay, 15–18% china stone and 33–35% flint.

Sanitary Fireclay. A type of sanitaryware made from a grogged fireclay body, which is covered with a white ENGOBE (q.v.), which in turn is covered with a glaze. A typical body composition is 60–80% fireclay, 20–40% grog. The engobe contains 5–15% ball clay, 30–50% china clay, 15–30% flint, 20–35% china stone, 0–10% feldspar; the proportions of china stone and feldspar vary inversely as one another.

Sanitaryware. The various types include: SANITARY EARTHEN-WARE (q.v.); VITREOUS CHINA SANITARYWARE (q.v.); SANITARY FIRECLAY (q.v.); VITREOUS-ENAMEL SANITARYWARE, e.g. baths.

Santorin Earth. A POZZOLANA (q.v.) from the Grecian island, Santorin. A quoted composition is (per cent): SiO_2, 64; Al_2O_3, 13; Fe_2O_3, 5·5; TiO_2, 1; CaO, 3·5; MgO, 2; alkalis, 6·5; loss on ignition, 4.

Sapphire. Single-crystal alumina; sapphire boules can be made by the VERNEUIL PROCESS (q.v.) and find use as bearings and thread guides.

Satin Glaze, Satin-vellum Glaze or Vellum Glaze. A semi-matt glaze, particularly for wall tiles, with a characteristic satin appearance. Such glazes are generally of the tin-zinc-titanium type.

Saturation Coefficient. The ratio of the amount of water taken up by a porous building material after it has been immersed in cold water for an arbitrary period (e.g. 24 h) to its water absorption, normally determined by boiling in water for 5 h; the ratio is usually expressed as a decimal fraction, e.g. 0·85. A low saturation coefficient generally indicates good frost resistance but the correlation is far from perfect.

Saucer Wheel. An abrasive wheel shaped like a saucer.

Saxon Slabs. A less common name for NORMAN SLABS (q.v.).

Scab. (1) A fault in the base-metal for vitreous enamelling; the 'scab' is a partially detached piece of metal (which may subsequently have been rolled into the metal surface) and is liable to cause faults in the applied enamel coating.

(2) A fault in glass caused by an undissolved inclusion of sodium sulphate; also known as SULPHATE SCAB and as WHITE-WASH.

Scalding. A term that has been used to describe the fault in the glost firing of pottery when glaze falls off the ware before it has fused; a cause is too great a difference between dimensional changes of body and applied glaze.

Scale. A fault, in glass or vitreous enamelware, in the form of an embedded particle of metal oxide or carbon.

Scaling. Preparation of the base-metal for vitreous enamelling by heating it (often in an atmosphere of acid fumes or of sulphur gases) to form a surface coating of oxide—'scale'—which is subsequently removed by PICKLING (q.v.).

Scallop. The rims and edges of pottery-ware are sometimes trimmed to give a scalloped effect, i.e. small segments are symmetrically removed from the edges before the ware is fired. Until 1955, when a scalloping machine was introduced, the process had always been carried out by hand.

Scandium Oxide. Sc_2O_3; m.p. $> 2400°C$; sp. gr. 3·86.

Scatter Coefficient. See COEFFICIENT OF SCATTER.

Scheidhauer and Giessing Process. See S.u.G. PROCESS.

Schellbach Tubing. A type of ENAMEL-BACK TUBING (q.v.) with a central blue line.

Schenck Porosimeter. Apparatus for the determination of PORE-SIZE DISTRIBUTION (q.v.) by the MERCURY PENETRATION METHOD (q.v.); it has been applied to the study of refractories. (H. Schenck and J. Cloth, *Arch. Eisenhüttenw.*, **27**, 421, 1956.)

Schlenkermann's Stone. A German FIRESTONE (q.v.); it contains about 90% SiO_2.

Schmidt Hammer. A device for the non-destructive testing of set concrete; it is based on the principle that the rebound of a steel hammer, after impact against the concrete, is proportional to the compressive strength of the concrete. (E. Schmidt, *Schweiz. Bauzeitung*, 15/7/50.)

Schnitzler's Gold Purple. A tin-gold colour, produced by a wet method; it has been used for the decoration of porcelain.

Schöne's Apparatus. An elutriator consisting of a tall glass vessel tapering towards the bottom, where water enters at a constant rate. Shöne's formula is: $V = 104·7(S-1)^{1.57}D^{1.57}$ where V is the velocity of water (mm/s) required to carry away particles of diameter D and sp. gr. S. (E. Schöne, *Uber Schlämmanalyse und einen neuen Schlämmapparat*, Berlin, 1867.)

Schuhmann Equation. An equation for the particle-size distribution resulting from a crushing process: $y = 100(x/K)^\alpha$, where y is the cumulative percentage finer than x, α is the distribution modulus, and K is the size modulus; α and K are both constants. (R. Schuhmann, *Amer. Inst. Min. Engrs.*, *Tech. Paper*, 1189, 1940) cf. GAUDIN'S EQUATION).

Schuller Process. See UP-DRAW PROCESS.

Schulze Elutriator. The original type of water elutriator; it has since been improved by H. Harkort (*Ber. Deut. Keram. Ges.*, **8**, 6, 1927).

Schurecht's Ratio. A term that has been used for SATURATION COEFFICIENT (q.v.); named from H. G. Schurecht (USA) who introduced this coefficient in his research on frost-resistance of terra cotta carried out at the National Bureau of Standards in 1926 but never published; the term 'Schurecht Ratio' was first applied by T. W. McBurney.

Scone. See SPLIT.

Scotch Block. One form of gas port in an open-hearth steel furnace; the distinguishing feature is that it is monolithic, being made by ramming suitably graded refractory material around a metal template.

Scotch Kiln. See SCOVE.

Scouring. The cleaning and smoothing of biscuit-fired ceramic ware by placing the ware in a revolving drum together with coarse abrasive material, e.g. PITCHERS (q.v.). This process is used in the making of bone china and of some other types of vitreous table-ware (in USA); when applied in the manufacture of electrical porcelain, the process is known (in UK) as RUMBLING.

Scouring Block. An abrasive block of ceramically-bonded SiC or Al_2O_3 used in the grinding of steel rolls. Such a block is 5–6 in. long and 2–3 in. 'square' (the actual section may be trapezoidal, roughly semicircular or of other special shape).

Scove or Scotch Kiln. An early type of up-draught intermittent kiln for the firing of bricks, etc. It was rectangular and consisted of side-walls and end-walls only, with fire-holes in each side-wall and WICKETS (q.v.) in each end wall. The top of the setting was covered with a PLATTING consisting of a layer of fired bricks, with ashes or earth above.

SCR Brick. A perforated clay building brick introduced in 1952 by the Structural Clay Products Research Institute, USA, whence its name. The brick is $11\frac{1}{2} \times 5\frac{1}{2} \times 2\frac{1}{16}$ in., with 10 holes $1\frac{3}{8}$ in. dia. and a $\frac{3}{4}$ in. square jamb-slot in one end. The weight is 8 lb.

Scrapings. OVERSPRAY (q.v.) removed from the sides and bottom of a spray-booth in vitreous enamelling; it is re-milled and used as a first coat.

Scrapping. The removal of excess body from a shaped piece of pottery-ware while the latter is still in or on the mould; (cf. FETTLING).

Scraps

Scraps. Excess body removed during the shaping of pottery-ware, together with any broken, unfired, pieces. Scraps are usually returned to the blunger for re-use.

Screen. A unit for size classification on an industrial scale (a SIEVE (q.v.) is for laboratory use). A screen consists of a wire mesh or a perforated metal plate; the usual types are REVOLVING SCREENS (q.v.) and VIBRATING SCREENS (q.v.). (For screen-printing see SILK-SCREEN PROCESS).

Screw Press. See FRICTION PRESS.

Scrub Marks. A surface blemish on glass bottles; see BRUSH MARKS.

Scuffing. The dull mark that sometimes results from abrasion of a glazed ceramic surface or of glass-ware.

Scull. See SKULL.

Scum; Scumming. (1) A surface deposit sometimes formed on clay building bricks. The deposit may be of soluble salts present in the clay and carried to the surface of the bricks by the water as it escapes during drying; it is then known as DRYER SCUM. The deposit may also be formed during kiln firing, either from soluble salts in the clay or by reaction between the sulphur gases in the kiln atmosphere and minerals in the clay bricks; it is then known as KILN SCUM (cf. EFFLORESCENCE).

(2) Undissolved batch constituents floating as a layer above the molten glass in a pot or tank furnace.

(3) Areas of poor gloss on a vitreous enamel; the fault may be due to the action of furnace gases, to a non-uniform firing temperature, or a film of clay arising from faulty enamel suspension.

Scurf. A hard deposit, mainly of crystalline carbon, formed by the thermal cracking of crude coal gas on the refractory walls of gas retorts and coke ovens. When the scurf is periodically removed, by burning off, there is a danger that the refractory brickwork of the retort or oven may be damaged by overheating.

Sea-water Magnesia. Sea-water contains approx. $0 \cdot 14\%$ Mg. It can be extracted by treatment with slaked lime or with lightly calcined dolomite:

$$MgCl_2 + Ca(OH)_2 \rightarrow Mg(OH)_2 + CaCl_2$$
$$MgSO_4 + Ca(OH)_2 \rightarrow Mg(OH)_2 + CaSO_4$$

When calcined dolomite is used as precipitant, the yield is almost doubled because the MgO in the calcined dolomite is also largely recovered. The precipitated $Mg(OH)_2$ is settled in tanks, filtered,

244

and then calcined or dead-burned to produce MgO. The first small-scale plant was put into operation in California in 1935. The first large-scale production was at West Hartlepool, England, in 1938; most of the magnesia needed for the production of basic refractories in Britain has since been derived from the sea.

Séailles Process. For the simultaneous production of alumina and portland cement from siliceous bauxite, or from an aluminous slag or clay. An appropriately proportioned batch is fired in a rotary kiln to give $5CaO.3Al_2O_3$ and $2CaO.SiO_2$; leaching of this product yields alumina and reburning of the residue with more lime yields a cement clinker. The process was introduced in 1928 by J. S. Séailles who later worked in collaboration with W. R. G. Dyckerhoff (whence the alternative name Séailles–Dyckerhoff Process), see their joint British Patent 545 149, 2/8/38.

Seal. See Ceramic-to-metal Seal; Glass-to-metal Seal; Metallizing; Sealing Glass.

Sealed Porosity. See under Porosity.

Sealing Glass. A glass that is suitable for use in sealing a glass envelope of an electronic valve, for example, to metal. The usual basis for the selection of such a glass is matching its thermal expansion and contraction with that of the metal over the range of temperature from that at which the seal is made to room temperature; however, a glass that is sufficiently soft (e.g. a lead glass) can accommodate considerable stress at a glass-metal seal by slowly yielding (see *Glass-to-Metal Seals* by J. H. Partridge: 1949).

Seam. See Joint Line.

Seat. A place prepared on the Siege (q.v.) of a glass-pot furnace for the support of a pot.

Seat Earth or Underclay. The material immediately beneath a coal seam, normally a fireclay but in some cases a silica rock (Ganister, q.v.).

Seating Block. A block of fireclay refractory shaped to support a boiler.

Secar. Trade-name: a pure calcium aluminate cement, suitable for use in making special refractory castables or shapes, made by Lafarge Aluminous Cement Co. Ltd. It is available in two grades; the purer (Secar 250) contains 70–72% Al_2O_3, 26–29% CaO, $< 1\%$ SiO_2 and $< 1\%$ Fe_2O_3.

Secondary Air. Air that is admitted to a kiln (that is being fired under oxidizing conditions) in an amount and in a suitable location to complete the combustion of the fuel initiated by the PRIMARY AIR (q.v.). In a kiln fired by solid fuel on a grate, the secondary air passes over the fuel bed and burns the combustible gases arising therefrom.

Secondary Clay. See SEDIMENTARY CLAY.

Secondary Crusher. A machine for the size-reduction of a feed up to 1 in. to a product passing about 8 mesh. This group of machines includes the finer types of jaw crusher and gyratory crusher, and also crushing rolls, hammer mills and edge-runner mills.

Seconds. Pottery-ware with small, not readily noticeable, blemishes (cf. FIRSTS, LUMP).

Sedimentary Clay or Secondary Clay. A clay that has been geologically transported from the site of its formation and re-deposited elsewhere. The English ball clays, for example, are secondary kaolins (cf. PRIMARY CLAYS).

Sedimentation. The settling out of solid particles from a liquid. Some deposits of clay, sand, etc., have been formed by sedimentation from lakes, estuaries and the sea. Sedimentation is used as a method of purifying clays, heavy impurities (e.g. iron compounds) settling out more quickly than the lighter particles of clay. Sedimentation is also a method for determining the particle size of clays and powders (cf. ELUTRIATION).

Sedimentation Volume. The volume occupied by solid particles after they have settled from suspension in a liquid. With most clays, the sedimentation volume depends on the degree of flocculation or deflocculation. Determination of the sedimentation volume of brick clays has been found to provide some indication of the fineness of the clay, its working moisture content, and drying shrinkage.

Seed. A fault, in the form of small bubbles, in glass. When near the surface of plate glass they sometimes become exposed, as minute depressions, during the polishing process; they are then known as BROKEN SEED.

Seger Cone. The first series of PYROMETRIC CONES (q.v.) was that of H. A. Seger (*Tonind. Zeit.*, **10**, 135, 168, 229, 1886); the standard cones of Germany are still known by his name. For table of equivalent temperatures see Appendix 2.

Seger Formula. A procedure introduced by H. A. Seger (*Collected Writings*, Vol. 2, p. 557), and still commonly used, for the representation of the composition of a ceramic glaze. The chemical composition is recalculated to molecular fractions and the constituent oxides are then arranged in three groups: the bases, which are made equal to unity; R_2O_3; RO_2. Example: $(0.3 Na_2O . 0.2 CaO . 0.5 PbO)$. $0.2 Al_2O_3 . (3.0SiO_2 . 0.7SnO_2)$.

Seger's Porcelain. A German porcelain compounded by H. A. Seger: 30% feldspar, 35–40% quartz, 30–35% kaolin. It is covered with a glaze prepared from 83·5 parts feldspar, 26 parts kaolin, 35 parts whiting, 54 parts flint. It is biscuit fired at a low temperature and glost fired at Cone 9.

Seger's Rules. A series of empirical rules put forward by the German ceramist H. A. Seger (*Collected Writings*, Vol. 2, p. 577) for the prevention of crazing and peeling. To prevent crazing, the body should be adjusted as follows: decrease the clay, increase the free silica (e.g. flint); replace some of the plastic clay by kaolin; decrease the feldspar; grind the flint more finely; biscuit fire at higher temperature. Alternatively, the glaze can be adjusted: increase silica and/or decrease fluxes; replace some SiO_2 by B_2O_3; replace fluxes of high equivalent weight by fluxes of lower equivalent weight. To prevent peeling, the body or glaze should be adjusted in the reverse direction.

Segmental Arch. A sprung arch having the contour of a segment of a circle.

Segmental Wheel. An abrasive wheel that has been built up from specially made segments of bonded abrasive; wheels up to 5 ft dia. can be made in this way.

Segregation. Partial re-separation of a previously mixed batch of material into its constituents; this can occur either as a result of differences in particle size or in density. Segregation is liable to occur in storage bins, on conveyors and in feeders during the dry or semi-dry processing of ceramic materials.

Seignette-electric. Seignette salt is the alternative name for Rochelle salt (Na-K tartrate). Crystals of this composition are markedly piezo-electric and were used, for this property, before titanate ceramics were introduced. The term 'seignette-electric' is still used in Western Europe and Russia to signify ferroelectric.

Selenium. Compounds of Sc are used for decolorizing glass; under other conditions this element will produce red coloured glass (see RUBY GLASS), pottery, or vitreous enamel.

Selvedge. The formed edge of a ribbon of rolled glass.

Semi-conducting Glaze. Porcelain insulators that are covered with a normal glaze are liable, particularly if the surface gets

Semi-conductor

dirty, to surface discharges which cause radio interference and may lead to complete flashover and interruption of the power supply. This can be largely prevented if the glaze is semi-conducting as a result of the incorporation of metal oxides such as Fe_2O_3, Fe_3O_4, MnO_2, Cr_2O_3, Co_3O_4, CuO or TiO_2; SiC and C have also been used as semi-conducting constituents.

Semi-conductor. Fundamentally, a material of intermediate energy gap between that of metals and that of dielectrics. Such materials have a negative temperature coefficient of resistance and are typified by some ceramic carbides (e.g. SiC) and variable-valency oxides. At very high temperatures most ceramics become semi-conducting.

Semi-continuous Kiln. A TRANSVERSE-ARCH KILN (q.v.) having only a single line of chambers, so that when the firing zone reaches one end of the kiln the process of fire travel must be re-started at the other end. Kilns of this type are uncommon.

Semi-dry Pressing. See DRY PRESSING.

Semi-muffle. See under MUFFLE.

Semi-porcelain. A trade term sometimes applied, both in the UK and the USA, to semi-vitreous tableware. The term is confusing and should not be used.

Semi-silica Refractory; Semi-siliceous Refractory. See SILICEOUS REFRACTORY; note, however, that ASTM – C27 defines a semi-silica refractory as containing $\not< 72\%$ SiO_2. PRE defines a semi-siliceous refractory as containing $< 93\%$ SiO_2 and $< 10\%$ $(Al_2O_3 + TiO_2)$; it is noted that PCE shall be considered as a further basis of classification.

Semi-vitreous or Semi-vitrified. These synonymous terms are defined in the USA (ASTM – C242) as signifying a ceramic whiteware having a water absorption between 0.5% and 10%, except for floor tiles and wall tiles which are deemed to be semi-vitreous when the water absorption is between 3% and 7%.

Sentinel Pyrometers. Small cylinders made from blended chemical compounds (non-ceramic) and so proportioned that they melt at stated temperatures within the range $220–1050°C$. They are used principally in controlling the heat treatment of metals and are available from The Amalgams Co. Ltd., Sheffield, England.

Sepiolite. $3MgO.4SiO_2.5H_2O$; the magnesian end-member of the series of clay minerals known as PALYGORSKITES (q.v.); it has been found in some of the Keuper Marls used for brickmaking in Central England.

Serpentine. A naturally occurring hydrated magnesium silicate, $3MgO.2SiO.2H_2O$; some of the Mg is commonly replaced by Fe. Serpentine is a common gangue mineral associated with chrome ore. Some types of serpentine are used as raw materials (together with dead-burned magnesite) for the manufacture of forsterite refractories.

Serrated Saddle. A refractory support for pottery-ware during kiln firing; this particular item is a rod of triangular section, its upper edge being serrated to facilitate the REARING (q.v.) of the ware.

SESCI Furnace. A rotary furnace, designed to burn low-volatile coal, and used for the melting of cast iron. A rammed siliceous refractory is generally used as lining material. The name derives from the initials of the French makers: Société des Entreprises Spéciales de Chauffage Industriel.

Setter. (1) A man who sets bricks or tiles in a kiln; (the man who sets pottery-ware is known as a PLACER).

(2) An item of KILN FURNITURE (q.v.); it is a piece of fired refractory material shaped so that its upper surface conforms to the lower surface of the piece of ware that it is designed to support during kiln firing.

Setting. (1) The arrangement of ware in a kiln. With pottery-ware, the setting generally consists of the individual pieces together with the kiln furniture that supports them; a setting of building bricks or refractories consists merely of the bricks themselves.

(2) A group of gas retorts or chambers within a bench, the group being heated independently of other groups of retorts in the same bench; also known as a BED.

(3) The process of hardening of a cement (see INITIAL SET and FINAL SET) or of a plaster.

Setting-up Agent. An electrolyte, e.g. K_2CO_3, $MgSO_4$, etc., added to a vitreous enamel slip or to slop glaze to flocculate the clay particles and thus to hold the coarser and heavier frit particles in suspension.

Settle Blow. The stage in the BLOW-AND-BLOW (q.v.) glass-making process when glass is forced into a finish or ring mould by air pressure.

Settle Mark. Any slight variation in the wall thickness of a glass container; sometimes called a SETTLE WAVE.

Sewer Brick. A specification for clay bricks intended for this use is given in ASTM – C32.

Sewer Pipe. An impervious clay pipe (usually glazed) for sewerage and trade-effluent disposal. Relevant British Standards are: B.S. 65, 539, 540, 1143, 1634. Relevant US specifications are: ASTM – C12, C13, C200, C261, C278, C301, C462, C463.

Sgraffito. A mode of decoration sometimes used by the studio potter: a coloured ENGOBE (q.v.) is applied to the dried ware and a pattern is then formed by scratching through the engobe to expose the differently coloured body. (From Italian *graffito*, to scratch.)

Shadow Wall. A refractory wall in a glass-tank furnace erected on the bridge cover; it may be solid or may have openings, its purpose being to screen the working end from excessive heat radiation. It is also sometimes known as a BAFFLE WALL or CURTAIN WALL but the latter term is by some authorities reserved for a suspended wall serving the same purpose.

Shaft. See STACK.

Shaft Kiln. A vertical kiln charged at the top and discharged at the bottom. If solid fuel is used it is fed in with the charge, but shaft kilns can also be fired with gas or oil by burners placed nearer to the bottom of the shaft. Such kilns are used for the calcination of flint, dolomite, fireclay, etc.

Shaft Mixer. See MIXER.

Shale. A hard, laminated, generally carbonaceous, clay. Colliery shales are frequently used as raw materials for building brick production.

Shale Cutter or Shale Planer. A mechanical excavator sometimes used for getting clay from deposits that are both hard and friable, and that will maintain a steep clay face. The main feature of the machine is an endless chain that carries a series of cutters which bear downwards on the clay, removing a layer about $\frac{3}{4}$ in. thick. The machine makes semi-circular sweeps into the clay face before being moved forward or sideways.

Shaping Block. A piece of wood for the preliminary shaping of glass on the blowing-iron before it is blown in a mould.

Sharon Quartzite. A quartzite occurring in Ohio, USA, and used in the manufacture of silica refractories. A quoted analysis is (per cent): SiO_2, 98·7; Al_2O_3, 0·3; Fe_2O_3; 0·3; alkalis, 0·3.

Sharp Finish. See FINISH.

Sharp Pitch. See under PITCH.

Shaw Kiln. A gas-fired chamber kiln; one feature is that a proportion of the hot gases pass beneath the kiln floor to diminish the temperature difference from top to bottom of the setting. The kiln was designed by Shaw's Gas Kiln Co. Ltd., in about 1925.

Shaw Process. A process for the precision casting of small metal components; its main feature is the use of silicon ester as the bond for the refractory powder, e.g. sillimanite, from which the mould is made. (N. Shaw, Brit. Pat., 716 394, 6/10/54 and later patents.)

Shearer Plastometer. See R & S PLASTOMETER.

Sheet Glass. Flat, transparent glass made by drawing or blowing.

Shell; Shelling. (1) The falling away of a 1–2 in. internal layer of refractory from the roof of an all-basic open-hearth steel furnace; the probable cause is the combined effect of flux migration, temperature gradient and stress. This form of wear is also known as SLABBING.

(2) A flake of glass chipped from the edge of glass-ware, or the hollow left by such a flake.

(3) The 'shell' of a hollow clay building block refers to the outer walls of the block.

Shellback, Shellbacked. An erroneous variant of SCHELLBACH (q.v.).

Shell Moulding. A method for the precision casting of metal in refractory moulds made from silica sand bonded with synthetic resin, sodium silicate or silicon ester; alumina and crushed high-silica (96% SiO_2) glass have also been used as the refractory component of these moulds.

Shell-stone. A CHINA STONE (q.v.) from Cornwall containing too much iron (as brown mica) for use as a flux in pottery glazes but of potential value in sewer-pipe glazes.

Shell Wall or Ring Wall. The wall of fireclay refractories that protects the outer steel casing of a HOT-BLAST STOVE (q.v.).

Shelving. The effect produced in the refractory lining of a glass-tank furnace by severe erosion of the horizontal joints between the tank blocks.

Shielding Glass. A protective glass for use in nuclear engineering; although it is transparent to visible light, it absorbs high-energy electromagnetic radiation. Such glasses contain a maximum proportion of oxides of heavy elements, e.g. PbO, Ta_2O_5, Nb_2O_5, WO_3.

Shillet. Local term for the Devonian slates used for brick-making in the Plymouth and Torquay areas of England; the word is a dialect form of 'shale'.

Shiners. Minute FISH SCALES (q.v.) that sparkle in reflected light; they are liable to occur in vitreous enamelware if the ground-coat is overfired.

Ship-and-Galley Tile. US term for a special FLOOR QUARRY (q.v.) with an indented anti-slip pattern on its face.

Shivering. See PEELING.

Shoe. A hollow refractory shape that is placed in the mouth of a glass pot and used for heating the BLOWING IRONS (q.v.).

Shop. (1) A department or room in a pottery, etc.

(2) A team of men operating a blowing or pressing process in a glass-works.

Shore Hardness Tester or Scleroscope. A procedure for the determination of the hardness of a surface by dropping a ball from a fixed height above the surface and noting the height of rebound. This technique was first proposed by A. F. Shore, an American, in 1906. Although primarily for the testing of the hardness of metals, it has also been applied to a limited extent in the testing of ceramics.

Shore-lines. A fault in vitreous enamelware similar in appearance to the marks left on a shore by receding water. The basic cause is the drying out of soluble salts. The cure is adjustment of the mill additions and of the drying process.

Short. Term applied to a clay body that has little workability, or to a glass that quickly sets.

Short Glazed. See STARVED GLAZE.

Shot. See SLUG.

Shoulder Angle. A special shape of ceramic wall tile (see Fig. 6, p. 307).

Shouldering. The splay at the top right-hand and bottom left-hand corners of a single-lap roofing tile.

Shraff. N. Staffordshire word for the waste (e.g. broken fired ware, broken saggars, old plaster moulds) from a pottery.

Shredder. See CLAY SHREDDER.

Shrend. US term for the process of making CULLET (q.v.) by running molten glass into water (cf. DRAG-LADLE).

Shuff. A clamp-fired STOCK BRICK (q.v.) that is of too poor a quality for use.

Shuttle Kiln. An intermittent bogie kiln consisting of a box-like structure with doors at each end and accommodating kiln cars (usually two in number). Pottery-ware is set on the refractory decks of the cars which are then pushed along rails into the kiln, displacing two other cars of fired ware from the kiln. The fired ware is taken from the displaced cars which are then re-set with more ware to be fired. The shuttle movement of the kiln cars is then repeated (cf. BOGIE KILN).

Sial. A borosilicate glass, of high chemical resistance and thermal endurance, made by the Kavalier Glassworks, Sazava, Czechoslovakia.

Side Arch. A brick with the two largest faces symmetrically inclined towards each other, see Fig. 1, p. 37.

Side-blown Converter. See CONVERTER.

Side Feather. See FEATHER BRICK.

Side Lap. The distance by which the side of a roofing tile overlaps the joint in the course of tiles next below.

Side Pocket. Alternative name for SLAG POCKET (q.v.) as applied to glass-tank furnaces.

Side Skew. A brick with one of the side faces completely bevelled at an angle of 60°. An arch can be sprung from such a brick. See Fig. 1, p. 37.

Siderite. Natural ferrous carbonate, $FeCO_3$, found as an impurity in some clays.

Siege. The refractory floor of a glass furnace, particularly of a pot furnace; (from French *siège*, a seat).

Siemens or Siemens-Martin Furnace. See OPEN-HEARTH FURNACE.

Sieve. This term is generally reserved for testing equipment; the corresponding industrial equipment is generally called a SCREEN (q.v.). There are several standard series of test sieves; those most frequently met with in the ceramic industry are:

British Standard sieves (conforming with B.S. 410)

USA Standard sieves (conforming with National Bureau of Standards LC-584; or ASTM – E11)

French Standard sieves (AFNOR NF 11-501)

German Standard sieves (DIN 4188)

Tyler sieves, in which the ratio between the mesh sizes of successive sieves in the series is $\sqrt{2}$; thus the areas of the openings of each sieve are double those of the next finer sieve. Institution of Mining and Metallurgy, London (I.M.M.-Sieves) (For a comparison of these series see Appendix 1).

Sigma Cement

Sigma Cement. Trade-name: a hydraulic cement made by mixing portland cement consisting only of particles $< 30\mu$ with 16–50% of an inert extender, e.g. limestone, basalt or flint, the particle size of which is 30–200μ. (S. Gottlieb, Brit. Pat., 580 291, 26/6/44.)

Signal Glass. Coloured glass for light signals on railways, roads, airfields and at sea; the glass must conform to a close specification (B.S. 1376).

Silcrete. A type of quartzite that is either cryptocrystalline or chalcedonic. It occurs, typically, at Albertinia, Mossel Bay, S. Africa; the porosity is low and the Al_2O_3 content is less than 0·5%.

Silex. A name sometimes applied to CHERT (q.v.), particularly to the trimmed blocks of this material, from Belgium, used in the lining of ball mills. For the milling of vitreous-enamel frits, silex blocks are used more in mills for ground-coats than for cover-coats.

Silica. Silicon dioxide, SiO_2. For the various forms in which silica occurs see CHALCEDONY, CHERT, COESITE, CRISTOBALITE, FLINT, KEATITE, QUARTZ, TRIDYMITE and VITREOUS SILICA.

Silica Fireclay. A US term defined as: A refractory mortar consisting of a finely ground mixture of quartzite, silica brick and fireclay.

Silica Gel. A form of silica produced by treatment of a solution of sodium silicate with acid and/or other precipitant. The dried gel is highly porous. It has found little use in the ceramic industry.

Silica Glass. See VITREOUS SILICA.

Silica Modulus. The ratio of $SiO_2 : (Al_2O_3 + Fe_2O_3)$ in a hydraulic cement. In portland cement this modulus usually lies between 2 and 3. A cement with a low silica modulus can be expected to have high early strength, but if this modulus is high the final strength will be the greater.

Silica Refractory. Defined in B.S. 3446 as a refractory material that, in the fired state, contains not less than 92% SiO_2. Refractories of this type were first made in 1822, by W. W. Young, from the quartzite of Dinas, S. Wales, and are still often referred to as Dinas bricks in Germany and Russia. They are characterized by high RuL (q.v.) but are sensitive to thermal shock at temperatures up to 600°C. Their principal use is in steel furnaces, coke-ovens, gas-retorts and glass-tank furnaces. Silica refractories for coke-ovens and gas retorts in the UK must comply with the specification laid down by the Gas Council in collaboration with the

Society of British Gas Industries and the British Coking Industry Association. The principal features of this specification are: $\ll 94\%$ SiO$_2$; PCE $\ll 31$; specific gravity $\gg 2\cdot345$; permanent linear change after 2 h at 1450°C $\gg 0\cdot50$; cold crushing strength $\ll 2500$ p.s.i. In the USA silica refractories are defined (ASTM – C416) as containing $\gg 1\cdot50\%$ Al$_2$O$_3$, $\gg 0\cdot20\%$ TiO$_2$, $\gg 1\%$ Fe$_2$O$_3$, $\gg 4\%$ CaO; they are classified on the basis of a 'flux factor' equal to the percentage Al$_2$O$_3$ plus twice the total percentage of alkalis; Type A has a flux factor of $0\cdot50$ or less, Type B includes all other silica refractories meeting the general part of the specification. (See also SUPER-DUTY SILICA REFRACTORY.)

Silica Sol. Colloidal silica in the form of a dispersion in water. The modern method of manufacture involves the passage of sodium silicate solution through an ion-exchanger. As made, the sols contain about 3% SiO$_2$ but the concentration can be increased to 20% by evaporation; to give stability, a concentrated sol must have a larger particle size or a stabilizer, e.g. NaOH, must be added. In the ceramic industry, silica sol has been used to a small extent as a bond.

Silicate Bond. (1) As applied to ceramics other than abrasives, this term means that the fired material consists of larger grains set in a matrix of a complex, usually glassy, silicate.

(2) In the abrasives industry this term means a bond consisting essentially of sodium silicate, the abrasive wheels so bonded being baked at about 250°C.

Silicate Cement. As used in chemical engineering, this term denotes an acid-proof cement, consisting of an inert powder and sodium silicate solution, for jointing chemical stoneware or acid-resisting bricks.

Siliceous Refractory. Defined in B.S. 3446 as a refractory material that, in the fired state, contains not less than 78% but less than 92% SiO$_2$, the remainder being essentially Al$_2$O$_3$. SEMI-SILICA and SEMI-SILICEOUS refractories are also comprised within this definition. The specification for refractory materials for the gas and coking industries laid down by the Gas Council in collaboration with the Society of British Gas Industries and the British Coking Industry Association subdivides siliccous refractories into three groups: Clay-bonded Siliceous 'B', with $\ll 88\%$ SiO$_2$; Clay-bonded Siliceous 'C', with $\ll 85\%$ SiO$_2$; Semi-siliceous Firebrick, with 78–85% SiO$_2$.

The PRE definition of a siliceous refractory is that the SiO$_2$

content shall not exceed 93%, a further quality specification being according to use.

Silicides. A group of special ceramic materials: see under the silicides of the following elements: Cr, Mo, Ti.

Silicon Borides. Two compounds have been reported: SiB_4, oxidation-resistant to 1370°C; SiB_6, m.p. 1950°C. A special refractory has been made by reacting Si and B in air, the product containing SiB_4 and Si in a borosilicate matrix; it is stable in air to at least 1550°C and has good thermal-shock resistance.

Silicon Carbide. SiC; dissociates at approx. 2250°C; sp. gr. 3·2; hardness > 9 Moh; thermal expansion (25–1400°C) $4·4 \times 10^{-6}$; thermal conductivity 90 Btu/h/ft²/in./°F, the value varying but little with temperature; excellent thermal-shock resistance. Electrically, SiC is a semi-conductor; the resistivity of the bonded material is largely determined by the nature of the bond. Major uses include abrasives, refractories and furnace heating elements. Self-bonded SiC, of almost zero porosity, is a special ceramic having a Knoop hardness of 2740 and a modulus of rupture of 25 000 p.s.i.; it finds considerable use as a wear-resistant material, particularly in paper-making machinery.

Silicon Ester. An organic silicate, e.g. ethyl silicate; it is used as a binder in cases when the addition of any fluxing material would be undesirable.

Silicon Nitride. Si_3N_4; there are two forms, both hexagonal, the α-form changing irreversibly to the β-form at approx. 1700°C; sublimation occurs at 1900°C. Si_3N_4 is made by forming (pressing, slip-casting, extruding) fine Si powder and nitriding it at 1350°C; the firing involves an expansion of about 1%. As a special ceramic, Si_3N_4 is strong, hard, and has good resistance to thermal shock and to attack by many chemicals and molten metals; thermal expansion, (20–1000°C), $2·5 \times 10^{-6}$. It is a good electrical insulator.

Silicone. Organo-silicon polymers some of which can be used up to a temperature of 300°C or even higher. Silicones have been used in the ceramic industry as mould-release agents, for the coating of glass-ware to increase its strength and chemical durability, and for spraying on brickwork to give it water-repellancy.

Silicosis. A lung disease caused by inhalation, over a long period, of siliceous dusts, particularly those containing a high proportion of free silica of a size between 1μ and 2μ. In the UK the dust conditions in ceramic factories are limited by the Pottery (Health and Welfare) Special Regulations of 1947 and 1950.

Silit. Trade-mark: a heating element consisting principally of SiC; to maintain a constant resistance for a long period, Si is included in the batch and the shaped rods are fired in a controlled atmosphere to cause some nitridation and/or carbonation. (Siemens–Planiawerke A.G., Germany.)

Silk-screen Process. A decorating method that can be applied to pottery, glassware or vitreous enamelware. The simplest silk-screen equipment consists of a frame over which is stretched silk bolting cloth, or fine wire gauze, of 125–150 meshes per inch. A stencil is then placed on the frame and varnish is applied to fill in those parts of the screen not covered by the design. Colour, dispersed in a suitable oil, is pressed through the open parts of the screen by means of a roller and the pattern of the stencil is thus reproduced on the ware. The stencil can be made photographically. A further development was the screen-printed collodion-film transfer, which has itself been improved to give screen-printed cover-coat transfers.

Sill. A horizontal course of brickwork under an opening, e.g. the window of a house or the door or port of a furnace.

Sillimanite. A mineral having the same composition (Al_2SiO_5) as kyanite and andalusite but with different physical properties. The chief sources are S. Africa and India. Sillimanite changes into a mixture of mullite and cristobalite when fired at a high temperature (1550°C); this change occurs without any significant alteration in volume (cf. KYANITE). The mineral is used as a refractory, particularly in the glass industry (cf. SILLIMANITE REFRACTORY).

Sillimanite Refractory. A refractory material made from any of the SILLIMANITE (q.v.) group of minerals. Such refractories generally contain about 60% Al_2O_3; they have a high RuL (q.v.) and good spalling resistance. Sillimanite refractories are much used in glass-tank furnaces and in kiln furniture for use in the firing of pottery-ware.

Silver Lustre. Because silver itself tends to tarnish, 'silver lustre' is, in fact, made from platinum, with or without the addition of gold.

Silver-marking of Glazes. Silver cutlery, or other relatively soft metal, will leave a very thin smear of metal on pottery-ware if the glaze is minutely pitted. A glaze may have this defective surface as it leaves the glost kiln, or it may subsequently develop such a surface as a result of inadequate chemical durability. The fault is also known as CUTLERY-MARKING.

Silvering. The formation of a thin film of silver on glass; such a film can be produced by treatment of the glass surface with an ammoniacal solution of $AgNO_3$ together with a reducing agent. A trace of copper is also often included.

Simax. A glass, of very high thermal endurance and good chemical resistance, made by the Kavalier Glassworks, Sazava, Czechoslovakia.

Simplex Kiln. A type of annular kiln in which two barrel-arch galleries are each divided by transverse walls into eight or nine chambers, each of which has grates at the corners for the hand-firing of solid fuel. This kiln can be used for the firing of facing bricks, roofing tiles, firebricks or (because the flue system permits each chamber to be isolated) blue engineering bricks.

Singer's Test. A rough test for glaze-fit proposed by F. Singer (*Sprechsaal*, **50**, 279, 1917); the glaze is placed in a dish of the biscuit ware and fired to its normal maturing temperature; when cold, the glazed dish is examined for faults.

Single-bucket Excavator. There are two principal types of this machine: the MECHANICAL SHOVEL (q.v.) and the DRAG-LINE (q.v.).

Single-fired Ware. See ONCE-FIRED WARE.

Single-lap Roofing Tile. This term, as defined in B.S. 1424 ('Clay Single-lap Roofing Tiles and Fittings') includes pantiles, double Roman tiles, flat interlocking tiles and Pooles tiles.

Single-screened Ground Refractory Material. A US term defined as: A refractory material that contains its original gradation of particle sizes resulting from crushing, grinding, or both, and from which particles coarser than a specified size have been removed by screening (cf. DOUBLE-SCREENED).

Single-toggle Jaw Crusher. A jaw crusher with one jaw fixed, the other jaw oscillating through an eccentric mounted near its top. This type of jaw crusher has a relatively high output and the product is of fairly uniform size.

Sinter. See SINTERING.

Sintered Filter. A filter made from sintered glass, sintered silica or unglazed ceramic. B.S. 1752 refers to sintered disk filters for laboratory use varying in maximum pore size from Grade 00 (250–500μ) to Grade 5 ($< 2\mu$). B.S. 1969 describes methods for the testing of these filters (cf. CERAMIC FILTER).

Sintered Glass. Glass-ware of controlled porosity used for filtration, aeration, etc.; it is made by carefully heating powdered glass so that the surfaces of the particles begin to melt and adhere to one another.

Sintering. The process of heating a compacted powder to a temperature lower than that necessary to produce a liquid phase but sufficiently high for solid-state reaction or intercrystallization to take place so that the fired body acquires strength (cf. REACTION-SINTERING; VITRIFICATION).

Siporex. Trade-name: A lightweight (30 lb/ft³) pre-cast concrete made from portland cement, fine sand, Al powder and water; the set aerated blocks are autoclaved. (Costain Concrete Co. Ltd., London.)

Sitter-up. An assistant TEASER (q.v.) in a glass-works.

Size. A solution used to treat the surface of pottery ware or of plaster moulds. Mould-makers' size is commonly a solution of soft soap. Decorators' size is traditionally based on boiled linseed oil; after its application, and after it has become tacky, the lithograph is applied and brushed down.

SK Porosity Test. A method for the determination of the porosity of aggregates; the principle is to fill the voids in turn with mercury, air and water. The method, primarily developed for the testing of iron ores, was proposed by H. L. Saunders and H. J. Tress at South Kensington, hence the name 'SK'; (*J. Iron Steel Inst.*, **152**, No. 2, 291P, 1945).

Skewback. A refractory block having an inclined face, or a course of such blocks forming the top of a wall, from which an arch or furnace roof may be sprung; also known as SPRINGER.

Skids. Refractory skids for use in pusher-type reheating furnaces in the steel industry have been made from sintered or fused alumina; they have a long service life and permit more uniform heating of the ingots or billets than do water-cooled metal skids.

Skimmer. A single-bucket excavator in which the bucket travels along the boom, which is kept almost horizontal during operation. This machine is sometimes used for removing OVERBURDEN (q.v.).

Skimmer Block. A special refractory block that is partly immersed in the molten glass in a tank furnace to prevent impurities from entering the feeder channel.

Skimming Pocket. One of the small recesses, on the sides of a tank furnace for the production of flat glass, by means of which

Skintling

impurities on the surface of the molten glass can be removed. The skimming pockets are usually located a short distance before the FLOATERS (q.v.).

Skintling. The setting of bricks in a kiln so that courses of bricks lie obliquely to the courses above and below.

Skittle Pot. A small POT (q.v.) for glass melting.

Skiving. US term for a finishing process, on a lathe, of partially-dried ceramic ware such as h.t. insulators, ribbed formers or sparking-plug insulators.

Skull. Metal remaining in a steel-casting ladle at the end of teeming; its removal often causes damage to the refractory lining of the ladle. The same term is used for the glass left in a ladle after most of the molten glass has been poured in glass-making; in the glass industry the usual spelling is SCULL.

Slab Glass. A block of optical glass resulting from preliminary shaping of CHUNK GLASS (q.v.).

Slabbing. (1) The fixing of ceramic tiles to fireplace surrounds, etc., to produce a prefabricated unit.

(2) A form of failure of refractories, also known as SHELLING (q.v.).

Slag. (1) The non-metallic fusion that floats on a metal during its extraction or refining.

(2) The product of reaction between fluxing materials and a refractory furnace lining.

Slag Cement. A mixture of granulated blast-furnace slag and lime, together with an ACCELERATOR (q.v.); very little is now made (cf. PORTLAND BLAST-FURNACE CEMENT).

Slag Line. The normal level of the slag/metal interface in the working chamber of a metallurgical furnace. The refractory lining of the furnace is liable to be severely eroded at this level owing to the improbability of chemical equilibrium between slag, metal and refractory.

Slag Notch. The hole in the refractory brickwork of the wall of the hearth of a blast furnace permitting molten slag to flow from the furnace as and when necessary. It is also sometimes known as the CINDER NOTCH.

Slag Pocket. A refractory-lined chamber at the bottom of the downtake of an open-hearth steel furnace, or of a glass-tank furnace, designed to trap slag and dust from the waste gases before they enter the regenerator.

Sleeper Block. See THROAT.

260

Sleeper Wall. In structural brickwork, a low wall (generally built with openings checkerwise) built to carry floor joists.

Sleeve. See ROD COVER.

Slicking. The effect on plastic clay of moving over its surface a smooth metal blade; the combined action of pressure and movement tends to bring water, soluble salts and fine clay particles to the surface. If slicking occurs as a result of movement of the knives or screw in a pug, it causes lamination as the affected clay surfaces do not readily knit together again.

Slide-off Transfer. A type of transfer for the decoration of pottery. The pattern is SILK-SCREEN (q.v.) printed on litho paper and then covered with a suitable plastic medium. Prior to use, the transfer is soaked in water so that the pattern, still firmly fixed to the plastic, can be slid off the paper and applied to the ware. During the subsequent decorating fire, the plastic coating burns away.

Sliding. The faulty draining of wet-process enamel by the slipping downwards of patches of enamel; one cause is over-flocculation of the enamel slip.

Sliding-bat Kiln. See PUSHED-BAT KILN.

Slinger Process. A method used in USA for the moulding of insulating refractories; the wet batch is thrown on to pallets by a rotary machine—a 'slinger'—to form a column on a belt conveyor; the column is then cut to shape, dried and fired.

Slip. A suspension in water of clay and/or other ceramic materials; normally a DEFLOCCULANT (q.v.) is added to disperse the particles and to prevent their settling out. In the whiteware industry, a slip is made either as a means of mixing the constituents of a body (in which case it is subsequently dewatered, e.g. by filter-pressing) or preparatory to CASTING (q.v.). In vitreous enamelling, a slip is used for application of the enamel to the ware by spraying or dipping.

Slip Casting. See CASTING.

Slip Glaze. US term for a glaze made essentially from a fusible clay.

Slip House. The department of a pottery factory where the clays and other constituents are dispersed in water and blended to produce the slip.

Slip Kiln. A heat resistant trough placed over a heated flue; water can be evaporated from slip placed on the trough. This method of dewatering has been replaced by filter-pressing.

Slip Trailing. See TRAILING.

Slipware. An early type of pottery (revived by studio potters) usually having a red body and a lead glaze, decorated with white or coloured slip by dipping, trailing or sgraffito. English slipware was made in the 17th and early 18th century, the chief centres being Staffordshire, Kent, Sussex and the West Country.

Slop Glaze. A suspension of glaze-forming materials prepared for application to ceramic ware, usually by dipping or spraying. The materials are kept in suspension by the presence of dispersed clay and by the high concentration of solids. A small amount (about 0·02%) of calcium chloride is also added to prevent the glaze from setting in the glaze tub and to act as a thickening agent.

Slop Moulding. The hand-moulding of building bricks by a process in which the clay is first prepared at a water-content varying from 20 to 30% depending on the clay. The wet clay is thrown into a wooden mould, pressed into the corners, and the top surface is finally struck smooth with a wet wooden stick. The filled mould is set on a drying floor until the clay has dried sufficiently to maintain its shape; the mould is then removed and the drying process is completed.

Slop Peck. The volume of slip that contains 20 lb of dry material.

Slop Weight. The weight (oz) per pint of a suspension of clay, flint, etc., in water.

Slotting Wheel. A thin abrasive wheel used for cutting slots or grooves; such wheels are usually made with an organic bond.

Sludge Pan. See TEMPERING TUB.

Slug. (1) A rough piece of prepared clay body sufficient for making one piece of ware, by throwing or by jiggering for example.

(2) A fault in a glass-fibre product resulting from the presence of non-fibrous glass; also called SHOT.

Slugged Bottom. A fault in a glass container characterized by the bottom being thick on one side and thin on the other; also called a WEDGED BOTTOM and (in USA) HEEL TAP.

Slump Test. (1) A rough test for the consistency of freshly-mixed concrete in terms of the subsidence of a truncated cone of concrete when upturned from a bucket; ASTM – C143 and B.S. 1881.

(2) A works' test for the consistency of vitreous enamel slip; for details see IRWIN SLUMP TEST.

Slushing. A method for the application of vitreous-enamel slip

262

to ware, particularly to small awkwardly-shaped items. The slip is applied by dashing it on the ware to cover all its parts, excess then being removed by shaking the ware. A slip of 'thicker' consistency than normal is required for this process.

Smalt or Powder Blue. A fused mixture of cobalt oxide, sand and a flux, e.g. potash. It is sometimes used as a blue pigment for the decoration of pottery or for the colouring of glass or vitreous enamel (cf. ZAFFRE).

Smear. In the glass industry the word has the special meaning of a surface crack in the neck of a glass bottle (cf. CHECK).

Smectite. Obsolete name for the clay mineral HALLOYSITE (q.v.).

Smoked. Glass or glaze discoloured by a reducing flame.

Snagging. Rough grinding with an abrasive wheel to remove large surface defects; wheels with an organic bond are generally used.

Snakeskin Glaze. A decorative effect obtainable on potteryware by the application of a glaze of high surface tension, e.g. 320 dynes/cm. A quoted glaze formula for the production of a snakeskin glaze maturing at 1140°C is:
(0·3 PbO; 0·3 MgO; 0·2 CaO; 0·2 ZnO). 1·5 SiO_2. 0·2 Al_2O_3.

Snap-header. See HALF-BAT.

Soaking. As applied to the firing of ceramic ware this term signifies the maintenance of the maximum temperature for a period to effect a desired degree of vitrification, chemical reaction and/or recrystallization.

Soaking Area. The part of a cross-fired glass-tank furnace between the GABLE WALL (q.v.) and the first pair of ports; also known as the FRITTING ZONE.

Soaking Pit. In the steel industry, a refractory-lined furnace for the reheating of ingots. In the glass industry, a furnace for bringing pots of glass to a uniform temperature.

Soap. See PUP.

Soapstone. A popular name for STEATITE (q.v.).

Soda Ash. Anhydrous sodium carbonate, Na_2CO_3. A major constituent of most glass batches. Together with sodium silicate it is used to deflocculate clay slips.

Sodium Aluminate. An electrolyte used as a mill addition in the preparation of acid-resisting vitreous enamel slips; 0·10–0·25% is generally sufficient.

Sodium Antimonate. $Na_2O.Sb_2O_5.\frac{1}{2}H_2O$. An opacifier for vitreous enamels and a source of antimony for yellow ceramic colours.

Sodium Cyanide

Sodium Cyanide. NaCN; used as a neutralizer in vitreous enamelling.

Sodium-line Reversal Method. A technique for the measurement of flame temperatures. If a black-body is viewed, by means of a spectroscope, through a flame that has been coloured by sodium, there will be some temperature of the black-body at which its brightness in the spectral region of the Na_D line will equal the brightness of the light transmitted in this region through the flame, plus the brightness of the Na_D lines from the flame itself. At this temperature the spectrum of the black-body as seen in the spectroscope is continuous; there is a reversal in contrast above or below this temperature. The method was first described by C. Féry (*Compt. Rend.*, **137**, 909, 1903).

Sodium Niobate. $NaNbO_3$; a compound believed to be ferroelectric and having potential use as a special electroceramic. The Curie temperature is 360°C.

Sodium Nitrate. $NaNO_3$; used as an oxidizing agent in some glass batches and enamel frits. The old name was Saltpetre.

Sodium Nitrite. $NaNO_2$; added in small quantities to vitreous enamels to prevent rust-spotting and tearing; 0·1–0·25% is usually sufficient.

Sodium Phosphate. The hexa-metaphosphate (generally under the proprietary name CALGON) is used to control the viscosity of slips in the pottery and vitreous enamel industries.

Sodium Silicate. Name given to fused mixtures in which the $Na_2O:SiO_2$ ratio normally varies from about 1:2 to 1:3·5; care should be taken in selecting the grade best suited to the purpose in view, e.g. as a deflocculant or as an air-setting bond for refractory cements, etc. More alkaline silicates are used for cleaning metal, prior to enamelling for example.

Sodium Sulphate. See SALT CAKE; SODIUM SULPHATE TEST.

Sodium Sulphate Test. A test claimed to indicate the resistance of a clay building material to frost action. The test-piece is soaked in a saturated solution of sodium sulphate and is then drained and dried; the cycle is repeated and the test-piece is examined for cracks after each drying. The principle underlying the test is that the stresses caused by the expansion of sodium sulphate as it crystallizes are, to some exent, similar to the stresses caused by water as it freezes. The test has also been used to reveal laminations present in bricks.

Sodium Tannate. Sometimes used as a deflocculant for clay

264

slips; the effect is marked, only a small proportion being required. The material used for this purpose is generally prepared from NaOH and tannic acid; the former should be in excess and the pH should be about 8–9.

Sodium Tantalate. $NaTaO_3$; a ferroelectric compound having the ilmenite structure at room temperature; the Curie temperature is approx. 475°C. Of potential interest as a special electroceramic.

Sodium Uranate. $Na_2O.2UO_3.6H_2O$; has been used as a source of uranium for URANIUM RED (q.v.).

Sodium Vanadate. $NaVO_3$; a ferroelectric compound having potential use as a special electroceramic. The Curie temperature is approx. 330°C.

Sofim–Fichter Kiln. A gas-fired open-flame tunnel kiln; its novel feature, when introduced in about 1935, was the design of the pre-mix burners. The name derives from the initial letters of Société des Fours Industriels et Métallurgiques, the designers of the original kiln, and Fichter (of Sarreguemines) who designed the burners.

Soft. As applied to a glass or glaze, this word means that the softening temperature is low; such a glass or glaze, when cold, is also likely to be relatively soft, i.e. of lower than average hardness, in the normal sense.

Soft-mud Process. A process for the shaping of building bricks from clay at a water content of about 35%. The prepared wet clay is fed into sanded moulds which are then shaken or jolted until the clay fills the mould; because of its THIXOTROPY (q.v.), after the jolting ceases the clay stiffens sufficiently for the bricks to maintain their shape. The process can form the basis of hand-making or, more commonly, it can be mechanized as in the BERRY MACHINE (Berry & Son, Southend-on-Sea, England) or in the ABERSON MACHINE (Aberson, Olst, Holland).

Soft-paste or Fritted Porcelain. A type of porcelain made from a body containing a glassy frit and fired at a comparatively low temperature (1100°C). The most famous soft-paste ware was that produced in the 18th century at the Sèvres factory in France, and at Chelsea, Derby, Bow, Worcester and Longton Hall in England.

Softening Point. Generally, the indefinite temperature at which a ceramic material begins to melt. The term has a definite meaning however, when referring to glass, namely the temperature at which the viscosity is $10^{7.6}$ poises; this viscosity corresponds to the temperature at which tubes, for example, can be conveniently bent

Solar Furnace

and was first proposed as a basis for definition by J. T. Littleton (*J. Soc. Glass Tech.*, **24**, 176, 1940). Also known as the 7·6 TEMPERATURE or the LITTLETON SOFTENING POINT.

Solar Furnace. A particular type of IMAGE FURNACE (q.v.).

Solarization. An effect of strong sunlight (or artificial ultra-violet radiation) on some glasses, causing a change in their transparency. Glasses free from arsenic and of low soda and potash contents are less prone to this defect.

Soldier Blocks. Refractory blocks set on end. In the glass industry the term is particularly applied to blocks that extend more than the depth of the glass in a furnace.

Sole. The refractory brickwork forming the floor of a coke-oven; as the charge of coal—which is subsequently transformed into coke—rests on this brickwork, and as the coke is pushed out of the oven by means of a mechanical ram, the sole is subjected to severe abrasion.

Sole-flue Port Brick. See RIDER BRICKS.

Solid Casting. See under CASTING.

Solid Solution. A crystalline phase, the composition of which can, within limits, be varied without the appearance of an additional phase. In Fig. 2 (p. 100), a solid solution of mullite is shown to exist ranging in composition from that of pure mullite 60 mol% Al_2O_3) to 63 mol% Al_2O_3. A complete series of solid solutions exists between forsterite (Mg_2SiO_4) and fayalite (Fe_2SiO_4); these solid solutions occur naturally as the mineral olivine and are important in the technology of forsterite refractories.

Solidus. In a binary equilibrium diagram without solid solutions (see left-hand side of Fig. 2, p. 100, representing the binary between silica and mullite), the solidus is the line showing the temperature below which a given composition, when in chemical equilibrium, is completely solid. At temperatures between the solidus and the LIQUIDUS (q.v.) a composition will contain both solid and liquid material. In a ternary system without solid solutions, the solidus is a plane.

Soluble Salts. Salts, particularly sulphates of Ca, Mg and Na, present in some clays. When the clay is dried these salts migrate to the surface. In the pottery industry this can cause trouble during glazing; in the wet process of body preparation, however, these salts are mostly eliminated during filter-pressing. If present in clay building materials, soluble salts can cause EFFLORESCENCE (q.v.) and SCUMMING (q.v.).

266

Solution Ceramics. A type of ceramic coating introduced by Armour Research Foundation, USA. In the original process (Brit. Pat., 776 443, 5/6/57) a solution containing a decomposable metal salt is sprayed on the hot surface that is to be coated. A subsequent development (Brit. Pat., 807 302, 14/1/59) refers to the application of a coating of vitreous enamel or thermoplastic resin to the surface that has been flame-sprayed with the solution; this is claimed to result in a vitreous-enamelled surface having improved resistance to thermal shock.

Sorel Cement. A cement, used particularly in monolithic flooring compositions, first made by Stanislaus Sorel (*Compt. Rend.*, **65**, 102, 1867). Calcined magnesia is mixed with a 20% solution of $MgCl_2$, interaction forming a strong bond of Mg oxychloride. In flooring compositions, various fillers are used, e.g. sawdust, silica, asbestos, talc; colouring agents are also added. The relevant British specification is B.S. 776; there are ASTM specifications for sampling and testing the constituent materials.

Sorting. The removal, from pottery taken from the glost kiln, of adhering bedding material and/or particles that have become detached from the kiln furniture; sorting is usually done with a small pneumatic tool.

Soundness. As applied to portland cement, this term refers to its volume stability after it has set (cf. UNSOUNDNESS).

Souring. (1) An alternative term for AGEING (q.v.).

(2) The storage for a short while of the moistened batch for making basic refractories; some magnesium hydroxide is formed and this acts as a temporary bond after the bricks have been shaped and dried. If souring is allowed to proceed too far, cracking of the bricks is likely during drying and the initial stages of firing. The high pressure exerted by modern brick presses generally gives sufficient dry-strength without the bricks being soured, and this process is therefore now generally omitted.

Spacer. The tapered section of a pug joining the barrel to the die; in this section beyond the shaft carrying the screw or blades the clay is compressed before it issues through the die.

Spalling. (1) The cracking of a refractory product or, in severe cases, the breaking away of corners or faces. The principal causes are: thermal shock, crystalline inversion, a steep temperature gradient, slag absorption at the working face with a consequent change in properties, or pinching due to inadequate

267

Spanish Tile

expansion allowance (see also HOT-PLATE SPALLING TEST; PANEL SPALLING TEST).

(2) The chipping of vitreous enamelware in consequence of internal stress.

Spanish Tile. A fired clay roofing tile that, in section, is a segment of a circle; the tile tapers along its length so that the lower end of one tile will fit over the upper end of the tile below (cf. ITALIAN TILES).

Spark Plug (US); Sparking Plug (UK). The older type of ceramic core for sparking plugs was made of ELECTRICAL PORCELAIN (q.v.); the modern type is sintered alumina. The cores are shaped by pressing, rotary tamping or injection moulding; they are fired at 1600–1650°C. The standard size is 14 mm (B.S. 45).

Spark Test. A method for the detection of pin-holes in a vitreous enamel surface by the discharge of an electric spark.

Sparrow-pecked. See STIPPLED.

Special Oxides. See OXIDE CERAMICS.

Specific Gravity. See APPARENT SPECIFIC GRAVITY; BULK SPECIFIC GRAVITY; TRUE SPECIFIC GRAVITY. When the term is used without qualification it may be assumed that true specific gravity is meant.

Specific Surface. The total surface area per unit weight of a powder or porous solid; in the latter case the area of the internal surfaces of the pores is included. The usual units are m^2/g. Methods of determination include GAS ADSORPTION (q.v.) and PERMEABILITY (q.v.). See also SURFACE FACTOR.

Speckled Ware. Vitreous enamelware having small particles of one colour uniformly scattered in an enamel background of a different colour.

Specular Gloss (45°). An operational definition of this property, relevant in the surface evaluation of glazed or enamelled surfaces, is as follows (ASTM – C346): the ratio of reflected to incident light, times 1000, for specified apertures of illumination and reception when the axis of reception coincides with the mirror image of the axis of illumination.

Spengler Press. Trade-name: a cam-operated, rotary-table, dry-press brickmaking machine. (Spengler-Maschinenbau G.m.b.H., Berlin.)

Spew. See DUST.

Sphene (or Titanite). $CaO.SiO_2.TiO_2$; m.p. 1386°C. This

268

mineral has been found in the ceramic colour known as CHROME-TIN PINK (q.v.).

Spider. (1) A metal unit with two or more radial arms; for example, a spider is used in the mouthpiece of a pug to hold a core or to break up laminations.

(2) The term is sometimes used as an alternative to CENTRE BRICK (q.v.).

(3) A star-shaped fracture in a vitreous enamelled surface.

Spiked Bottom. One form of bottom for a CONVERTER (q.v.); it is of monolithic refractory material, air passages being formed by spikes which pierce the bottom while the refractory material is being rammed into place (cf. TUYERE BLOCK BOTTOM and TUBE BOTTOM).

Spinel. A group of minerals having the same general structure and formula—$R^{2+}O . R_2^{3+}O_3$, where R^{2+} is a divalent metal (Mg, Fe^{2+}, Zn, Co, Ni, etc.) and R^{3+} is a trivalent metal (Fe^{3+}, Cr, Al). The type mineral, to which the term 'spinel' refers if not qualified, is magnesium aluminate, $MgO . Al_2O_3$; this mineral is sometimes synthesized, by firing a mixture of the constituent oxides at a high temperature, for use as a refractory material. Another important group of spinels, known collectively as FERRITES (q.v.), are used as FERROMAGNETICS (q.v.). For other spinels see CHROMITE, HERCYNITE, PICROCHROMITE, MAGNESIO-FERRITE.

Spissograph. An automatic machine for checking the time of set of cement or mortar. (*Bull. ASTM*, No. **170,** 79, 1950.)

Spit-out. Pin-holes or craters sometimes occurring in glazed non-vitreous ceramics while they are in the decorating kiln. The The cause of this defect is the evolution of water vapour, adsorbed by the porous body via minute cracks in the glaze.

Split. (1) A brick of the shape of a STANDARD SQUARE (q.v.) split down its length, see Fig. 1, p. 37; commonly a $9 \times 4\frac{1}{2}$ in. brick with a thickness of $1\frac{1}{4}$, $1\frac{1}{2}$ or 2 in. (Also known as a SCONE.)

(2) A glass bottle containing $6\frac{1}{2}$ fl. oz (for mineral water).

(3) A crack that penetrates glass-ware, as distinct from a CHECK or VENT (q.v.).

Split Rock. US term for the grade of FIRESTONE (q.v.) that is most suitable for use as a refractory.

Spodumene. Natural lithium alumino-silicate, $Li_2O . Al_2O_3 . 4SiO_2$; it occurs in S. Rhodesia, USA and Brazil. When fired the α form changes irreversibly to the β form with a considerable expansion; the β-form is a constituent of some special ceramic bodies of zero thermal expansion.

Sponge, Sponging. The smoothing out, with a moistened sponge, of slight surface blemishes on pottery-ware before it is dried.

Spongy Enamel. Vitreous enamelware that is faulty because of local high concentrations of bubbles.

Spout. A refractory block shaped to carry molten glass, usually to a forming machine; see also FEEDER SPOUT and POT SPOUT.

Spray Drying. The process of dewatering a suspension, e.g. clay slip, by spraying the suspension into the top of a heated chamber, the dried powder being removed from the bottom.

Spray Sagging. The appearance of wavy lines on the vertical surface of vitreous enamelware after it has been sprayed and before the slip coating is dried. The fault is caused by incorrect adjustment of the fluidity and thixotropy of the slip.

Spray Welding. A process for the localized repair of cracks in the refractory brickwork of a gas retort or coke oven; a refractory powder is carried by a stream of oxygen into an oxy/coal-gas flame so that the powder fuses on the brickwork to seal the selected damaged area on which it is projected. This technique was first used by T. F. E. Rhead, City of Birmingham Gas Dept. (*Trans. Inst. Gas Engrs.*, **81,** 403, 1931) (cf. AIR-BORNE SEALING).

Spreader Block. See DEFLECTING BLOCK.

Sprigging. The decoration of pottery vases, etc., by affixing clay figures, usually classical, to form a bas relief; the figures are pressed separately from the vase and are made to adhere to the ware by means of clay slip (cf. EMBOSSING).

Springer. See SKEWBACK.

Sprue. In vitreous enamelling this term is sometimes applied to the frit that builds up at the tap-hole of a frit-kiln. From time to time pieces fall into the quenched frit and cause trouble in milling because of their hardness.

Sprung Arch; Sprung Roof. An arch, or furnace roof, the brickwork of which is supported solely by SKEWBACKS (q.v.); (cf. SUSPENDED ARCH or ROOF).

Spur. An item of KILN FURNITURE (q.v.) in the form of a tetrahedron with concave faces and sharp points on which the ware is set; see Fig. 4, p. 158.

Square. (1) See STANDARD SQUARE.

(2) A SQUARE of roofing tiles is 100 ft^2 as laid.

Square Pitch. See under PITCH.

Squeegee Process. See SILK-SCREEN PROCESS. The term SQUEEGEE OIL is used, particularly in the US industry, for the mixture of oils used to suspend a ceramic or enamel colour for silk-screen printing. Similarly, the term SQUEEGEE PASTE is used for the mixture of oils, colours and flux used in this process of decoration.

Squint. A building brick with one end chamfered on both edges so that the brick can be used at an oblique quoin.

SR Fusion-cast Refractory. A French fusion-cast refractory, e.g. glass-tank block, made by a process that largely eliminates the cavities that occur as a result of shrinkage while the block cools from the molten state. The block is made by a so-called high-filling process (SR = Sur-Remplis) and is then heat-treated.

Stack or Shaft. The upper part of a blast furnace, widening from the THROAT (q.v.) at the top to the LINTEL (q.v.) at the bottom. This part of the furnace is lined with fireclay refractories selected for their resistance to abrasion and to the disintegrating action of the carbon monoxide present in the stack atmosphere.

Staffordshire Blues. See under ENGINEERING BRICKS.

Staffordshire Cones. PYROMETRIC CONES (q.v.) made by Harrison & Sons Ltd., Stoke-on-Trent, England. For squatting temperatures (see Appendix 2).

Staffordshire Kiln. A particular design of transverse-arch kiln; it is fired from the top into combustion spaces in the setting. Such kilns are used for firing building bricks. The original design was patented by Dean, Hetherington & Co. in 1904.

Stahlton System. Prefabricated building elements made from hollow clay blocks, a row of which is united by prestressed wires and cement filling; each reinforcing wire is pre-stressed individually and anchored singly. The elements are used in floors and ceilings. The name derives from the German words *Stahl* (steel) and *ton* (clay).

Stain. (1) A coloured imperfection on the surface of ceramic ware.

(2) An inorganic colouring material used in the preparation of under-glaze and on-glaze pottery colours, for colouring pottery bodies and glazes, or for decorating the surface of glassware.

Stamping. A process for the application, by hand or by machine, of decoration to pottery-ware; a rubber stamp with a sponge backing is used. Stamping is particularly suitable for the application of BACK STAMPS (q.v.) and for some forms of gold decoration.

Standard Black. A ceramic colour; a quoted composition is 30 parts Co_2O_3, 56 parts Fe_2O_3, 48 parts Cr_2O_3, 8 parts NiO and 31 parts Al_2O_3.

Standard Square. In the UK this term refers to bricks of the following sizes:

(1) Common Clay Building Bricks (B.S. 657): $8\frac{5}{8} \times 4\frac{1}{8} \times 2\frac{7}{8}$ in. and $8\frac{5}{8} \times 4\frac{1}{8} \times 2\frac{5}{8}$ in.

(2) Refractory Bricks (B.S. 3056): $9 \times 4\frac{1}{2} \times 3$ in. and $9 \times 4\frac{1}{2} \times 2\frac{1}{2}$ in.; the ISO recommended size is $230 \times 114 \times 64$mm.

Standard Surface Factor. See SURFACE FACTOR.

Stanford Joint. A joint, for sewer-pipes, consisting of tar, sulphur and ground brick or sand; the proportions vary according to the nature of the tar, which should contain sufficient pitch to ensure that the joint will set.

Stannic Oxide. See TIN OXIDE.

Star Dresser. A tool for the DRESSING (q.v.) of abrasive wheels; it consists of a series of star-shaped metal cutters separated by washers, freely mounted on the spindle of a holder. Also sometimes known as a HUNTINGTON DRESSER.

Star Mark. A defect (star-shaped fracture) liable to occur in vitreous-enamel cover-coats if the ware is roughly handled as it is being placed on the pointed supports used in the firing process; the marks result from fracture of the enamel coating.

Starred Glaze. Term sometimes applied to a glaze that has partially devitrified, star-shaped crystals appearing on the surface; the cause may be SULPHURING (q.v.).

Starved Glaze. The poor quality of the glazed surface of ceramic ware if insufficient glaze has been applied; also sometimes called SHORT GLAZED.

Steam Curing. The rapid CURING (q.v.) of pre-cast concrete units; this can be done at high pressure in an AUTOCLAVE (q.v.) or at atmospheric pressure in chambers or tunnels.

Steam Shovel. See MECHANICAL SHOVEL.

Steam Tempering. The treatment of clay (particularly brick-clay) with steam to develop its plasticity for the shaping process and, as an additional advantage, to reduce the time required to dry the shaped bricks. The clay generally reaches a temperature of 70–80°C. (Also known as HOT PREPARATION.)

Steam Test. See under AUTOCLAVE.

Stearates. Organic compounds used as a lubricant in the dry-pressing of ceramics; the stearates used are those of Al, Ca, Mg and Zn.

Steatite. Massive form of the mineral talc, $Mg_3Si_4O_{10}(OH)_2$. The chief sources are in USA, France, Italy, India, Austria and Norway. The natural rock can be machined and the shaped parts fired for use as electroceramics. A far greater proportion is ground and shaped in the usual way with a clay bond to produce low-loss electroceramics. Steatite is also used as a raw material for cordierite electroceramics and refractories.

Steger's Crazing Test. A method for the assessment of the 'fit' between a ceramic body and a glaze by measuring any deformation, on cooling, of a thin bar glazed on one side only. (W. Steger, *Ber. Deut. Keram. Ges.*, **9,** 203, 1928; more precise details were given in a later paper—*ibid.*, **11,** 124, 1930.)

Stemware. A general term for wine glasses, etc., having stems. The stem may be pulled or drawn from the bowl (PULLED STEM or DRAWN STEM), or it may be made separately (STUCK SHANK).

Stent. Cornish term for the mixture of sand and stone occurring with china clay in the china-clay rock.

Sticking Up. The process of joining together, by means of SLIP (q.v.), the various parts of items of pottery-ware that cannot readily be made in one piece, e.g. putting the knob on a tureen cover, or spouts on teapots.

Stiff-mud Process. The equivalent US term for the WIRE-CUT PROCESS (q.v.).

Stiff-plastic Process. A process of brickmaking by mechanical presses; the clay is prepared to a moisture content of about 12% and the shaped bricks can be set direct in the kiln without preliminary drying. This is particularly common in brickworks sited on the clays and shales of the Coal Measures.

Stillage. A small platform on which shaped clayware may be placed to facilitate its handling within the factory—to and from a dryer, for example (cf. PALLET).

Stilt or Wedge-stilt. An item of KILN FURNITURE (q.v.) in the form of a three-pointed star; the three arms are SADDLES (q.v.) with points. See Fig. 4, p. 158.

Stippled. (1) A surface texture on a clay facing brick produced by the rapid impingement on the green brick of a head, carried on a reciprocating arm, fitted with steel spikes; this texture is also called SPARROW-PECKED.

Stock Brick

(2) A mottled decoration on pottery-ware or on vitreous enamel-ware, produced by applying colour with a sponge or brush.

Stock Brick. Originally a term localized to S.E. England and meaning a clay building brick made by hand on a 'stock', i.e. the block of wood that defined the position of the mould on the moulding table. Stock bricks are now machine-made. The London Stock Brick is a yellow brick of rough texture.

Stoichiometric. A chemical compound, or a batch for synthesis, is said to be stoichiometric when the ratio of its constituents is exactly that demanded by the chemical formula. Of more interest in the field of special ceramics are NON-STOICHIOMETRIC (q.v.) compounds.

Stoke. A unit of kinematic viscosity:
1 Stoke = 1 poise ÷ (density of the fluid).

Stoker. A mechanical device for feeding solid fuel into a kiln firebox or through the roof of a Hoffmann-type kiln.

Stokes's Law. The force required to move a sphere of radius r at uniform velocity v through a medium of viscosity η is equal to $6\pi\eta vr$. This law is applied in the determination of particle size by sedimentation, elutriation or centrifugation. For a particle of density d_1 settling in a medium of density d_2, g being the acceleration due to gravity, when equilibrium is reached the following equation applies:

$$r = \left[\frac{9v\eta}{2(d_1 - d_2)g} \right]^{\frac{1}{2}}$$

Stone. A crystalline inclusion present as a fault in glass; stones may result from incomplete reaction of particles of batch or from the pick-up of small fragments of the refractory lining of the pot or furnace in which the glass is melted. The most common constituents of stones are carnegieite, corundum, cristobalite, mullite, nephelite, tridymite and zirconia. See also CHINA STONE.

Stoner. A workman in a glass-works whose job is to smooth the rims of the ware; also called a FLATTER.

Stoneware. A vitreous, but opaque, type of ceramic ware. The body contains a naturally-vitrifying clay, e.g. a stoneware clay or a suitable ball clay; sometimes a non-plastic constituent and a flux are added. Stoneware may be once-fired or it may be biscuit fired at about 900°C followed by glost-firing at 1200–1250°C. At the present day stoneware is produced on a commercial scale chiefly as cooking-ware and some tableware, for which purposes its high strength and freedom from crazing are valuable; on a smaller scale, stoneware is much favoured by studio potters. See also CHEMICAL STONEWARE.

Stoning. The removal, by means of a RUBBING STONE (q.v.), of excrescences from enamelware during the stages of manufacture. The need for stoning may arise as a result of splashes of enamel slip from the spray-gun, or of non-uniform draining of the slip.

Stopper. (1) The round-ended refractory (fireclay or graphite) shape that terminates the stopper-rod assembly in a steel-casting ladle and controls the rate of flow of metal through the nozzle.

(2) A movable refractory shape for the control of the flow of glass in the channel leading from a glass-tank furnace to a revolving pot of a suction machine.

(3) A refractory closure for the mouth of a covered glass-pot or for the working-hole outside an open pot.

An alternative term with meanings (2) and (3) is GATE.

Stopping. The filling up of any cracks in biscuit pottery-ware prior to the glost firing.

Stormer Viscometer. A rotating cylinder viscometer of a type that has found considerable use in USA for the determination of the viscosity and thixotropy of clay slips. (*Industr. Engng. Chem.*, **1,** 317, 1909.)

Stove. See HOT-BLAST STOVE.

Stove Fillings. The special fireclay refractory shapes used as checker bricks in a HOT-BLAST STOVE (q.v.). Normally, these stoves operate at a max. temperature (at the top of the stove) of about 1200°C; the top 15 ft or so of fillings are therefore built with 40–42% Al_2O_3 refractories, with 35–37% Al_2O_3 refractories below. Recently, interest has been shown in the possibility of still higher hot-blast temperatures and trials have been made with 50–65% Al_2O_3 stove fillings for the upper courses.

St.P. Abbreviation for STRAIN POINT (q.v.).

Strain Disk. A glass disk internally stressed to give a calibrated amount of birefringence; the disk is used as a comparative measure of the degree of annealing of glass-ware.

Strain-line. Alternative term for the vitreous-enamel fault more generally known as HAIR-LINE (q.v.).

Strain Point. The temperature corresponding to the lower end of the annealing range and defined in USA (ASTM – C336) as the temperature at which the viscosity of a glass is $10^{14.5}$ poises; abbreviated to St.P.

Strainer Core. A porous refractory shape for use in foundries to control the flow of metal and to keep slag and sand inclusions out of the casting. High thermal-shock resistance is required.

Strand Count. The thickness of a strand of glass fibres expressed as the number of 100 yd lengths per pound weight.

Strass. A glass of high lead content and great brilliancy used for artificial jewelry; named from its 18th century German inventor, Josef Strasser.

Straw Stem. The slender hollow stem of a wine-glass.

Stretched-membrane Theory. See MEMBRANE THEORY OF PLASTICITY.

Stretcher. A brick with its length parallel to that of the wall in which it has been laid (cf. HEADER).

Striae. A fault, in glass, in the form of fine CORDS (q.v.); also known as VEIN.

Striking. (1) The development of opacity or colour in glass-ware by a heat-treatment process, e.g. by the formation of a colloidal dispersion, within the glass, of a small amount of Cu, Ag or Au.

(2) The smoothing of the wet clay surface in a mould by means of a wooden or metal rod, as in the hand-moulding of special clay building bricks.

(3) A fault sometimes encountered with enamel colours on pottery-ware, the colour becoming detached from the ware during firing; a common cause is lack of control during HARDEN-ING-ON (q.v.).

(4) Setting out the first mould and the profile tool for the shaping of ware on a JIGGER (q.v.).

String. A fault in glass with the appearance of a straight or curled line; the usual cause is slow solution of a large sand grain or other coarse material.

String Course. A distinctive, usually projecting, course in a brick wall; its purpose is aesthetic.

String Dryer. A tunnel-type dryer, particularly for building bricks, that is operated intermittently; in the early stages of drying, the exits and exhaust ducts are closed so that a high humidity is built up. This type of dryer has been used particularly in Texas, USA.

Stripper. A person employed in the pottery industry to remove the dried ware from the plaster moulds.

Strontium Boride. SrB_6; m.p. 2235°C; sp. gr., 3·42.

Strontium Oxide. SrO; m.p. 2430°C; sp. gr. 4·7; thermal expansion (20–1200°C) $13·9 \times 10^{-6}$.

Strontium Stannate. $SrSnO_3$; sometimes used as an additive to titanate bodies, one result being a decrease in the Curie temperature.

Strontium Titanate. $SrTiO_3$; used alone or in combination with $BaTiO_3$ as a ceramic dielectric. The power factor at low frequency is high. The Curie temperature is approx. -260°C. The dielectric constant is 230–250.

Strontium Zirconate. $SrZrO_3$; m.p. 2700°C; sp. gr. 5·48. Sometimes used in small amounts (3–5%) in ceramic dielectric bodies, one effect being to lower the Curie temperature.

Structon. A unit of atomic structure in an amorphous solid, such as glass, proposed by M. L. Huggins (*J. Phys. Chem.*, **58**, 1141, 1954) and defined as 'A single atom (or ion or molecule) surrounded in a specified manner by others'; examples for a sodium silicate glass would be Si(4O), O(2Si,Na), etc. The concept aimed to reconcile the randomness of the long-range structure of a glass with the relative regularity in the short-range structure. The term has not achieved popularity (see VITRON).

Structural Clay Tile. In USA this term is defined (ASTM – C43) as a hollow burned-clay masonry building unit with parallel cells; the equivalent term in the UK is 'Hollow Clay Block'.

Structural Spalling. A form of SPALLING (q.v.) defined in USA as: The spalling of a refractory unit caused by stresses resulting from differential changes in the structure of the unit.

Structure. As applied to abrasive wheels, this term refers to the proportion and distribution of the abrasive grains in relation to the bond. In USA the term GRAIN SPACING is also used.

Stub. An abrasive wheel that has been used until its diameter has been so much reduced by wear that it is no longer serviceable.

Stuck Shank. See STEMWARE.

Stuck Ware. Pottery-ware that has stuck to the kiln furniture during the glost firing and is therefore waste. The fault may be caused by careless placing, by the presence on the ware of too much glaze, or by firing at a temperature that is too high for the glaze being used, which therefore becomes too fluid.

Stupid. An extrusion machine in which the clay is forced through the die by means of a piston; such machines are now rare.

Submerged Wall. A wall of refractory material below the level

Substance

of the molten glass in a tank furnace and separating the melting zone from the refining zone.

Substance. In the glass industry this word means the thickness of glass sheets in oz/ft².

Suck-and-Blow Process. A method of shaping glass-ware, the PARISON (q.v.) is made by sucking molten glass into a mould, the final shape being subsequently produced in a BLOW-MOULD (q.v.). The process is known in USA as VACUUM-AND-BLOW.

Sucking. Loss of volatile oxides, particularly lead oxide, from a glaze by volatilization during glost firing in unglazed saggars or adjacent to non-vitreous kiln furniture, the vapour being 'sucked' into the porous refractory. The fault is prevented by washing the insides of saggars with glaze or, in saggarless firing, by the use of kiln furniture of low porosity.

Suction Pyrometer or High-Velocity Thermocouple. An instrument for the determination of the temperature of moving gases when it differs considerably from that of their surroundings; the hot gases are drawn rapidly past the junction of a fine-wire noble-metal thermocouple. Such an instrument is used, for example, in the determination of the temperature of hot kiln gases passing through a setting of relatively cool bricks.

Suffolk Kiln. An early type of up-draught intermittent kiln. The fireboxes were below the kiln floor which was perforated for the upward passage of the hot gases.

S.u.G Process. A German method for the shaping of highly-grogged fireclay refractories; the bond-clay is added as a slip and shaping is by a compressed-air rammer. The name derives from the originators: Scheidhauer und Giessing A.G. (Brit. Pat., 262 383, 24/6/26).

Sugar. A fault on lead crystal glass resulting from inadequate control during acid polishing and revealed as crystallites on the surface of the glass.

Sugar Test. A quality test for cement; see MERRIMAN TEST.

Sugary Cut. Undue roughness of the edge of flat glass resulting from faulty cutting.

Sullivan Process. A process for the treatment of sheet steel prior to vitreous enamelling; it is based on the spraying of a suspension of NiO on the metal to give a deposit of about 0·1–1·0 g/ft². (J. D. Sullivan, US Pat., 2 940 865, 14/6/60).

Sulphite Lye. A by-product of the paper industry used as a cheap temporary bond, e.g. for silica refractories; it has also been used as a plasticizer in making building bricks. It generally contains 50–70% of the ligno-sulphonate of Na, Ca or NH_4, together with 15–30% of a mixture of sugars; the ash content of the NH_4 type is very low but that of the Ca or Na types may amount to 30%.

Sulpho-aluminate Cement. A hydraulic cement made by grinding a mixture of high-alumina cement and gypsum (or anhydrite) (Lafarge, Brit. Pat., 317 783, 29/7/29).

Sulphuring. A surface bloom or dulling of glazed ceramic ware resulting from attack by SO_3 in the kiln, and more particularly by sulphuric acid condensed on the ware in cooler parts of the kiln in the early stages of firing. The sulphur compounds may originate in the fuel or in traces of sulphur compounds (e.g. pyrite) present in the ware itself. The fault is most common in glazes containing calcium or barium; glazes containing lead are usually more fluid and, although they may absorb more sulphur as sulphate, this may not result in a visible fault.

Sunburner. A faulty piece of hand-blown glass-ware characterized by excessive local thickness.

Sunderland Splatter. A characteristic effect found on some of the old lustre pottery made in Sunderland, England; after the lustre had been applied it was 'splattered' with a second medium which caused the formation of irregular patches on the base lustre.

Super-duty Fireclay Brick. Defined in ASTM – C27 as a fireclay refractory having a P.C.E. $\lessdot$ 33 and an after-contraction $\gtrdot 1\%$ when tested at 1600°C. There are three types: Regular, Spall Resistant, Slag Resistant. Limits are set for panel spalling loss, modulus of rupture and bulk density.

Super-duty Silica Refractory. In the UK the National Silica Brickmakers' Association has defined this type of refractory as a SILICA REFRACTORY (q.v.) containing not more than 0·5% Al_2O_3 and with a total of Al_2O_3 plus alkalis not exceeding 0·7%.

Superheater. A refractory-lined chamber in a water-gas plant; it is filled with checkers and ensures completion of the decomposition of the oil vapours begun in the CARBURETTOR (q.v.).

Super Staffordshire Kiln. A STAFFORDSHIRE KILN (q.v.) modified in design so that the fuel is charged on to a grate instead of within the setting of bricks. Kilns of this type can be used for temperatures up to about 1200°C.

Supersulphated Cement. A hydraulic cement made by finely grinding a mixture of 80–85% granulated blast furnace slag,

Surface Combustion

10–15% calcined gypsum and 5% portland cement (or lime). Its principal feature is resistance to attack by water containing dissolved sulphates.

Surface Combustion. That fuel gases burn more readily when brought into contact with a hot surface was first demonstrated by W. A. Bone and R. V. Wheeler (*Phil. Trans.*, A, **206**, 1, 1906). The principle has been applied to furnace design, combustion taking place on the incandescent surface of refractory material; in one surface-combustion device, the fuel gas is passed through a porous refractory burner, 'flameless combustion' occurring over the surface of the refractory.

Surface Factor. A factor used in the ceramic industry to indicate the fineness of a powder. It is calculated from the equation:

$$S = \frac{6}{G} \left(\frac{W_1}{d_1} + \frac{W_2}{d_2} + \frac{W_3}{d_3} + \ldots \right)$$

where G is the specific gravity of the powder; W_1, W_2, etc., are the fractional weights of material whose average diameters are d_1, d_2, etc.; $W_1 + W_2 + \ldots = 1$.

Surface Tension. The effect of internal molecular attraction acting at the surface of a liquid to cause the surface to behave as though it were a tense skin. Some surface tension values are (dynes/cm): paraffin (kerosene), about 20 at 20°C; water, 74·2 at 10°C; glasses in their working temperature range, 200–300.

Suspended Arch, Roof or Wall. An arch, furnace roof, or wall, in which some or all of the bricks are suspended by metal hangers from a steel framework. In an arch or roof, the object is to relieve the bricks from the mechanical stress resulting from the thrust of the arch; in a suspended furnace wall, the usual purpose is to permit refractory brickwork to be inclined inwards without danger of it falling into the furnace.

Suspending Agent. A material, such as clay, used to keep a vitreous-enamel or glaze in suspension so that it can be conveniently used for application to the ware by dipping or spraying.

Svedala Kiln. A chamber kiln with a dryer built above it; designed in Scandinavia for use in the building brick industry.

Swab Test. An electrical test, carried out with low voltage, for exploration of such defects in vitreous enamelware as pin-holes or other discontinuities.

Sweet. A term applied to glass that is easy to shape.

Swept Valley. A form of roof tiling designed to cover a re-entrant corner of a roof without any sharp valley; careful cutting of the roofing tiles is necessary to ensure a symmetrical finish.

Swindell–Dressler Kiln. See DRESSLER KILN.

Swing Press. A hand-operated screw press sometimes used, for example, in the shaping of a small quantity of special-shaped wall or floor tiles.

Syphon. A special type of refractory block for the bath of a glass-tank furnace.

T

T_a. See under REFRACTORINESS-UNDER-LOAD.

T_e. (1) The temperature at which the electrical resistivity is 1 megohm/cubic cm. The T_e value of some ceramics is: porcelain, 350°C; cordierite, 780°C; steatite, 840°C; zircon, 870°C; forsterite, 1040°C; alumina, 1070°C.

(2) A temperature associated with REFRACTORINESS-UNDER-LOAD (q.v.).

Taber Abraser. A device for assessing the abrasion resistance of a surface; the principle is contact with loaded abrasive wheels which are rotated against the surface to be tested. It has been used for testing vitreous enamel and floor-tile surfaces. (Taber Instrument Corp., 111 Goundry St., North Tonawanda, NY, USA.)

Tabular Alumina. Corundum (> 99·5% Al_2O_3) supplied by Aluminum Company of America: T-60 grade, for refractories, contains < 0·25% Na_2O; T-61, for electroceramics, contains < 0·05% Na_2O.

Tailings. The oversize material retained on a screen; normally the tailings are returned to the grinding plant for further grinding but in some cases the tailings contain a concentration of impurities and are discarded.

Talc. See STEATITE.

Talwalker-Parmelee Plasticity Index. This index, based on the results of tests on a clay in shear with a specially designed apparatus, is the ratio of the total deformation at fracture to the average stress beyond the proportional limit. (T. W. Talwalker and C. W. Parmelee, *J. Amer. Ceram. Soc.*, **10**, 670, 1927.)

Tamping. The shaping of a semi-dry powder, e.g. of refractory material, in a mould by repeated blows delivered mechanically on the top mould-plate (cf. JOLT MOULDING and RAMMING).

Tank Block. A refractory block used in the lower part of a glass-tank furnace. These blocks are normally made of sillimanite, mullite or corundum; they are frequently made by electrofusion of the refractory, which is then cast in a mould to form a highly crystalline, virtually non-porous, block which is very resistant to attack by the molten glass.

Tank Furnace. A furnace for the continuous process of glass melting and refining; it is usually gas-fired and regenerative. The glass is melted in the bath of the furnace, which is lined with refractory blocks.

Tannin. A complex organic compound of C, H and O produced by metabolism in trees and plants. Sodium tannate is used to some extent as a deflocculant for clay slips.

Tantalum Borides. Several borides are known, including the following. TaB_2; m.p. 3200°C; sp. gr. 12·5; thermal expansion, $5·5 \times 10^{-6}$. TaB; m.p. 2400°C; sp. gr. 14·3. Ta_3B_4; melts incongruently at 2650°C; sp. gr. 13·6.

Tantalum Carbides. Two carbides exist; TaC, m.p. 3800°C; Ta_2C, m.p. 3400°C. The monocarbide has the following properties: sp. gr. (theoretical), 14·5 g/ml; hardness, 1950 (K100); modulus of rupture, 30 000 p.s.i. at 25°C and 17 000 p.s.i. at 2000°C.

Tantalum Nitrides. Two nitrides are known: TaN, m.p. 3090° $\pm$ 50°C; Ta_2N, which loses nitrogen at 1900°C.

Tapered Wheel. An abrasive wheel that is thicker at the hub than at the circumference.

Taphole. A small passage through the refractory lining of the hearth-wall of a metallurgical furnace, e.g. a blast furnace or open-hearth steel furnace.

Taphole Clay. The plastic refractory material used to plug the tap-hole of a blast furnace generally consists of a mixture of fire-clay and grog together with carbon in the form of coal or coke (in USA pitch is also sometimes used). The range of chemical composition of a number of British taphole clays is (per cent): SiO_2, 40–60; Al_2O_3, 22–27; loss on ignition, 8–18. The material is adjusted to have a moisture content of 15–18% and is forced into the taphole by means of a 'gun'.

Taphole Gun. A gun-shaped device used to force TAPHOLE

282

CLAY (q.v.) into the TAPHOLE (q.v.) of a blast furnace after it has been tapped.

Tappit Hen. A 3-quart wine bottle.

Tea-dust Glaze. An opaque stoneware glaze, greenish in colour from reduced iron compounds, sometimes used by studio potters.

Tea-pot Ladle. A ladle, lined with refractory material and used in foundries for the transfer of molten iron or steel; its distinctive feature is the refractory dam below which the metal passes to reach the ladle spout; the refractory dam prevents the outflow of slag.

Tear. An open crack in glass-ware (cf. CHECK).

Tearing. Cracks (usually partially healed) in the cover-coat of vitreous enamelware. The cause is high drying shrinkage resulting, most commonly, from too-fine grinding, too wet a slip, or too heavy an application of slip. The amount of electrolyte in the slip is important; the presence of sodium nitrite is beneficial.

Teaser. A glass-worker who controls the temperature of a pot or tank-furnace to that required for feeding the batch; also called a FOUNDER.

Tecalamit System. Trade-name: an impulse system for the oil-firing (from the top or side) of ceramic kilns; the impulses are controlled by a compressed-air device.

Technate. A less common name for DIDYMIUM (q.v.).

Teem. To pour molten metal (or glass) from a ladle (or pot).

Tegula. See ITALIAN TILES.

Telefunken Process. See under METALLIZING.

Temper. The residual stress in annealed glass-ware as measured by comparison with STRAIN DISKS (q.v.). See also TEMPERING.

Temperature. For conversion table from Fahrenheit to Centigrade, and vice versa, see Appendix 3.

7·6 Temperature. See SOFTENING POINT.

13·0 Temperature. See ANNEALING POINT.

Temperature-gradient Furnace. A laboratory electric-resistance furnace in which the heating element is wound round the furnace tube in such a manner that there is a steady temperature gradient along the axis of the furnace. Such a furnace is useful in the ceramic laboratory in that it will expose a long test-piece placed within the furnace to different temperatures at different points, so that the effect of various firing temperatures can be studied in a single operation.

Tempered Glass. See TOUGHENED GLASS.

Tempering. The mechanical treatment of moistened clay, or body, to disperse the water more uniformly and so to improve the plasticity; tempering is usually done in a solid-bottom edge-runner mill with the mullers raised slightly above the pan bottom.

Tempering Tub. A combined pan and vertical pug mill for the preparation of clay for brickmaking. The mixing pan is about 7 ft dia., a central vertical shaft carrying the mixing blades; the shaft continues downward as the shaft of the pug mill. In some districts of England this equipment is known as a SLUDGE PAN.

Tenmoku. A glaze containing iron compounds sometimes used by studio potters; it is lustrous black except where it is thinner and has oxidized to a red colour.

Terra Cotta. Unglazed fired clay building blocks and moulded ornamental building components. In England, the use of architectural terra cotta reached its peak in the 19th century (cf. FAIENCE).

Terra Sigillata. A type of pottery having a fine-textured red body and a glossy surface, usually with a raised decoration. The Latin name derives from the pastilles made from certain clay deposits in Lemnos and Samos (hence the alternative name Samian ware) and impressed with the sacred symbol of Diana; these pastilles were used medicinally.

Tertiary Air. Preheated air introduced into the waste-gas flue of a kiln firing under reducing conditions, e.g. a blue-brick kiln; its purpose is to burn the combustible matter in the gases leaving the kiln chamber, thus helping to minimize smoke emission from the stack (cf. PRIMARY AIR and SECONDARY AIR).

Tessha. A version of the TENMOKU (q.v.) glaze; it is described by Bernard Leach as 'more metallic and broken'.

Test Cone. A test-piece cut or moulded from a sample of refractory material that is to be tested for refractoriness or P.C.E. (q.v.). The shape is that of a pyramid on a triangular base; the dimensions vary according to which national specification is being followed.

Tetracalcium Aluminoferrite. See BROWNMILLERITE.

Texture. (1) The physical property of a ceramic product determined by the shapes and sizes of the pores and the grading of the solid constituents. The texture can to some extent be evaluated in terms of porosity and permeability; additional information is provided by pore size measurement and the total internal surface area (as measured by gas adsorption, for example). (2) The general surface appearance of a clay facing brick.

Tg Point. See TRANSFORMATION POINTS.

Thénard's Blue. A dark blue colour for use under glaze in pottery decoration; it consists of approx. four parts cobalt oxide to five parts alumina.

Therm. A unit of heat equivalent to 100 000 Btu.

Thermal Analysis. A technique for mineral identification that is usually elaborated into DIFFERENTIAL THERMAL ANALYSIS (q.v.).

Thermal Barrier. The zone where the temperature is highest in the melting end of a glass-tank furnace.

Thermal Conductivity. The quantity of heat transmitted through a material in unit time, per unit temperature gradient along the direction of flow, and per unit of cross-sectional area. The thermal conductivity of crystalline ceramics of low porosity decreases with increasing temperature; that of porous ceramics having a glassy bond generally increases with temperature. Some typical room-temperature values (Btu/ft^2/h/°F/in.) are: Silicon carbide, 100; fireclay refractory, 8; insulating refractory, 2·5.

Thermal Diffusivity. A measure of the rate of change of temperature at a point in a material when heat is applied or removed at another point, e.g. the rate of rise of temperature at the centre of a furnace wall when the furnace is heated. The thermal diffusivity is measured as the ratio of the thermal conductivity of the material to the product of the bulk density and the specific heat.

Thermal Endurance. The ability of a piece of glass-ware to resist thermal shock. An attempt to assess this in terms of the physical properties of the glass is offered by the WINKELMANN AND SCHOTT EQUATION (q.v.).

Thermal Expansion. The reversible increase in dimensions of a material when it is heated. Normally, the linear expansion is quoted, either as a percentage or as a coefficient, in either case over a stated temperature range; for example, the thermal expansion of a silica refractory may be quoted as $1·2\%$ or as 12×10^{-6}, between 0° and 1000°C.

Thermal Expansion Factors for Glass. Factors that have been proposed from time to time for the calculation of the coefficient of linear thermal expansion of a glass on the assumption that this is an additive property. The factors are inserted in the equation $\lambda(= 10^8 \Delta l/l\Delta T) = a_1 p_1 + a_2 p_2 + \ldots a_n p_n$ where p_1, etc., are the

Thermal Shock

percentages of the oxides and a_1, etc., are the factors in the following table:

	Winkelmann & Schott (1)	English & Turner (2)	Gilard & Dubrul (3)	Hall (4)
SiO_2	2·67	0·50	0·4	—
B_2O_3	0·33	−6·53	−4+0·1p	2·0
Na_2O	33·33	41·6	51−0·33p	38·0
K_2O	28·33	39·0	42−0·33p	30·0
MgO	0·33	4·5	0	2·0
CaO	16·67	16·3	7·5 +0·35p	15·0
ZnO	6·0	7·0	7·75−0·25p	10·0
BaO	10·0	14·0	9·1 +0·14p	12·0
PbO	10·0	10·6	11·5 −0·05p	7·5
Al_2O_3	16·67	1·4	2	5·0

1. A. Winkelmann and O. Schott, *Ann. Physik.*, **51**, 735, 1894.
2. S. English and W. E. S. Turner, *J. Amer. Ceram. Soc.*, **10**, 551, 1927; ibid., **12**, 760, 1929.
3. P. Gilard and L. Dubrul, *Verres Silicates Ind.*, **5**, 122, 141, 1934.
4. F. P. Hall, *J. Amer. Ceram. Soc.*, **13**, 182, 1930.

Thermal Shock. Sudden heating or cooling: the stresses set up by the differential expansion or contraction between the outside and inside of a thermally-shocked ceramic may cause it to crack.

Thermal Spalling. SPALLING (q.v.) caused by stresses resulting from non-uniform dimensional changes of a brick or block produced by a difference in temperature.

Thermalite Ytong. Trade-name: A lightweight concrete made from portland cement, sand, and pulverized fuel ash; these are well mixed with water and a small proportion of aluminium powder is then added. This causes gas bubbles to form. Blocks of the cellulated material are then autoclaved. The material is a development of the Swedish material YTONG (q.v.). (Thermalite Ytong Ltd, Lea Marston, Sutton Coldfield, England.)

Thermistor. A material having a high temperature coefficient of electrical resistance. Ceramic thermistors having a negative temperature coefficient (the usual type) are made from mixtures of the oxides of Mn, Fe, Ni, Co, Cu and U; the batch is shaped and fired to low porosity. The electrical resistance of $BaTiO_3$ to which

286

rare earths, Bi or Th have been added has a high positive temperature coefficient. The properties of thermistors make them useful in various instruments and controllers.

Thermocouple. A device for the measurement of temperature on the basis of the electric current generated when the junction between two dissimilar conductors is heated. The common types of thermocouple, with their maximum temperature of use (when protected) are: Copper/Constantan, 400°C; Iron-Constantan, 1000°C; Chromel/Alumel, 1200°C; Platinum/Platinum-Rhodium, 1700°C. Thermocouples with Tungsten/Iridium, or with Silicon-Carbide/Carbon, elements have been used for special purposes at > 2000°C.

Thermodyne Test. A test to determine the durability of optical glass in contact with moist air. Freshly broken or optically polished surfaces, sealed off in a flask together with a quantity of water, are subjected to a series of temperature cycles, each of 2 h duration, from 15 to 60°C in air saturated with water vapour for a period of 12 days. (W. M. Hampton, *Proc. Physical Soc.*, **54**, 391, 1942.)

Thermoscope. See PYROSCOPE.

Thimble. (1) An item of kiln furniture, Fig. 4, p. 158; it is conical, hollow, and with a projection at the base to support the pottery-ware being fired; thimbles are inserted into one another so that a bung of ware can be built up.

(2) A refractory shape, usually resembling the letter L, used to stir optical glass in a pot.

Thimble-Bat. A refractory BAT (q.v.) of a type used in the firing of pottery; it is perforated to hold the ends of THIMBLES (q.v.).

Thin Section. A section of material that has been prepared, by grinding to extreme thinness (about 30μ), for examination by transmitted light under a polarizing microscope. For details of the technique see such books as *Thin Section Mineralogy of Ceramic Materials* by G. R. Rigby, 2nd Ed. 1953, and *Microscopy of Ceramics and Cements* by H. Insley and D. Frechette, 1955.

Thirting. Cutting ball clay into blocks about 9 in. cube.

Thivier Earth. A siliceous hydrated iron oxide from Thivier, 19 miles N.E. of Perigueux, France. A quoted composition is: 83% SiO_2, 10% Fe_2O_3, 2% Al_2O_3, 1% CaO, 1% alkalis, 3% loss on ignition. It has been used as a red colour for pottery decoration.

Thixotropy. The property, exhibited by many clay slips for

287

Thorium Borides

example, of becoming more viscous when left undisturbed but more fluid when stirred.

Thorium Borides. Two borides are known: ThB_4 (grey) and ThB_6 (deep red). The more attention has been paid to the tetraboride, the properties of which are: m.p. $> 2,200°C$ (but oxidizes slowly above 1000°C); thermal expansion, $5·9 \times 10^{-6}$ (20–1000°C); sp. gr. 8·45 g/ml; modulus of rupture (20°C), 20 000 p.s.i. Some properties of ThB_6 are: m.p. 2200°C; sp. gr. 7·1.

Thorium Carbides. Two carbides are known: ThC, m.p. 2625°C; ThC_2, m.p. 2655°C. These special carbides are of potential interest in nuclear engineering.

Thorium Nitride. Three nitrides have been reported: ThN, Th_2N_3 and Th_3N_4.

Thorium Oxide or Thoria. ThO_2; m.p. 3300°C; sp. gr. 9·69; thermal expansion (20–1000°C) $9·3 \times 10^{-6}$.

Thorium Sulphides. Three sulphides have been reported: Th_4S_7, Th_2S_3 and ThS. Crucibles made of these sulphides have been used as containers for molten Ce.

Thorpe's Ratio. A formula suggested by Prof. T. E. Thorpe in 1901 for assessing the probable solubility of a lead frit: the sum of the bases expressed as PbO divided by the sum of the acid oxides expressed as SiO_2 should not exceed 2.

Thread Guide. Porcelain thread guides are satisfactory for use with cotton, wool, or silk; man-made fibres, e.g. rayon and nylon, are more abrasive and sintered alumina or synthetic sapphire thread guides are used.

Throat. (1) The part of a blast furnace at the top of the stack.

(2) The zone of decreased cross-section found between the port area and the furnace chamber in some designs of open-hearth steel furnace.

(3) The submerged passage connecting the melting end to the working end of a glass-tank furnace; the refractory blocks forming the sides of the throat are known as THROAT CHEEKS, SLEEPER BLOCKS or DICE BLOCKS, the refractories for the top are the THROAT COVER.

Throwing. The method of shaping pottery hollow-ware in which a ball of the prepared body is thrown on a revolving potter's wheel, where it is centred and then worked into shape with the hands. The process is now chiefly used by studio potters, although a small amount of high-class commercial pottery is still made in this way.

Thwacking. The process by which clay pantiles are given their final curved shape. When partially dry, the tiles are placed on a wooden block of the correct curvature and beaten to that contour by means of a bevelled piece of wood; any distortion of the tile caused by unequal shrinkage during the preliminary drying is thus corrected.

Tickell Roundness Number. An index of the shape of a particle in terms of the ratio of the actual area of the projection of the grain to the area of the smallest circumscribing circle; (F. G. Tickell, *The Examination of Fragmental Rocks*, 1931).

Tiering. See TORCHING.

Tiger Eye. A particular type of AVENTURINE (q.v.) glaze.

Tile. See FLOOR TILE, ROOFING TILE, WALL TILE. In the USA the word 'Tile' also denotes a hollow clay building block and is used in such terms as 'Load-Bearing Tile', 'Partition Tile', the meaning usually being self-evident. The term 'Drain Tile', in the USA, signifies an unglazed, clay, field-drain pipe.

Tile Hanging. The process of fixing roofing tiles (as distinct from wall tiles) on a vertical outside wall.

TI-LOC Process. Trade-name: a process for the treatment of steel prior to enamelling; it is claimed to improve adherence and to eliminate the need for a ground-coat. (Strong Mfg. Co., Sebring, Ohio, USA.)

Tin Oxide. Two oxides exist but the more common is stannic oxide, SnO_2, which is used as an opacifier and in the preparation of colours, e.g. CHROME-TIN PINK (q.v.). Properties making this oxide of some interest as a special ceramic are its high thermal conductivity and low thermal expansion ($3 \cdot 8 \times 10^{-6}$); it is sensitive to reducing atmospheres.

Tin-Vanadium Yellow. See VANADIUM YELLOW.

Tinsel. Very thin glass that has been crushed and silvered for use as a decorative material (cf. GLASS FROST).

Titanate Ceramics. A group of electroceramic materials generally based on the compound BARIUM TITANATE (q.v.) but often with the addition of other titanates, zirconates, stannates or niobates. These ceramics are notable for their high dielectric constant (up to, and even exceeding, 10 000 compared with a value of 5–10 for the more common ceramic materials); because of this property they find use in capacitors. Titanate ceramics are also used where piezoelectric properties are needed, i.e. in transducers.

Titania. See TITANIUM OXIDE.

Titanite

Titanite. See SPHENE.

Titanium Boride. See TITANIUM DIBORIDE.

Titanium Carbide. TiC; m.p. 3140°C; sp. gr. 4·25, thermal expansion (25–800°C) $7·4 \times 10^{-6}$. A hard refractory compound; it has found use as the ceramic component of some cermets.

Titanium Diboride. TiB_2; m.p. 2920°C with excellent oxidation-resistance up to 1000°C; sp. gr. (theoretical), 4·5; Knoop hardness (K100), 2710; transverse strength at 2000°C, 35 000 p.s.i.; thermal expansion, $9·7 \times 10^{-6}$ (200–1800°C); inert to molten Al and Zn but not to ferrous alloys. It has been used in rocket nozzles, as a constituent of cermets, and for boats for the evaporation of Al (for thin-film coatings).

Titanium Dioxide. See TITANIUM OXIDE.

Titanium Nitride. TiN; a special refractory material (m.p. 2950°C). It can readily be produced from $TiCl_4$ and NH_3.

Titanium Oxide. TiO_2; used as an opacifier, particularly in vitreous enamels, and as a constituent of some ceramic colours. Titania and titanate electroceramics, for use in the radio-frequency field, are based on this oxide and its compounds. Titania occurs in three crystalline forms: ANATASE, BROOKITE and RUTILE (see under each mineral name).

Titanium Silicide. Ti_5Si_3; 2120°C; sp. gr. 4·2. This special ceramic has good resistance to high-temperature oxidation but not to thermal shock.

Tobermorite. A hydrated calcium silicate approximating in composition to $5CaO.6SiO_2.5H_2O$. Tobermorite gel is the principal cementing compound in hardened portland cement.

Toggle Press. See MECHANICAL PRESS.

Tong Outcrop Clay. A fireclay associated with the Better Bed coal, Yorkshire, England. The raw clay contains about 65% SiO_2, 22% Al_2O_3, 1·5% Fe_2O_3 and 1·6% alkalis.

Tongue. See MIDFEATHER.

Tool Tips. Ceramic tool-tips for use in the machining of metals are of two types: carbide, e.g. tungsten carbide; oxide, e.g. corundum. It is important that the size of the individual corundum crystals in an oxide tool tip should be small ($< 10\mu$); the presence of a small amount of MgO inhibits crystal growth. Important advantages of oxide tool tips is their retention of strength, hardness and wear-resistance at high temperatures; these factors result in long life and clean cutting even at cutting speeds higher than normal.

290

Tooling. The rubber or plastic bag used in ISOSTATIC PRESSING (q.v.).

Toothing. In structural brickwork, bricks left projecting at the temporary termination of a wall so that future extensions can be bonded in.

Top-fired Kiln. A kiln fired by feeding coal or oil through apertures in the roof. The typical kiln of this type is the Hoffmann annular kiln, but the Monnier kiln provides an example of a top-fired car-tunnel kiln.

Top-hat Kiln. An intermittent kiln of a type sometimes used in the firing of pottery. The ware is set on a refractory hearth, or plinth, over which a box-shaped cover is then lowered.

Top Pouring. See DIRECT TEEMING.

Topaz. $Al_2SiO_4(OH,F)_2$. Occurs in economic quantities in Australia, Brazil, Nigeria and USA. After it has been calcined, the material has a composition similar to that of mullite and it has been used to a small extent, either alone or mixed with calcined kyanite, for making high-alumina refractories.

Torching. The pointing, with cement or mortar, of the underside of a tiled roof; also sometimes known as TIERING.

Torkret Process. German process for spraying refractory patching material on the walls of steel-furnace ladles.

Tornebohm's Minerals. See ALITE; BELITE; CELITE; and FELITE.

Torsion Viscometer. An instrument much used for works' control of the viscosity and thixotropy of clay slips. It consists of two concentric cylinders, the slip occupying the space between them; the inner cylinder is suspended from a torsion wire and is released from a position equivalent to a 360° twist in the wire. The overswing gives a measure of viscosity; comparison of this degree of overswing with that after a specified lapse of time provides a measure of the thixotropy of the slip.

Tortoiseshell. A decorative effect produced on a lead glaze by dusting metal oxides (MnO_2, CoO or CuO) over the surface and firing.

Totanin. Trade-name: an ammonium-based SULPHITE LYE (q.v.).

Toughened Glass. Glass that has been rapidly cooled so that the surface layers are in compression; the thermal and mechanical endurance are increased and, if the glass does break, it will shatter into small, granular, fragments rather than into large and dangerously jagged pieces.

Tourmaline

Tourmaline. A complex alumino-borosilicate that will often contain small amounts of iron and alkalis. When heated, tourmaline loses water at 150–750°C; B_2O_3 is evolved at 950–970°C. Tourmaline is an accessory mineral in the granite from which the Cornish china clays are derived.

Towing. The smoothing, generally on a powered wheel, of the outer edge of dried pottery flatware before it is fired. Tow is commonly used for the purpose, but sandpaper or a profile scraper are also sometimes used.

Trace Hole. A small horizontal passage or flue left in a setting of bricks in a kiln to facilitate the movement of hot gases.

Trailing. A method of slip decoration sometimes used by the studio potter: a pattern is formed on the ware by means of a viscous slip squeezed through a fine orifice, e.g. a quill.

Transducer. A device for the direct transformation of electrical energy into mechanical energy. PIEZOELECTRIC (q.v.) ceramics are used for this purpose.

Transfer Glass. US term for optical glass that has been cooled in the pot in which it was melted (cf. ROLLED GLASS).

Transfer Ladle or Hot-metal Ladle. A large ladle lined with refractory material (usually fireclay bricks) for the transport of molten pig-iron from a blast furnace to a hot-metal mixer or to a steelmaking furnace.

Transfer Printing. An intaglio process of decoration, particularly applicable to pottery-ware; a single-coloured pattern is transferred directly from a printing plate or roller by means of thin paper. The colour used is generally dispersed in linseed oil; soft-soap is the traditional size for the transfer paper but various synthetics have also been used (cf. LITHOGRAPHY).

Transfer Ring. See HOLDING RING.

Transformation Points. Temperatures at which the coefficient of thermal expansion of a glass changes. For any one glass, there are normally two such points known respectively as the Tg point and the Mg point: the Tg point is the first temperature at which there is a sudden change in expansion when the glass is heated at 4°C/min; the Mg point is the temperature at which the thermal expansion curve reaches a maximum and is usually equal to the softening temperature.

Translucency. The ability to transmit light is an important feature of bone china and of most porcelain. The translucency of these materials agrees reasonably well with Lambert's Law: the

292

ratio of the intensity of the emergent light to that of the incident light is an exponential function of the thickness of the ware and of a constant, the latter depending on the nature of the material. Translucency can be measured by an apparatus comprising a standard light source and a photoelectric cell.

Transverse-arch Kiln. An ANNULAR KILN (q.v.) that is divided into a series of chambers by fixed walls (hence the alternative name CONTINUOUS CHAMBER KILN). The axis of the arched roof of each chamber is transverse to the length of the kiln. This type of kiln finds use in the heavy-clay industry.

Transverse Strength. See MODULUS OF RUPTURE.

Trass. A product of the partial decomposition of volcanic ash; it is often consolidated as a result of infiltration of calcareous or siliceous solutions. Trass is used in POZZOLANA (q.v.).

Trébuchon–Kieffer Annealing Schedule. A procedure for annealing the glass components of electron tubes; it is based on annealing at the transformation temperature for 20 min followed by cooling at a rate dependent on the nature of the glass and its thickness. (G. Trébuchon and J. Kieffer, *Verres Réfract.*, **4**, 230, 1950; *Glass Industry*, **32**, 240, 1951.)

Tremolite. $2CaO.5MgO.8SiO_2.H_2O$. Tremolite is sometimes present in steatite rocks, rendering them unsuitable for use as a ceramic raw material.

Triangle Bar. One type of metal support for vitreous enamelware during firing.

Triaxial Test. A method of testing clay in which the test-piece, in a plastic state, is enclosed in a rubber envelope and is then subjected to uniform hydrostatic pressure while it is also being loaded axially. A stress/deformation diagram is plotted.

Tricalcium Aluminate. $3CaO.Al_2O_3$; melts incongruently at 1535°C. This compound is present in portland cement.

Tricalcium Disilicate. $3CaO.2SiO_2$ (see RANKINITE).

Tricalcium Pentaluminate. A compound, $3CaO.5Al_2O_3$, formerly believed to be present in high-alumina hydraulic cement. It is now known that a melt of this composition consists of a mixture of $CaO.2Al_2O_3$ and $CaO.Al_2O_3$, the latter compound being responsible for the hydraulic properties.

Tricalcium Silicate. $3CaO.SiO_2$; dissociates at approx. 1900°C to form CaO and $2CaO.SiO_2$. This compound is the principal cementing constituent of portland cement, small quantities of MgO and Al_2O_3 usually being present in solid solution.

Tridymite

Tricalcium silicate is also present in some stabilized dolomite refractories.

Tridymite. SiO_2; sp. gr. 2·28. According to the classical research of C. N. Fenner (*Amer. J. Sci.*, **36**, 331, 1913) tridymite is the form of silica that is stable between 870 and 1470°C; he considered that there are three crystalline varieties changing reversibly, one into another, as follows:

$$\alpha - \text{tridymite} \underset{\rightleftharpoons}{117°C} \beta_1 - \text{tridymite} \underset{\rightleftharpoons}{163°C} \beta_2 - \text{tridymite}$$

More recent research has indicated a further inversion at about 250°C and has suggested that tridymite can be produced only in the presence of foreign ions, which enter the crystal lattice and cause disorder in the structure.

Trief Process. A process for making concrete with PORTLAND BLAST-FURNACE CEMENT (q.v.) first proposed by a Belgian, V. Trief (Brit. Pat., 673 866, 11/6/52; 674 913, 2/7/52). The slag is wet ground and fed as a slurry to a concrete mixer together with portland cement and aggregate.

Trimmers. See FITTINGS.

Triple Round Edge. A type of wall tile (see Fig. 6, p. 307).

Tripoli. A sedimentary rock consisting essentially of silica and having a porous and friable texture. A principal use is as an abrasive.

Trommel. See REVOLVING SCREEN.

Tropenas Converter. See CONVERTER.

Truck Chamber Kiln. See BOGIE KILN.

True Density. A term used when considering the density of a porous solid, e.g. a silica refractory. It is defined as the ratio of the mass of the material to its TRUE VOLUME (q.v.). (Sometimes referred to as POWDER DENSITY.)

True Porosity. See under POROSITY.

True Specific Gravity. The ratio of the mass of a material to the mass of a quantity of water that, at 4°C, has a volume equal to the TRUE SOLID VOLUME of the material at the temperature of measurement.

True Volume. A term used in relation to the density and volume of a porous solid, e.g. a brick. It is defined as the volume of the solid material only, the volume of any pores being neglected.

Trumpet or Bell. A fireclay-refractory funnel placed at the top of the assembly of GUIDE TUBES (q.v.) to receive molten metal from the nozzle of a ladle in the Bottom Pouring (q.v.) of steel.

Tube Bottom. One form of bottom for a Converter (q.v.); it is made of monolithic refractory material, the air passages being lined with copper tubes (cf. Spiked Bottom and Tuyere Block Bottom).

Tube Furnace. A type of furnace, particularly for vitreous enamelling, heated by tubes of heat-resisting metal in which gas is burned.

Tube Mill. A ball mill having a cylinder longer than usual, this usually being sub-divided internally so that the material to be ground passes from one compartment to the next, the grinding media in successive compartments being appropriately smaller than in preceding compartments.

Tubelining. A process of decoration, particularly for wall tiles requiring 'one-off' designs. Lines of coloured slip are added to the tile by squeezing it from a rubber bag through a narrow tube; the tile is then fired and the pattern between the raised lines is filled-in with various colours and refired; alternatively, the colours can be applied to the unfired tile and the once-fired process used.

Tuckstone; Tuckwall. A shaped refractory block fitting above the tank blocks of a glass furnace. The general purpose of the tuckstones is to protect the top of the tank blocks from the furnace gases and, in some types of tank furnace, to act as a seal between the tank blocks and the side- and end-walls. The course of tuck-stones is sometimes called the Tuckwall.

Tundish. A rectangular trough lined with fireclay refractories and with one or more refractory nozzles in its base. Tundishes are sometimes used between the ladle and the ingot moulds in the teeming of steel.

Tungsten Borides. Data have been reported on three com-pounds. W_2B; dissociates at 1900°C; sp. gr. 16·7. WB exists in two crystalline forms: α-WB, sp. gr. 16·0; β-WB, sp. gr. 15·7 (both forms melt at approx. 2400°C). W_2B_5; m.p. 2200°C; sp. gr. 13·1.

Tungsten Carbide. WC; m.p. 2865°C; hardness, 2100 (K100); sp. gr. (theoretical), 15·7 g/ml; modulus of rupture, 50 000–80 000 at 25°C. The principal use of this carbide is in cutting tools. There is also a ditungsten carbide: W_2C; m.p. 2855°C; sp. gr. 17·2.

Tuning-fork Test (for Glaze-Fit). A test-piece is made by joining, with clay slip, two bars of the extruded pottery body to a short piece of the same material. The test-piece is biscuit fired and the outer faces are then glazed. The test-piece is placed in a furnace and fired so that the glaze matures; it is then allowed to cool, while still in the furnace, and any relative movement of the two ends of the 'tuning fork' is measured by a micrometer telescope. From this measurement the magnitude of any stress in the glaze can be calculated. The test was devised by A. M. Blakely (*J. Amer. Ceram. Soc.*, **21**, 243, 1938).

Tunnel Dryer. A continuous dryer through which shaped clay-ware can be transported on cars; it is controlled so that the humidity is high at the entrance and low at the exit.

Tunnel Kiln. A continuous kiln of the type in which ware passes through a stationary firing zone near the centre of the kiln. In the most common type of tunnel kiln the ware is placed on the refractory-lined deck of a car, a continuous series of loaded cars being slowly pushed through a long, straight, tunnel.

Turbidimeter. An instrument for determining the concentration of particles in a suspension in terms of the proportion of light absorbed from a transmitted beam. An instrument of this type designed for particle-size analysis is the WAGNER TURBIDIMETER (q.v.).

Turbine Blades. See GAS TURBINE.

Tuscarora Quartzite. An important source of raw material for silica refractories occurring in Pennsylvania, USA. A typical analysis is (per cent): SiO_2, 97·8; Al_2O_3, 0·9; Fe_2O_3, 0·7; alkalis, 0·4.

Tuyere. A tube or opening in a metallurgical furnace through which air is blown as part of the extraction or refining process. In a blast furnace the tuyeres are water-cooled metal tubes which pass through the refractory lining of the BOSH (q.v.) (French word meaning 'a tube').

Tuyere Block Bottom. One form of bottom for a CONVERTER (q.v.). The passages for the air blast are separate pre-formed tuyeres each having several holes; these tuyere blocks are interspersed with solid refractory blocks, the whole bottom then being finished by ramming refractory material into any spaces.

Twaddell Degrees (°Tw). A system for denoting the specific gravity of a liquid:

$$\text{Degrees Twaddell} = (\text{sp. gr.} - 1) \times 200$$

The specific gravity of solutions of sodium silicate, for example, is often quoted in this form; each Twaddell degree corresponds to a sp. gr. interval of 0·005. This scale is named after William

Twaddell who, in Glasgow in 1809, made a hydrometer with this scale to the design of Charles Macintosh (of rainwear fame).

Tweel Block. A type of refractory block used in the glass industry for such purposes as protection of a newly-set pot, the construction of a furnace door or damper, or control of the flow of molten glass. (From French *tuile*, a tile.)

Twin-plate Process. A process for the simultaneous grinding and polishing of both faces of a continuously-produced ribbon of glass; the complete flow-line is nearly 1300 ft long. The process was introduced by Pilkington Bros. Ltd, England, in 1952.

Tyler Sieve. See under SIEVE; for mesh sizes see Appendix 1.

U

U-type Furnace or Hair-pin Furnace. A furnace for the firing of vitreous enamelware, which is carried along a U-shaped path so that ware enters and leaves the furnace at adjacent points.

U-value. A unit of heat transmission used in heat-loss calculations for buildings and defined as: Btu transmitted per ft^2 per h per °F difference in air temperature between the two faces of the wall under consideration. The walls of a house should have a U-value of 0·20 or less.

Udden Grade Scale. A scale of sieve sizes introduced by J. A. Udden (*Augustana Library Pub.*, No. 1, 1898). The basic opening is 1 mm, the scale above and below being a geometrical series with a ratio of 2 (above 1 mm) and $\frac{1}{2}$ (below 1 mm).

Ulexite. A boron mineral approximating in composition to $Na_2O.2CaO.5B_2O_3.16H_2O$. Ulexite occurs in Chile and Argentina. Trials have been made with this mineral as a flux in ceramic glazes.

Ultimate Analysis. The chemical analysis reported in terms of constituent oxides, as distinct from the RATIONAL ANALYSIS (q.v.), which is in terms of the minerals actually present.

Ultrasonic Equipment. The word 'ultrasonic' signifies vibration at a *frequency* greater than the maximum audible frequency, and should not be confused with 'supersonic', which signifies a *velocity* greater than that of sound. Ultrasonic vibrations can be generated by piezoelectric ceramics, by magnetostrictive devices, or by 'whistles' in which there is a steel blade vibrated by a high-pressure jet of liquid. Ultrasonic equipment has been used in the

Ultra-violet Absorbing Glass

ceramic industry for the dispersion of clay slips, for metal clean-ing prior to vitreous enamelling, and for flaw detection, particu-larly in large electrical porcelain insulators.

Ultra-violet Absorbing Glass. Glasses can be made to absorb U.V. light, while transmitting visible light, by the inclusion of CeO_2 in the batch. Other elements absorbing U.V. light include Cr, Co, Cu, Fe, Pb, Mn, Nd, Ni, Ti, U and V.

Ultra-violet Transmitting Glass. For high transmittance of U.V. light, a glass must be free from Fe, Ti and S. Phosphate glasses and some borosilicate glasses have good U.V. transmittance. Uses include special windows and germicidal lamps.

Umber. A naturally-occurring hydrated iron oxide occasionally used as a colouring agent for pottery decoration.

Underclay. See SEAT EARTH.

Undercloak. A layer, of plain clay tiles for example, between the laths and the roof tiling proper at the VERGE (q.v.) of a tiled roof.

Undercutting. Faulty cutting of flat glass resulting in an edge that is oblique to the surface of the glass.

Under-glaze Decoration. Decoration applied to pottery before it has been glazed. Because it is finally covered by the glaze, such decoration is completely durable, but because the subsequent glost firing is at a high temperature the range of available colours for under-glaze decoration is limited.

Under-ridge Tile. A roofing tile for use at the top of a tiled roof. Such tiles are shorter than standard roofing tiles and are used to complete the roof along the ridge beneath the RIDGE TILES (q.v.).

UNI. Abbreviation for *Unificazione Italiana*; the prefix to the identification number of an Italian standard specification.

Unsoundness. As applied to portland cement, this term refers to slow expansion after the cement has set. The principal causes of this fault are the presence of free CaO, excess MgO, or excess sulphates.

Up-draught Kiln. An intermittent kiln in which the combustion gases pass from the fireboxes through the setting and thence through one or more chimneys in the roof. Such kilns are in-efficient and are now little used.

Up-draw Process. The continuous vertical drawing of glass rod or tubing from an orifice; to produce tubing, the rod is drawn around a refractory cone. (This process has also been called the SCHULLER PROCESS, or WOOD'S PROCESS.)

Uphill Teeming. See Bottom Pouring.

Upright. See Post.

Uptake. See Downtake.

Uranium Borides. Three borides are known: UB_2, UB_4 and UB_{12}. The most attention has been paid to the tetraboride, the properties of which are: m.p. $> 2100°C$ (but oxidizes rapidly above 600°C); sp. gr. 9·38 g/ml; thermal expansion, $7·1 \times 10^{-6}$ (20–1000°C); modulus of rupture (20°C), 60 000 p.s.i.; electrical resistivity, 3×10^{-5} ohm.cm.

Uranium Carbide. Two carbides exist, UC (m.p. $2590° \pm 50°C$) and UC_2; the powdered carbides can be made, by ceramic processes, into nuclear fuel elements.

Uranium Nitride. UN; m.p. $2650° \pm 100°C$.

Uranium Oxide. The important oxides of uranium are UO_2, U_3O_8 and UO_3. The dioxide (m.p. 2880°C) is used as a nuclear fuel element. Uranium oxide has been used to produce red and yellow glazes and ceramic colours.

Uranium Red. A ceramic stain for coloured glazes suitable for firing temperatures up to 1000°C. Increasing the uranium oxide content strengthens the colour from orange, through red to tomato red. The glaze should be basic, preferably 0·5 mol SiO_2, 0·1–0·2 mol Al_2O_3, 0·1 mol K_2O and the remaining bases chiefly PbO; B_2O_3 should not be present in significant quantity.

Uranium Silicide. β-USi_2 has a highly anisotropic thermal expansion.

Uvarovite. $3CaO . Cr_2O_3 . 3SiO_2$; the colouring agent in Victoria Green (q.v.).

V

V-bricks. A series of perforated clay building bricks designed by the Building Research Station, England, in 1959–60; the name derives from the fact that the perforations are Vertical.

V-draining. See Double-draining.

Vacuity. The expansion space left above the liquid in a closed glass container.

Vacuum-and-Blow Process. See Suck-and-Blow Process.

Vacuum Firing. A process for the firing of special types of ceramic either to prevent oxidation of the ware or to reduce its porosity. Vacuum firing is used, for example, in the firing of dental porcelain to produce teeth of almost zero porosity.

Vacuum Mixer. A machine for the simultaneous de-airing and moistening of dry, prepared clay as it is fed to a pug. In the original design (L. Walker, *Claycraft*, **25**, 76, 1951) the clay fell as a powder through a vertical de-airing chamber where water was added as a fine spray; from the bottom of the de-airing chamber the moist, de-aired, clay passed into a pug. There have been several developments of this principle.

Vacuum Pug. A PUG (q.v.) with a vacuum chamber in which the clay is de-aired before it passes into the extrusion chamber.

Vallendar Clay. A clay from Westerwald, Germany, used in the vitreous enamel industry.

Valley Tiles. Specially-shaped roofing tiles for use in the 'valley' where two roof slopes meet; these tiles are made to fit into the angle. They lap and course in with the normal tiling.

Vanadium Borides. Several borides have been reported, including the following: VB_2; m.p. $2400°C$; sp. gr. $5·0$; the thermal expansion is highly anisotropic. VB; m.p. $2250°C$; sp. gr. $5·6$. V_3B_4; melts incongruently at $2300°C$; sp. gr. $5·5$.

Vanadium Carbide. VC; m.p. $2830°C$; sp. gr. $5·36$.

Vanadium Yellow or Vanadium-Tin Yellow. A ceramic colour produced by the calcination, at about $1000°C$, of a mixture of $10–20\%$ V_2O_5 (as ammonium metavanadate) and $80–90\%$ SnO_2. A stronger yellow results if a small amount of TiO_2 is added. These colours can be used in most glazes and either SnO_2 or zircon can be used as opacifier.

Vanadium-Zirconium Blue (or Turquoise). See ZIRCONIUM–VANADIUM BLUE.

Vanal. Trade-name: Hagenberger–Schwalb A.G., Hettenleidelheim/Pfalz, Germany. A coating for the protection of refractories against slag attack developed by A. Staerker (*Ber. Deut. Keram. Ges.*, **28**, 390, 1951; **29**, 122, 1952). It contains vanadium and is claimed to prevent slags from wetting the refractory.

Vane Feeder. A device for feeding dry ground clay from a hopper to a tempering machine or mixer, for example. Vanes fixed to a horizontal shaft at the base of the hopper rotate to discharge the material.

Varistor. A material having an electrical resistance that is sensitive to changes in applied voltage. A typical example is the varistor made from a batch consisting of granular silicon carbide, mixed with carbon, clay and water; the shaped components are fired at 1100–1250°C in H_2 or N_2. Varistors are used in some types of telephone equipment.

VE. Abbreviation for VITREOUS ENAMEL (q.v.).

Vebe Apparatus. A device developed by the Swedish Cement Association for the measurement of the consistency of concrete. It is a slump test in which the consistency is expressed in degrees, the value being obtained by multiplying the ratio of the volume of the test-piece after vibration to that before vibration by the number of vibrations required to cause the test-piece to settle.

Vegard's Law. States that in a binary system forming a continuous series of solid solutions, the lattice parameters are linearly related to the atomic percentage of one of the components. This law has been applied, for example, in the study of mixed spinels of the type formed in chrome-magnesite refractories. (L. Vegard, *Z. Physik*, **5,** 17, 1921.)

Vein. See STRIAE.

Vein Quartz. An irregular deposit of quartz, often of high purity, intruded into other rocks. In consequence of its mode of formation, vein quartz usually contains occluded gas bubbles and is unsuitable as a raw material for silica refractories.

Vello Process. A method for the production of glass tubing; molten glass flows vertically through an annular orifice; the central refractory pipe within the orifice is hollow and rotates. The process is considerably faster than the DANNER PROCESS (q.v.).

Vellum Glaze. See SATIN GLAZE.

Veneered Wall. A wall having a facing (of faience panels, for example) which is attached to the backing but not in a way to transmit a full share of any imposed load; the veneer and the backing do not exert a common action under load (cf. FACED WALL).

Vent. See CHECK.

Verge. The gable edge of a tiled roof. At the verge the roofing tiles are edge-bedded, preferably on a single or double undercloak of plain tiles. This form of undercloak gives a neat appearance to the verge and slightly inclines the verge tiles so that rainwater is turned back on to the main roof.

Vermiculite. A group name for certain biotite micas that have been altered hydrothermally. When rapidly heated to 800–950°C vermiculite exfoliates as the combined water is expelled; the

Verneuil Process

volume increase is about 15%. Large deposits of vermiculite occur in USA, Transvaal, Uganda, Australia and the Ural Mountains of Russia. The exfoliated material is used as loose-fill insulation and in the manufacture of vermiculite insulating bricks. Standards for the loose-fill are laid down in B.S. 1785 and in US Govt. Fed. Spec. HH-I-00585. Vermiculite bricks are made by bonding the graded material with about 30% of clay, shaping, and firing at 1000°C. The bulk density of such bricks is 30–40 lb ft³; thermal conductivity 1·5–2·0 Btu/ft²/h/in./°F. They can be used up to 1100°C.

Verneuil Process. A method for the production of ceramic BOULES (q.v.) by feeding powdered material, e.g. Al_2O_3, into an oxy-hydrogen flame. (A. Verneuil, *Ann. Chim. Phys.*, **3**, 20, 1904.)

Vibrating Ball Mill; A BALL MILL (q.v.) supported on springs so that an out-of-balance mechanism can impart vibration to the mill, usually in the vertical plane and typically at about 1500 cycles/minute. Advantages over the ordinary ball mill are increased rate of grinding (particularly with very hard materials), lower energy consumption per ton of product, and less wear.

Vibrating Screen. A screen, set at an angle of 25–35° to the horizontal, and vibrated by an eccentric, a cam or hammer, an out-of-balance pulley, or an electro-magnet. When operating with 10-mesh cloth on damp clay, such a screen should have an output of about 0·5 ton/h/ft². The screening efficiency is generally 80–85%.

Vibro-energy Mill. Trade-name: a VIBRATING BALL MILL (q.v.) designed to oscillate both horizontally and vertically, the vertical motion being of small but sufficient amplitude to prevent the charge from becoming tightly packed. (W. Podmore Ltd., and W. Boulton Ltd., Stoke-on-Trent, England.)

Vicat Needle. An instrument for evaluating the consistency of cement in terms of the depth of penetration of a 'needle' of standard shape and under a standard load; it was designed by L. J. Vicat, a Frenchman, in the early 19th century. Details of dimensions and method of use are given in B.S. 12 (cf. GILLMORE NEEDLE).

Vickers Hardness. An indentation test in which a diamond pyramid is used. The diamond is loaded mechanically by a lever, the application of the load being hydraulically controlled to give the correct duration. Some use has been made of this form of test in studying the hardness of glasses and glazes.

Victoria Green. A bright green ceramic colour. A typical batch composition is: 38% $K_2Cr_2O_7$, 20% $CaCO_3$, 22% CaF_2, 20% SiO_2. This batch is calcined, washed free from soluble chromates, and ground. The colouring agent is stated to be uvarovite $(3CaO . Cr_2O_3 . 3SiO_2)$.

Vignetting. The decoration of a glass surface by firing-on a metal or other suitable powder; the surface is first coated with sodium silicate solution and the powder is then dusted on and fired in.

Vinsol Resin. Trade-name: a thermoplastic powder used as an AIR-ENTRAINING (q.v.) agent in the mixing of concrete. (Hercules Powder Co. Ltd., London.)

Viscometer. The commonest types of viscometer, as used in the ceramic industry, depend on measurement of the flow of the test-liquid through an orifice or of the drag on one of a pair of concentric cylinders when the other is rotated, the test-liquid occupying the space between the cylinders.

Viscosity. The viscosity of a true ('Newtonian') liquid is the ratio of the shearing stress to the rate of shear, of which the viscosity itself is independent; this is not true of clay slips, vitreous-enamel slips, or glaze suspensions, all of which exhibit THIXO-TROPY (q.v.).

Vitreous. This term, meaning 'glassy', is applied to ceramic ware that, as a result of a high degree of vitrification (as distinct from sintering) has an extremely low porosity. In the USA the term is defined (ASTM – C242) as generally signifying that the ware has a water absorption below 0·5%, except for floor tiles, wall tiles and low-tension electrical porcelain which are considered to be vitreous provided that the water absorption does not exceed 3%.

Vitreous-china Sanitaryware. Defined in B.S. 3402 as: A strong high-grade ceramic ware used for sanitary appliances and made from a mixture of white-burning clays and finely ground minerals. After it has been fired at a high temperature the ware will not, even when unglazed, have a mean value of water absorption greater than 0·5% of the ware when dry. It is coated on all exposed surfaces with an impervious non-crazing vitreous glaze giving a white or coloured finish. A typical batch for this type of body is: 20–30% ball clay, 20–30% china clay, 10–20% feldspar, 30–40% flint, 0–3% talc; sometimes nepheline syenite is used instead of feldspar.

Vitreous Enamel

Vitreous Enamel. Defined in B.S. 232 as: An inorganic glass which is fused on to a metal article in the form of a relatively thin coating and provides protection against corrosion. The equivalent US term is PORCELAIN ENAMEL (q.v.).

Vitreous Silica. A glass made from silica; it may contain numerous small bubbles, in which case it is translucent; when free from bubbles it is transparent. An important property is extremely low thermal expansion, hence a high resistance to thermal shock; vitreous silica tubes, etc., find considerable use in chemical engineering.

Vitreous Slip. A US term defined (ASTM – C242) as a slip coating matured on a ceramic body to produce a vitrified surface.

Vitrification. The progressive partial fusion of a clay, or of a body, as a result of a firing process or, in the case of a refractory material, of the conditions of use in a furnace lining. As vitrification proceeds the proportion of glassy bond increases and the apparent porosity of the fired product becomes progressively lower. The VITRIFICATION RANGE is the temperature interval between the beginning of vitrification of a ceramic body and the temperature at which the body begins to become deformed.

Vitrified. See VITREOUS.

Vitrified Wheel. An abrasive wheel made from a batch consisting of abrasive grains and a ceramic bond formed by kiln firing at 1200–1300°C. Over half the abrasive wheels currently produced are of this type.

Vitrite. The black glass used in the caps of electric lamps.

Vitroceramic. One of several terms proposed for the type of ceramic product formed by the controlled devitrification of a glass; see DEVITRIFIED GLASS.

Vitron. A unit of atomic structure, particularly in silica glass, proposed by L. W. Tilton (*J. Res. Nat. Bur. Stand.*, **59**, 139, 1957). Its basis is a pentagonal ring of five SiO_4 tetrahedra; these rings can be built up into three-dimensional clusters but only to a limited extent because of increasing distortional stress; a cluster of the pentagonal SiO_4 rings is a VITRON. Its most important property, as a basis for the understanding of the properties of glass, is its fivefold symmetry which precludes the formation of crystals. (cf. STRUCTON).

Vogel's Red. A pure ferric oxide produced by precipitating ferrous oxalate which is then calcined. It has been used as a basis for some 'iron' colours on porcelain.

304

Volclay. A sodium bentonite from Wyoming, USA. (Trade name.)

VPB Kiln. A kiln for the firing of building bricks; it consists of two groups of chambers in which the fire travel follows a zig-zag course. The bricks are set on refractory bats and are put into, and subsequently drawn from, the kiln without the workmen having to enter the hot chambers. The name is from the initials of the inventor, V. P. Bodin (*French Pat.*, 1, 156, 918, 22/5/58).

Vycor. A highly siliceous glass introduced by the Corning Glass Co. USA, in 1938. A borosilicate glass is first made and this solidifies in two phases, one of which is soluble in dilute acid and is thus removed, leaving a highly siliceous skeleton. The porous ware is then heated at about 1000°C; it shrinks considerably and non-porous high-silica glass (96% SiO_2) is produced The name is derived from Viking and Corning.

W

Wad. An extruded strip or rod of fireclay (with or without the addition of fine grog) used in the firing of pottery to seal the joints between saggars, or to level the supporting surfaces of saggars in a bung.

Wad Box. A simple hand-extrusion device for producing cylindrical fireclay WADS (q.v.) for use during the setting of bungs of saggars. A wad box was also used in the old method of pressing handles and similar shapes of pottery ware; used for this purpose, the device was sometimes called a DOD BOX.

Wadhurst Clay. A Cretaceous clay used for brickmaking in parts of Kent and Sussex.

Waechter's Gold Purple. A colour, of various shades, that has been used in the decoration of porcelain.

Waelz Furnace. A rotary furnace used particularly for the calcination of non-ferrous ores; chrome-magnesite linings have been used in Waelz furnaces producing ZnO.

Wafers. Small sheets of electroceramic material (e.g. $BaTiO_3$), 0·001–0·01 in. thick, for use in electronic equipment, particularly in miniature capacitors, transistors, resistors and other circuit components.

Wagner Turbidimeter. Apparatus for the determination of the fineness of a powder by measurement of the turbidity, at a

Waist

specified level and after the lapse of a specified time, of a suspension of particles that are settling by gravity according to Stokes' Law. The method was proposed by L. A. Wagner (*Proc. ASTM*, **33**, Pt. 2, 553, 1933).

Waist. See BELLY.

Walker Vacuum Mixer. See VACUUM MIXER.

Walking-beam Kiln. A tunnel kiln of unusual type, the ware (set on bats) being moved through the kiln in steps by a mechanism that alternately lifts the bats and sets them down further along the kiln.

Wall Tiles and Fittings. Glazed wall tiles (which are also used in fireplace surrounds) are made by a highly mechanized dust pressing process from white or buff bodies; to reduce the firing contraction and MOISTURE EXPANSION (q.v.) these bodies often contain lime compounds. Wall tiles are made in a wide variety of colours and glazes; a satin or matt glaze is often used. Some of the shapes of tiles and fittings are shown in Fig. 6. (Note: in USA the terms 'Tile' and 'Wall Tile' denote HOLLOW CLAY BLOCKS (q.v.).)

Warpage. A test for warpage of vitreous-enamelled flatware is given in ASTM – C314

Wash. See REFRACTORY COATING.

Wash-back. See under WASH-MILL.

FIG. 6. (*opposite*) WALL TILES AND FITTINGS: UK NAMES AND SYMBOLS (the latter, in brackets, follow the name).

1. Plain Tile. 2. Round Edge Tile (RE). 3. Round Edge External Corner (REX). 4. Round Edge Reveal (RER). 5. Round Edge Opposite (REO). 6. Round Edge External Reveal (REXR). 7. Triple Round Edge (TRE). 8. Round Edge Tile, Long side (REL). 9. Round Edge External Corner, Left Hand (HREX). 10. Round Edge External Corner, Right Hand (REXH). 11. Round Edge Opposites, Long Sides (REOL). 12. Round Edge Tile, Short Side (RES). 13. Round Edge External Reveal, Curb Corner, (REXR). 14. Curb Bend. 15. Internal Angle Bead. 16. External Angle Bead. 17. Internal Bird's Beak. 18. External Bird's Beak. 19. Internal Coconut Piece. 20. External Coconut Piece. 21. External Shoulder Angle. 22. Internal Shoulder Angle. 23. Cove Skirting. 24. External Angle to Cove Skirting. 25. Internal Angle to Cove Skirting.

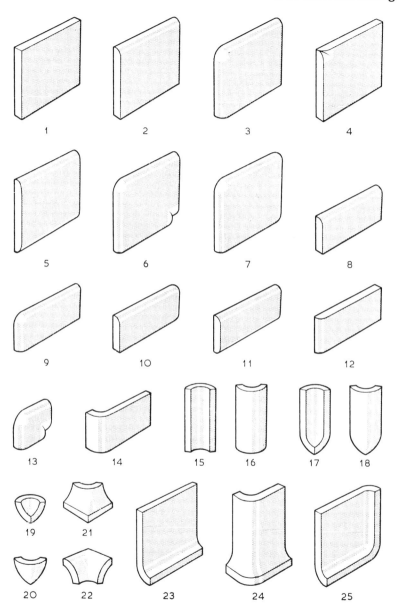

1   2   3   4

5   6   7   8

9   10   11   12

13   14   15   16   17   18

19   21   23   24   25

20   22

Washbanding. A form of pottery decoration, usually on-glaze, in which a thin layer of colour is applied over a large surface of the ware by means of a brush.

Washboard. Unintentional waviness on the surface of glassware; also known as LADDERS.

Wash-gate. See under WASH-MILL.

Washing Off. Removing printing paper from pottery ware that has been decorated by the transfer process.

Wash-mill. A large tank fitted with stirrers (known as HARROWS or WASH-GATES) for the cleaning of the impure surface clays used in the manufacture of STOCK BRICKS (q.v.). From the wash-mill the clay slurry, together with a slurry of any lime or chalk that is to be added, is pumped into a settling tank known as a WASH-BACK.

Waste-heat Dryer. A dryer for clayware that derives its heat from the cooling ware in the kilns. Such dryers are common in the brick industry and, even though the fuel consumption in the kilns may be slightly increased, this method of heat utilization results in an overall economy in fuel. Hot-floors, chamber dryers or tunnel dryers can be operated on this principle and the necessary heat can be derived from intermittent, annular or tunnel kilns.

Waster. A brick, structural or refractory, that is defective as drawn from the kiln; wasters in the refractories industry are crushed and re-used as grog.

Water Absorption. The weight of water absorbed by a porous ceramic material, under specified conditions, expressed as a percentage of the weight of the dry material. This property is much quoted when referring to structural clay products; the apparent porosity is more commonly quoted for refractories and whitewares. The two properties are related by the equation:

Apparent Porosity = Water Absorption × Bulk Density

(Note: In the USA, the term 'Absorption' is preferred to 'Water Absorption').

Waterfall Process. A method for the application of glaze materials to a ceramic body by mechanically conveying the ware through a continuously flowing (recirculated) vertical stream of the glaze suspension. The process is used in the glazing of wall tiles.

Waterford Glass. Cut or gilded glass made in the Waterford district of Ireland and characterized by a slight blueness resulting from the presence of a trace of cobalt.

Water Gain. See BLEEDING.

Water Glass. Popular name for soluble grades of SODIUM SILICATE (q.v.).

Water-line. A defect in vitreous enamelware in the form of a line marking the limit of water penetration from wet beading enamel into the unfired enamel coating.

Water Mark or Water Spot. (1) A shallow depressed spot sometimes appearing as a defect in vitreous enamelware.

(2) During transfer-printing on pottery, a water-mark may form if a drop of water dries on the ware, leaving a deposit of soluble salts.

Waterproofing. (1) Concrete can be made more waterproof by surface treatment of the set concrete or by the addition of an integral waterproofer. For surface treatment a solution of sodium silicate or of a silicofluoride may be used; silicones, drying oils and mineral oils are also sometimes employed. Integral waterproofers include calcium chloride solution and/or various stearates.

(2) Silicones have been recommended for the waterproofing of brickwork.

Water-smoking Period. The stage in the firing of heavy-clayware and of fireclay refractories when the mechanically-held water in the clay is being evolved, i.e. the temperature range 100–250°C. The temperature distribution in the setting should be such as to prevent condensation of any of this water vapour, and the temperature should not be raised further until all the water has been evolved.

Water Spot. See WATER MARK.

Water Streak. A fault in vitreous enamelware arising from drops of water running down the ware, while it is being dried, and partially removing the enamel coating. The obvious cause is the use of a slip that is too wet, when water-streaks are liable to occur in any sharp angle of the ware; condensation of drops of water on parts of the ware during the drying process is another cause.

Water Test. See FLOC TEST.

Watkin Heat Recorders. Cylindrical pellets ($\frac{3}{8}$ in. high, $\frac{1}{4}$ in. dia.) made of a blend of ceramic materials and fluxes so proportioned that, when heated under suitable conditions, they will fuse at stated temperatures. They are numbered from 1 (600°C) to 59 (2000°C) and were introduced by H. Watkin (Stoke-on-Trent, England) in 1899; they are now available from Harrison & Sons, Ltd., Stoke-on-Trent.

Wave. An optical defect in glass caused by uneven glass distribution or by STRIAE (q.v.).

Weald Clay. A Cretaceous clay, often variable in composition even within the same clay-pit, used for brickmaking in parts of Surrey, Kent and Sussex.

Weathering. (1) The preparation, particularly of clay, by exposure to the weather for a long period. This helps to oxidize any pyrite present, rendering it soluble, so that this and other soluble impurities are to some extent leached out; the water content also becomes more uniform and agglomerates of clay are broken down with a consequent increase in plasticity.

(2) The attack of glass or other ceramics by the atmosphere.

Web. One of the clay partitions dividing a hollow building block into cells.

Webb Effect. The increase in volume of a pottery slip as deflocculation proceeds. (H. W. Webb, *Trans. Brit. Ceram. Soc.,* **33,** 129, 1934.)

Wedge Pyrometer. An instrument for the approximate measurement of high temperatures. It depends on a wedge of coloured glass, the position of which is adjusted until the source of heat is no longer visible when viewed through the glass; movement of the wedge operates a scale calibrated in temperatures.

Wedge-stilt. See STILT.

Wedged Bottom. See SLUGGED BOTTOM.

Wedging. (1) A procedure for preparing clay or a clay body by hand: the lump of clay is repeatedly thrown down on a workbench; between each operation the lump is turned and sometimes cut through and rejoined in a different orientation. The object is to disperse the water more uniformly, to remove lamination and to remove air.

(2) A fault in dust-pressed tiles if the powder is not charged to a uniform depth in the die.

Wedgwood Pyrometer. A device for the determination of high temperatures on the basis of the approximate relationship between the contraction of clay test-pieces and the temperature to which they have been exposed; this pyrometer was introduced by Josiah Wedgwood (*Phil. Trans.,* **72,** 305, 1782) and now forms the basis of the BULLERS' RING (q.v.).

Weibull's Theory. A statistical theory of the strength of materials proposed by W. Weibull (*Ing. Vetenskaps Akad. Hand.,* No. 151, 1939); its basic postulate is that the probability of fracture of a solid body depends on the volume under stress and on the stress distribution. This theory has been applied in studies of the strength of ceramic materials and their resistance to the stresses induced by thermal shock.

Well. The term sometimes used for the lowest part of a blast furnace, i.e. the part in which molten iron collects. This part of the blast furnace is lined with aluminous firebricks or carbon blocks.

Well-hole Pipe. One of the short fireclay pipes that were used to carry the flame upwards from the well-hole in the bottom of a BOTTLE OVEN (q.v.).

Wentworth Grade Scale. An extension of the UDDEN (q.v.) scale of sieve openings proposed by C. K. Wentworth (*J. Geol.*, **30**, 377, 1922).

Westerwald Clay. A refractory clay occurring in an area E. of the Rhine between Coblenz and Marburg. These clays are of Oligocene origin and vary widely in composition from highly siliceous to clays that contain (raw) over 35% Al_2O_3.

Westlake Process. An automatic method for making glass-ware in paste moulds; the process closely imitates hand-making and was invented in 1916 by the Westlake European Machine Co., Toledo, USA.

Wet Pan. An EDGE-RUNNER MILL (q.v.) used for grinding relatively wet material in the refractories and structural clay-wares industries. The bottom has slotted grids with a proportion of solid plates on which the mullers can grind.

Wet Pressing. Alternative term in USA for PLASTIC PRESSING (q.v.).

Wet Process. (1) The method of blending the constituents of a whiteware body in the form of SLIPS (q.v.).

(2) The process of portland cement manufacture in which the limestone (or chalk) and clay are fed to the kiln as a slurry.

(3) The method of applying vitreous enamel as a slip, usually by spraying.

Wet-rubbing Test. A test to determine the degree of attack of a vitreous-enamelled surface after an acid-resistance test; (see ASTM – C282).

Wetting Off. The severing of a hand-made glass bottle from the blow-pipe by means of a fine jet of water.

Wheelabrator. A shot-blasting machine of a type used for cleaning castings prior to vitreous enamelling. This equipment has also been adapted to the testing of refractory bricks for abrasion resistance (see *Trans. Brit. Ceram. Soc.*, **50**, 145, 1951).

Whelp. A refractory brick of the same thickness and breadth as a standard square but of greater length, e.g. $12 \times 4\frac{1}{2} \times 3$ in. (see Fig. 1, p. 37).

Whirler. (1) A piece of tableware that has warped slightly during drying and/or firing; in consequence, such ware will 'whirl' on its foot if spun on a flat surface.

(2) A turntable used for checking the symmetry of a model in pottery making, or for the hand-making of a SAGGAR (q.v.).

Whirlering. The plaster moulds for bone-china hollow-ware are often revolved on a turntable while they are being filled with slip; this is known as 'whirlering' and the object is to prevent WREATHING (q.v.).

White Cement. Portland cement made from non-ferruginous raw materials, i.e. chalk (or low-iron limestone) and china clay. The Fe_2O_3 content is $< 1\%$.

White Flint. See FLINT GLASS.

White Ground-coat. Term sometimes used for a white vitreous enamel of high opacity used for one-coat application.

White Lead. Basic lead carbonate, $2PbCO_3 \cdot Pb(OH)_2$. Used to some extent in USA as a glaze constituent.

White Spot. A fault sometimes occurring in pottery colours, e.g. in chrome-tin pink and in manganese colours. It is caused by evolution of gas during firing, the glaze subsequently flowing over the crater left by the gas bubble without carrying with it sufficient colour to match the surrounding area.

White's Test. A method for the detection of free lime, for example in portland cement or dolomite refractories. A few mg of the powdered sample is placed on a glass microscope-slide and wetted with a solution of 5g phenol dissolved in 5 mg nitrobenzene with the addition of two drops of water. Micro-examination ($\times 80$) will reveal the formation of long birefringent needles if free CaO is present. (A. H. White, *Industr. Engng. Chem.*, **1**, 5, 1909.)

Whiteware. A general term for all those varieties of pottery that usually have a white body, e.g. tableware, sanitary ware and wall tiles. See also CERAMIC WHITEWARE, which has an ASTM definition.

Whitewash. (1) Local term for EFFLORESCENCE (q.v.) or SCUM (q.v.) on bricks.

(2) A fault in glass; see SCAB.

Whiting. Finely-ground cretaceous chalk, $CaCO_3$. British Whiting is 97–98% pure and practically all finer than 25 μ. It is used as a source of lime in pottery bodies and glazes, and to a small extent in glasses and vitreous enamels.

Wicket. A wall built of refractories to close an opening into a kiln or furnace; it is of a temporary nature, serving as a door, for example, in intermittent or annular kilns.

Wiegner Sedimentation Tube. Apparatus for particle-size analysis by sedimentation from a suspension in a tube (or cylinder) of relatively large diameter; the rate of sedimentation is indicated by the movement of the meniscus in a narrow side-tube, joined to the large tube near the base of the latter and itself containing the dispersing liquid free from solid particles. It was designed by G. Wiegner (*Landw. Vers.-Stat.*, **91**, 41, 1918); a later development was the KELLY SEDIMENTATION TUBE (q.v.).

Wilkinite. A particular type of bentonitic clay; it has been used as a suspending agent for glazes.

Wilkinson Oven. A pottery BOTTLE OVEN (q.v.) designed so that the hot gases rise through the bag-walls and the central well-hole, and descend between the saggars to leave the kiln through flue-openings in the floor mid-way along radii. (A. J. Wilkinson, Brit. Pat. 4356, 20/3/1890.)

Willemite. Zn_2SiO_4; formed in crystalline glazes that are loaded with ZnO.

Williams' Plastometer. A parallel-plate compression apparatus designed by I. Williams (*Industr. Eng. Chem.*, **16**, 362, 1924) for the testing of rubber; it has since been used quite considerably in the testing of clay.

Williamson Kiln. A tunnel kiln of the combined direct-flame and muffle type designed by J. Williamson (*Trans. Brit. Ceram. Soc.*, **27**, 290, 1928) and first used, in England, for the firing of wall-tiles and sewer-pipes. This kiln differed from earlier tunnel kilns in that the hot combustion gases passed across, rather than along, the kiln.

Willow Blue. Cobalt blue diluted with white ingredients such as ground silica; a quoted recipe is 40% cobalt oxide, 40% feldspar and 20% flint.

Willow Pattern. This well-known pseudo-Chinese scene was first engraved, in 1780, for the decoration of pottery-ware, by Thomas Minton for Thomas Turner of the Caughley Pottery, Shropshire, England.

Winchester. A straight-sided glass bottle of the type used for transporting laboratory liquids; the two British Standard Winchesters have volumes of 80 and 90 fl. oz respectively, i.e. approx. $2\frac{1}{4}$ and $2\frac{1}{2}$ litres. (Brit. Stand. 830.)

Winchester Cutting. A method of splay-cutting roofing tiles for the top course(s) in the vertical tiling of exterior walls, so that the final course of tiles is perpendicular to the overhanging verge.

Wind Ridge. A type of RIDGE TILE (q.v.).

Winkelmann and Schott Equation. An equation proposed by A. Winkelmann and O. Schott (*Ann. Phys. Chem.*, **51**, 730, 1894) for assessing the thermal endurance (F) of glass-ware on the basis of the tensile strength (P), modulus of elasticity (E), coefficient of linear thermal expansion (α), thermal conductivity (K), specific heat (C) and specific gravity (S):

$$F = \frac{P}{\alpha E}\sqrt{\frac{K}{SC}}$$

See also THERMAL EXPANSION FACTORS FOR GLASS.

Winning. The combined process of getting (i.e. excavating) and transporting a raw material such as clay to a brickworks or stockpile.

Wire-cut Process (UK); Stiff-mud Process (USA). The shaping of bricks by extruding a column of clay through a die, the column being subsequently cut to the size of bricks by means of taut wires.

Wired Glass. Flat glass that has been reinforced by the incorporation of wire mesh. One use is as a 'fire-stop': whereas, in a fire, ordinary window panes crack, fall out and allow flames to spread, wired glass will crack but hold together.

Witherite. Natural BARIUM CARBONATE (q.v.).

Wohl Block. A hollow clay building block designed for the construction of walls of various thicknesses and for use in ceilings. (E. Wohl, *Ziegelindustrie*, **2**, 99, 1949.)

Wollastonite. Calcium metasilicate, $CaSiO_3$; m.p. 1544°C; thermal expansion $5 \cdot 9 \times 10^{-6}$ (25–700°C). There are two forms: the natural, β, form, and the α-wollastonite that is produced when the β-form is heated above 1100°C. There are economic deposits in New York State, Finland and USSR. Wollastonite is used in some pottery bodies (particularly wall tiles), and in special low-loss electroceramics.

Wonderstone. A popular name (particularly in S. Africa) for PYROPHYLLITE (q.v.).

Wood's Glass. A special glass that transmits ultra-violet but is almost opaque to visible light; such a glass was first made by Prof. R. W. Wood, John Hopkins University, USA, and was used for invisible signalling during World War I.

Woods Hole Sediment Analyzer. A method of particle-size analysis based on measurement of pressure changes resulting from sedimentation in a suspension of the particles in water. It is applicable to coarse silts and fine gravels, and permits 150 tests to be made in a day. The instrument was designed by Woods Hole Oceanographic Institution (*J. Sed. Petrology*, **30**, 490, 1960).

Wood's Process. See UP-DRAW PROCESS.

Woodhall-Duckham Kiln. See ROTARY-HEARTH KILN.

Wool-drag. A fault in GROUND LAYING (q.v.) resulting from accidental smearing of the colour.

Worcester Shape. A tea- or coffee-cup having the general shape of a plain cylinder, rounded sharply near the bottom, which has a broad but shallow foot. For specification see B.S. 3542.

Work-board. A board, about 6 ft long and 9 in. wide, on which pottery-ware may be placed and carried from one process to the next.

Work Speed. A term relating to the process of grinding with abrasive wheels. In surface grinding, the work speed is the rate of table traverse, usually expressed in ft/min. In centreless, cylindrical and internal grinding, the work speed is the rate at which the object being ground (the 'work') revolves; this may be expressed either in rev/min or in ft/min.

Workability. An evasive synonym for PLASTICITY (q.v.) introduced a generation ago when it became clear that the structural and physical factors causing plasticity in clay were not known; the term is still much used.

Working End. Part of a glass-tank furnace: (1) in such a furnace as used for container glass, the term signifies the compartment following the melting end and separated from it by the bridge; (2) the end from which the glass is withdrawn in a glass-tank furnace that has no bridge.

Working Mould. See under MOULD.

Working Range. The temperature range within which glass is amenable to shaping; at higher temperatures the glass is too fluid (viscosity less than about 10^3 poises) and at lower temperatures too rigid (viscosity above 10^7 poises). For comparative purposes, the Working Range is equated to a viscosity range from 10^4 to $10^{7.6}$ poises (see SOFTENING POINT).

Wreathing. A fault that sometimes occurs on the inside of cast whiteware as a slightly raised crescent or snake-like area; it is probably a result of orientation of the plate-shaped clay particles and can usually be prevented by increasing the viscosity and the thixotropy of the casting slip.

X

Xerography. A method of copying designs by means of a photo-conductive plate and an electrostatic powder. It has been applied in the production of transfers for the decoration of glass and ceramic ware.

Y

Yellow Ware. US term for CANE WARE (q.v.).

Young's Modulus. The MODULUS OF ELASTICITY (q.v.) in tension or compression. Some typical values for this property are: earthenware, 6×10^6 p.s.i.; electrical porcelain, $10–12 \times 10^6$ p.s.i.; sintered alumina, 40×10^6 p.s.i.

Ytong. A cellular (lightweight) concrete made in block form from shale and lime, and subsequently hardened by autoclave treatment. It was first produced in 1918, by A. Eriksson in Sweden. (cf. THERMALITE YTONG).

Yttrium-Iron Garnet. $Y_3Fe_2(FeO_4)_3$; it can be synthesized by heating a mixture of the oxides at 800–1000°C. This compound is characterized by exceptionally narrow ferrimagnetic resonance line widths, and is of interest in microwave applications.

Yttrium Oxide. Y_2O_3; m.p. 2410°C; sp. gr. 4·84; thermal expansion (25–1400°C) $9·3 \times 10^{-6}$. An oxide extracted from MONAZITE (q.v.) though not itself, strictly, a rare earth. Yttrium-iron garnets are used as ferromagnetics. A well-vitrified special ceramic can be produced by firing the compacted oxide at 1800°C, up to which temperature the cubic form is stable.

Z

Zachariasen's Theory. See under GLASS.

Zaffre. A roasted mixture of cobalt ore and sand, formerly used as a blue colouring material for pottery and glass (cf. SMALT).

Zahn Cup. An orifice-type viscometer; it has been used for the determination of the viscosity of glaze suspensions. (E. A. Zahn, *Chem. Industries*, **51**, (2), 220, 1942.)

Zebra Roof. A type of roof for basic O.H. steel furnaces, the feature being alternate rings of chrome-magnesite and of silica refractories, hence the name from the dark and light stripes across the roof. The Zebra Roof was introduced in 1947 with a view to combining the merits of the two types of refractory; by 1952 there were 300 such roofs in service in USA alone, but the Zebra Roof has now been displaced by the all-basic roof.

Zeissig Green. An underglaze colour that has been used for pottery decoration. It is made by calcining a mixture of 10 parts barium chromate, 8 parts whiting and 5 parts boric acid.

Zeta Potential. The electrical potential ζ set up by the double layer of charged ions in a colloidal solution (such as a clay slip) where it is close to a solid surface:

$$\zeta = 4\pi ed/D$$

where e = density of charge on the surface of the particles; d = thickness of the double layer; D = dielectric constant of the liquid medium. The zeta potential is of significance in slip casting.

Zig-zag Kiln. A TRANSVERSE-ARCH KILN (q.v.) with staggered dividing walls, the fire-travel thus being forced to follow a zig-zag path. Such kilns find use in the firing of structural clay products.

Zinc Aluminate (Gahnite). $ZnAl_2O_4$; m.p. 1950°C. This spinel, when made from industrial-grade oxides, has a P.C.E. > 1900°C and RuL > 1700°C. Russian experiments indicate that it can be used as a refractory lining for electric furnaces melting Al, Zn, Pb or Sn. It is rapidly attacked by alkalis.

Zinc Oxide. ZnO; sp. gr. 5·65: sublimes at 1800°C. Used in glasses, glazes, enamels, and more recently in special ferro-magnetic ceramics.

Zircon. Zirconium orthosilicate, $ZrO_2 . SiO_2$. The principal source is along the most easterly part of the coast of Australia, on both sides of the border between Queensland and New South

Zircon Porcelain

Wales, where it occurs abundantly as beach sands. When heated, zircon dissociates at 1600–1800°C into SiO_2 and ZrO_2; it is nevertheless used as a refractory, in the lining of aluminium-melting furnaces for example. Zircon is also used as an opacifier in vitreous enamels and glazes, and as a constituent of special electrical porcelains.

Zircon Porcelain. An electroceramic made from a batch consisting of 60–70% zircon, 20–30% flux and 10–20% clay; the flux may be a complex Ca-Mg-Ba-Zr silicate or other alkaline-earth composition. Zircon porcelain has high mechanical strength and good thermal-shock resistance over a wide temperature range; electrically, it is a low-loss material.

Zirconia. Zirconium oxide, ZrO_2; m.p. 2690°C; sp. gr. 5·56. There are two forms: monoclinic, stable up to about 1200°C; tetragonal, stable from 1200°C to the m.p. The inversion is accompanied by a reversible linear contraction of about 2·5% which causes pure zirconia ware to spall. Heating ZrO_2 with small amounts of cubic oxides, e.g. CaO or MgO, causes these oxides to enter into solid solution in the ZrO_2, which then itself assumes a cubic structure and remains stable—thus becoming more resistant to thermal shock. Fused zirconia is the only refractory material that is suitable for the making of kiln furniture for the firing of titanate electroceramics.

Zirconium Borides. ZrB_2; m.p. approx. 3050°C; sp. gr. 6·1. A hard, refractory, chemically resistant material. When shaped by ceramic processes it may find use in nozzles and in the casting of alloys. It has a high electrical conductivity; the thermal expansion is $4·0 \times 10^{-6}$ (25–700°C). A less well-known compound is ZrB_{12}; m.p. 2680°C; sp. gr. 3·6. The diboride has been used in rocket nozzles and in thermocouple sheaths.

Zirconium Carbide. ZrC; m.p. 3530°C; theoretical density, 6·7 g/ml. This special refractory material has been used in the manufacture of a flame deflector operating at very high temperatures.

Zirconium-Iron Pink. A ceramic stain suitable for the colouring of a variety of glazes maturing at 1220–1280°C. This firing range is greater than that permissible with chrome-tin pink or with chrome-alumina pink.

Zirconium Nitride. ZrN; m.p. 2980 ± 50°C; sp. gr. 7·32.

Zirconium Oxide. See ZIRCONIA.

Zirconium Phosphate. Normal zirconium phosphate, ZrP_2O_7, has a reversible inversion at 300°C and at 1550°C dissociates into zirconyl phosphate, $(ZrO)_2P_2O_7$, with loss of P_2O_5 as vapour. Zirconyl phosphate is stable up to about 1600°C and has a very low thermal expansion -1×10^{-6} (20–1000°C).

Zirconium Silicate. See ZIRCON.

Zirconium-Vanadium Blue (or Turquoise). A pigment, for use in ceramic glazes, introduced by Harshaw Chemical Co. (US Pat., 2 441 447, 11/5/48; Brit. Pat. 625 448, 28/6/49). The composition is (parts by wt.): ZrO_2, 60–70; SiO_2, 26–36; V_2O_5, 3–5. Alkali must also be present, e.g. 0·5–5% Na_2O. In the absence of alkali a green colour is produced.

Zirkite. A term used in the mineral trade for the natural, impure, zirconia ore (baddeleyite) that occurs in Brazil.

APPENDIX 1

Sieve Sizes

The following table has been arranged for quick comparison between one standard series of sieves and another. The nominal apertures of the first column are, for the most part, those of the British Standard sieves; where the aperture of a sieve in another series differs slightly from that given in the first column, the exact aperture is printed in brackets.

Nominal Aperture (mm)	Mesh Number					
	Britain	USA	France	Germany	Tyler	IMM
5·00		4 (4·76)	38	5·0	4 (4·70)	
4·00		5	37	4·0	5 (3·96)	
3·35	5	6 (3·36)			6 (3·33)	
3·15			36	3·15		
2·80	6	7 (2·83)			7 (2·79)	
2·50			35	2·50		5 (2·54)
2·40	7	8 (2·38)			8 (2·36)	
2·00	8	10	34	2·00	9 (1·98)	
1·68	10	12			10 (1·65)	
1·60			33	1·60		8 (1·57)
1·40	12	14 (1·41)			12	
1·25			32	1·25		10 (1·27)
1·20	14	16 (1·19)			14 (1·17)	
1·00	16	18	31	1·00	16 (0·99)	12 (1·06)
0·85	18	20 (0·84)			20 (0·83)	
0·80			30	0·80		16 (0·792)
0·71	22	25 (0·707)			24 (0·701)	
0·63			29	0·63		20 (0·635)
0·60	25	30 (0·595)			28 (0·589)	
0·50	30	35	28	0·50	32 (0·495)	
0·42	36	40	27 (0·40)	0·40 (0·40)	35 (0·417)	30
0·355	44	45 (0·354)			42 (0·351)	
0·315			26	0·315		40 (0·317)
0·30	52	50 (0·297)			48 (0·295)	
0·25	60	60	25	0·25	60 (0·246)	50 (0·254)
0·21	72	70	24 (0·20)	0·20 (0·20)	65 (0·208)	60 (0·211)
0·18	85	80 (0·177)			80 (0·175)	70
0·15	100	100 (0·149)	23 (0·16)	0·16 (0·16)	100 (0·147)	80 (0·157)
0·14						90 (0·139)
0·125	120	120	22	0·125	115 (0·124)	100 (0·127)
0·105	150	140	21 (0·100)	0·10 (0·10)	150 (0·104)	120 (0·107)
0·09	170	170 (0·088)		0·09	170 (0·088)	150 (0·084)
0·075	200	200 (0·074)	20 (0·080)	0·08 (0·080)	200 (0·074)	
0·063	240	230	19	0·063	250 (0·061)	200
0·053	300	270	18 (0·050)	0·05 (0·050)	270	
0·045	350	325 (0·044)		0·045	325 (0·043)	
0·040			17	0·04		
0·037		400			400 (0·038)	

APPENDIX 2

Nominal Temperature (°C) Equivalents of Pyrometric Cones.

CONE No.	BRITISH (Stafford-shire) 4°C/min	GERMAN (Seger)	AMERICAN (Orton)		
			Large 1°C/min	Large 2½°C/min	Small 5°C/min
022	600	600	585	600	630
022A	625	—	—	—	—
021	650	650	602	614	643
020	670	670	625	635	666
019	690	690	668	683	723
018	710	710	696	717	752
017	730	730	727	747	784
016	750	750	767	792	825
015	790	—	790	804	843
015A	—	790	—	—	—
014	815	—	834	838	870
014A	—	815	—	—	—
013	835	—	869	852	880
013A	—	835	—	—	—
012	855	—	866	884	900
012A	—	855	—	—	—
011	880	—	886	894	915
011A	—	880	—	—	—
010	900	—	887	894	919
010A	—	900	—	—	—
09	920	—	915	923	955
09A	—	920	—	—	—
08	940	—	945	955	983
08A	950	940	—	—	—
07	960	—	973	984	1008
07A	970	960	--	—	—
06	980	—	991	999	1023
06A	990	980	—	—	—
05	1000	—	1031	1046	1062
05A	1010	1000	—	—	—
04	1020	—	1050	1060	1098

Appendix 2

Nominal Temperature (°C) Equivalents of Pyrometric Cones. (Cont.).

CONE No.	BRITISH (Stafford-shire) 4°C/min	GERMAN (Seger)	AMERICAN (Orton) Large 1°C/min	AMERICAN (Orton) Large 2½°C/min	AMERICAN (Orton) Small 5°C/min
04A	1030	1020	—	—	—
03	1040	—	1086	1101	1131
03A	1050	1040	—	—	—
02	1060	—	1101	1120	1148
02A	1070	1060	—	—	—
01	1080	—	1117	1137	1178
01A	1090	1080	—	—	—
1	1100	—	1136	1154	1179
1A	1110	1100	—	—	—
2	1120	—	1142	1162	1179
2A	1130	1120	—	—	—
3	1140	—	1152	1168	1196
3A	1150	1140	—	—	—
4	1160	—	1168	1186	1209
4A	1170	1160	—	—	—
5	1180	—	1177	1196	1221
5A	1190	1180	—	—	—
6	1200	—	1201	1222	1255
6A	1215	1200	—	—	—
7	1230	1230	1215	1240	1264
7A	1240	—	—	—	—
8	1250	1250	1236	1263	1300
8A	1260	—	—	—	—
8B	1270	—	—	—	—
9	1280	1280	1260	1280	1317
9A	1290	—	—	—	—
10	1300	1300	1285	1305	1330
10A	1310	—	—	—	—
11	1320	1320	1294	1315	1336
					P.C.E. Cones 2½°C/min
12	1350	1350	1306	1326	1337
13	1380	1380	1321	1346	1349

Nominal Temperature (°C) Equivalents of Pyrometric Cones. (Cont.).

CONE No.	BRITISH (Stafford-shire) 4°C/min	GERMAN (Seger)	AMERICAN (Orton) Large 1°C/min	Large 2½°C/min	P.C.E. Cones 2½°C/min
14	1410	1410	1388	1366	1398
15	1435	1435	1424	1431	1430
16	1460	1460	1455	1473	1491
17	1480	1480	1477	1485	1512
18	1500	1500	1500	1506	1522
19	1520	1520	1520	1528	1541
20	1530	1530	1542	1549	1564
23	—	—	1586	1590	1605
26	1580	1580	1589	1605	1621
27	1610	1610	1614	1627	1640
28	1630	1630	1614	1633	1646
29	1650	1650	1624	1645	1659
30	1670	1670	1636	1654	1665
31	1690	1690	1661	1679	1683
31½	—	—	1685	1700	1699
32	1710	1710	1706	1717	1717
32½	—	—	1718	1730	1724
33	1730	1730	1732	1741	1743
34	1750	1750	1757	1759	1763
35	1770	1770	1784	1784	1785
36	1790	1790	1798	1796	1804
37	1825	1825	—	—	1820
38	1850	1850	—	—	1835*
39	1880	1880	—	—	1865*
40	1920	1920	—	—	1885*
41	1960	1960	—	—	1970*
42	2000	2000	—	—	2015*

*10°C/min

APPENDIX 3

Conversion Table from Fahrenheit to Celsius (Centigrade) Degrees and vice versa.

The number to be converted is in the middle of each set of three columns; if this number is °F, the equivalent °C is found in the left-hand column; if the number is °C, the equivalent °F is found in the right-hand column.

°C		°F	°C		°F
−18	0	32	138	280	536
−12	10	50	143	290	554
−7	20	68	149	300	572
−1	30	86	154	310	590
4	40	104	160	320	608
10	50	122	166	330	626
16	60	140	171	340	644
21	70	158	177	350	662
27	80	176	182	360	680
32	90	194	188	370	698
38	100	212	193	380	716
43	110	230	199	390	734
49	120	248	204	400	752
54	130	266	210	410	770
60	140	284	216	420	788
66	150	302	221	430	806
71	160	320	227	440	824
77	170	338	232	450	842
82	180	356	238	460	860
88	190	374	243	470	878
93	200	392	249	480	896
99	210	410	254	490	914
104	220	428	260	500	932
110	230	446	266	510	950
116	240	464	271	520	968
121	250	482	277	530	986
127	260	500	282	540	1004
132	270	518	288	550	1022

°C		°F	°C		°F
293	560	1040	527	980	1796
299	570	1058	532	990	1814
304	580	1076	538	1000	1832
310	590	1094	543	1010	1850
316	600	1112	549	1020	1868
321	610	1130	554	1030	1886
327	620	1148	560	1040	1904
332	630	1166	566	1050	1922
338	640	1184	571	1060	1940
343	650	1202	577	1070	1958
349	660	1220	582	1080	1976
354	670	1238	588	1090	1994
360	680	1256	593	1100	2012
366	690	1274	599	1110	2030
371	700	1292	604	1120	2048
377	710	1310	610	1130	2066
382	720	1328	616	1140	2084
388	730	1346	621	1150	2102
393	740	1364	627	1160	2120
399	750	1382	632	1170	2138
404	760	1400	638	1180	2156
410	770	1418	643	1190	2174
416	780	1436	649	1200	2192
421	790	1454	654	1210	2210
427	800	1472	660	1220	2228
432	810	1490	666	1230	2246
438	820	1508	671	1240	2264
443	830	1526	677	1250	2282
449	840	1544	682	1260	2300
454	850	1562	688	1270	2318
460	860	1580	693	1280	2336
466	870	1598	699	1290	2354
471	880	1616	704	1300	2372
477	890	1634	710	1310	2390
482	900	1652	716	1320	2408
488	910	1670	721	1330	2426
493	920	1688	727	1340	2444
499	930	1706	732	1350	2462
504	940	1724	738	1360	2480
510	950	1742	743	1370	2498
516	960	1760	749	1380	2516
521	970	1778	754	1390	2534

°C		°F	°C		°F
760	1400	2552	993	1820	3308
766	1410	2570	999	1830	3326
771	1420	2588	1004	1840	3344
777	1430	2606	1010	1850	3362
782	1440	2624	1016	1860	3380
788	1450	2642	1021	1870	3398
793	1460	2660	1027	1880	3416
799	1470	2678	1032	1890	3434
804	1480	2696	1038	1900	3452
810	1490	2714	1043	1910	3470
816	1500	2732	1049	1920	3488
821	1510	2750	1054	1930	3506
827	1520	2768	1060	1940	3524
832	1530	2786	1066	1950	3542
838	1540	2804	1071	1960	3560
843	1550	2822	1077	1970	3578
849	1560	2840	1082	1980	3596
854	1570	2858	1088	1990	3614
860	1580	2876	1093	2000	3632
866	1590	2894	1099	2010	3650
871	1600	2912	1104	2020	3668
877	1610	2930	1110	2030	3686
882	1620	2948	1116	2040	3704
888	1630	2966	1121	2050	3722
893	1640	2984	1127	2060	3740
899	1650	3002	1132	2070	3758
904	1660	3020	1138	2080	3776
910	1670	3038	1143	2090	3794
916	1680	3056	1149	2100	3812
921	1690	3074	1154	2110	3830
927	1700	3092	1160	2120	3848
932	1710	3110	1166	2130	3866
938	1720	3128	1171	2140	3884
943	1730	3146	1177	2150	3902
949	1740	3164	1182	2160	3920
954	1750	3182	1188	2170	3938
960	1760	3200	1193	2180	3956
966	1770	3218	1199	2190	3974
971	1780	3236	1204	2200	3992
977	1790	3254	1210	2210	4010
982	1800	3272	1216	2220	4028
988	1810	3290	1221	2230	4046

°C		°F	°C		°F
1227	2240	4064	1443	2630	4766
1232	2250	4082	1449	2640	4784
1238	2260	4100	1454	2650	4802
1243	2270	4118	1460	2660	4820
1249	2280	4136	1466	2670	4838
1254	2290	4154	1471	2680	4856
1260	2300	4172	1477	2690	4874
1266	2310	4190	1482	2700	4892
1271	2320	4208	1488	2710	4910
1277	2330	4226	1493	2720	4928
1282	2340	4244	1499	2730	4946
1288	2350	4262	1504	2740	4964
1293	2360	4280	1510	2750	4982
1299	2370	4298	1516	2760	5000
1304	2380	4316	1521	2770	5018
1310	2390	4334	1527	2780	5036
1316	2400	4352	1532	2790	5054
1321	2410	4370	1538	2800	5072
1327	2420	4388	1543	2810	5090
1332	2430	4406	1549	2820	5108
1338	2440	4424	1554	2830	5126
1343	2450	4442	1560	2840	5144
1349	2460	4460	1566	2850	5162
1354	2470	4478	1571	2860	5180
1360	2480	4496	1577	2870	5198
1366	2490	4514	1582	2880	5216
1371	2500	4532	1588	2890	5234
1377	2510	4550	1593	2900	5252
1382	2520	4568	1599	2910	5270
1388	2530	4586	1604	2920	5288
1393	2540	4604	1610	2930	5306
1399	2550	4622	1616	2940	5324
1404	2560	4640	1621	2950	5342
1410	2570	4658	1627	2960	5360
1416	2580	4676	1632	2970	5378
1421	2590	4694	1638	2980	5396
1427	2600	4712	1643	2990	5414
1432	2610	4730	1649	3000	5432
1438	2620	4748			

327